GaAs Integrated Circuits

GaAs Integrated Circuits

Edited by

Joseph Mun

MACMILLAN PUBLISHING COMPANY

NEW YORK

First published in 1988 by
BSP Professional Books

Distributed in the United States by Macmillan
Publishing Company
866 Third Avenue, New York,
NY 10022

Printed and bound in Great Britain

Library of Congress Cataloging-in-Publication Data

GaAs integrated circuits.
    Bibliography: p.
    1. Integrated circuits. 2. Gallium arsenide
semiconductors. I. Mun, Joseph.
TK7874.G33   1988   621.381′73   87–60605
ISBN 0–02–948821–4

Contents

Contributors' Biographies

Thomas R. Apel received the BS degree in Physics and Mathematics from Loras College in 1976 and the MSEE degree from the University of Wisconsin (Madison) in 1978. He is currently section head of the Power Circuits Section of the Torrance Research Center of Hughes Aircraft Company where he is responsible for power amplifier development. He is also responsible for several high efficiency monolithic power amplifier development efforts. *(Chapter 4)*

Etienne Delhaye received his degree in Electrical Engineering from Ecole Nationale Supérieure des Télécommunications de Bretagne, Brest, France, in 1983. He joined Laboratoires d'Electronique et de Physique Appliquée (LEP) in 1983 to work on GaAs digital ICs and more particularly on frequency dividers and arithmetic multipliers. He received a PhD in microelectronics from the University of Grenoble, France, in 1986. He is currently working on the design of GaAs large scale digital integrated circuits. *(Chapter 2)*

Thierry Ducourant received his degree in Electrical Engineering from l'Institut Supérieur d'Electronique du Nord, Lille, France, in 1982. In 1983 he joined LEP (Laboratoires d'Electronique et de Physique Appliquée), to work on ultra-high-speed GaAs ICs and more particularly on the design and measurements of comparators, A to D converters and static RAMs. M. Ducourant holds five patents on circuit design. *(Chapter 2)*

Allen Firstenberg is director of the Information Sciences Function for Rockwell International Science Center. Prior to joining Rockwell in January, 1978, Allen Firstenberg was highly successful in the technical management and design of a number of aerospace systems. Among his program successes were Voyager '77, an imaging system that acquired photos and data from Jupiter, Saturn and their satellites, and Viking '75, a Mars lander that acquired photos and evaluated both Mar's soil and atmosphere. Allen Firstenberg received his BS degree in Electrical

Engineering from the University of California at Los Angeles (UCLA) under academic scholarship in 1963. He has authored and co-authored over 30 technical papers and is holder of two patents in the areas of image processing and data compression. *(Chapter 8)*

Bertrand Gabillard received his degree in Electrical Engineering from l'Institut Supérieur d'Electronique du Nord, Lille, France, in 1978. He received a PhD in electronics from the University of Lille, France, in 1981. In 1981, he joined LEP (Laboratoires d'Electronique et de Physique Appliquée), to work on high-speed digital GaAs IC designs and, more particularly, on the design of static RAM memories and frequency dividers. *(Chapter 2)*

Stephen J. Harrold graduated with a degree in Applied Physics and Electronics from Durham University in 1978. He was awarded a PhD in Engineering Science at Exeter University, in 1982. Dr Harrold was with STC Technology Ltd for three years before joining the University of Manchester Institute of Science and Technology in the IC Design and Test Centre which provides an IC design facility for Science and Engineering Research Centre funded users. *(Chapter 6)*

William A. Hughes received a BSc in Physics from Manchester University, UK in 1983. In 1983 he joined GEC Research working on direct write electron beam lithography of FETs. He became group leader of FET and HEMT based digital logic in 1985. He is also involved in the analogue application of HEMTs. He is presently completing his PhD thesis on 'Design and Fabrication of HEMTs' with Leeds University. *(Chapter 7)*

Donald T. J. Hurle is an Individual Merit Deputy Chief Scientific Officer at the Royal Signals and Radar Establishment, Malvern, England where he has worked for the last 27 years on the theory and practice of the growth of electronic materials. From 1976 to 1986 he was chairman of the GaAs Consortium which co-ordinated materials research in GaAs in the UK. He has BSc, PhD and DSc degrees from the University of Southampton and is currently Visiting Professor of Physics and Mathematics at the University of Bristol. *(Chapter 1)*

Naoki Kato received a BS degree in Applied Physics from Waseda University and an MS degree from Tokyo Metropolitan University in 1973 and 1976, respectively. He received a PhD in Electronics from Waseda University in 1986. He joined the Electrical Communications Laboratories, Nippon Telegraph and Telephone Corporation in 1976 and he is currently managing a GaAs R&D processing line. Dr. Kato is a member of the IEEE, the Institute of Electronics and Communication Engineers of Japan, and the Japan Society of Applied Physics. *(Chapter 3)*

Yutaka Matsuoka received his BS and MS degrees in Physics from the Tokyo Institute of Technology in 1974 and 1976 respectively. In 1976, he joined the Musashino Electrical Communication Laboratory, Nippon Telegraph and Telephone Public Corporation, Tokyo, Japan. Since 1982, he has been engaged in the development of GaAs LSIs. He is now a research engineer at the NTT Atsugi Electrical Communication Laboratories. Mr. Matsuoka is a member of the Institute of Electronics and Communication Engineers of Japan and the Japan Society of Applied Physics. *(Chapter 3)*

Joseph Mun graduated with a BSc degree in Electrical Engineering from the University of Aston in 1968 and with an MSc in Microwave Engineering from University College London in 1969. He joined STL in 1969. In 1977 Mr Mun initiated the GaAs MESFET research at STL and has been responsible for all aspects of its development leading to a wide range of integrated circuits. He is manager of the GaAs Integrated Circuits Laboratory and he is currently extending his interest to opto-electronic integrated circuits. Mr Mun is a member of the Institution of Electrical Engineers and the Institute of Physics. *(Editor)*

John A. Phillips joined STC Technology Ltd in 1978 having gained a first class honours degree in Electronics at the University of Southampton. From 1982 he was leading a team responsible for the design, testing and packaging of GaAs ICs. He is currently working on the application of high-speed integrated circuits for defense systems. *(Chapter 6)*

Ali A. Rezazadeh received a PhD degree in Solid-State Physics from the University of Sussex, UK, in 1982 where he continued his research as research fellow, in the characterization of defects in GaAs. In 1983, he joined the GEC Research Laboratories, Hirst Research Centre, Wembley, England, where he worked in the III–V Compound Semiconductor Laboratory in the research and development of heterojunction bipolar transistors. His particular interests include physics and technology of GaAs/GaAlAs HBT devices and logic circuits. He is now the group leader of the HBT technology. *(Chapter 7)*

Marc Rocchi received his degree in Electrical Engineering and Solid State Physics from l'Ecole Supérieure d'Electricité de Paris, in 1972. In 1976, he joined LEP (Laboratoires d'Electronique et de Physique Appliquée) to work on high-speed digital GaAs ICs. He is currently group leader at LEP in charge of the design and fabrication of GaAs discrete devices and ICs. *(Chapter 2)*

Sven A. Roosild received his BA degree in Engineering and Applied Physics from Harvard University in 1960 and his MA in Physics from Boston University in 1966. In August 1980, Sven Roosild joined DARPA's Defense Sciences Office as Program Manager in charge of R&D on solid state electronic materials and devices. He initiated the DARPA low power, radiation hard GaAs LSI Pilot Line programs that are designed to establish the first production capability for LSI complexity GaAs integrated circuits. In August 1984, Mr Roosild was made Assistant Director for Electronic Sciences in the Defense Sciences Office of DARPA. *(Chapter 8)*

James M. Schellenberg received the BS and MS degrees in Electrical Engineering from Fresno State University, Fresno, California in 1969 and Johns Hopkins University, Baltimore, Maryland, in 1973, respectively. Mr. Schellenberg is currently head of the FET Circuits Department of the Torrance Research Center of Hughes Aircraft Company. Mr. Schellenberg is the author of numerous technical papers on FET amplifiers and holds six US patents. He is a member of the Microwave Theory and Techniques Society and the Electron Devices Society of the IEEE. *(Chapter 4)*

James A. Turner received his BSc (Hons) in Physics from Sheffield University in 1960. He joined Plessey Research Caswell in the same year and from 1960 to 1966 he carried out studies of silicon and GaAs bipolar transistors. He became Department Manager of Microwave Device Technology in 1982. He became Senior Research Executive in 1986 with responsibilities of technical co-ordination of microwave and opto technologies and liaison between R&D and manufacturing. In recognition of Mr. Turner's pioneering work on GaAs FETs he was awarded the Member of the Order of the British Empire (MBE) by the Queen in 1981. *(Chapter 5)*

Colin C. E. Wood received a PhD from Nottingham University in 1970 in Physical Chemistry. In 1978, he joined Cornell University as a founder member of the National Sub-micron Facility and a Senior Research Associate of the Electrical Engineering Department. In 1982, he joined Hirst Research Centre as Manager of Compound Semiconductor Research Laboratory, and is currently Chief Scientist of the Compound Semiconductor Laboratory. In 1985 Dr. Wood received the Patterson Medal from the Institute of Physics. *(Chapter 7)*

Kimiyoshi Yamasaki received his BE, ME, and PhD degrees in Electronic Engineering from the University of Tokyo in 1975, 1977, and 1980 respectively. In 1980, he joined Electrical Communication Laboratories, NTT, Tokyo, Japan working on the development of GaAs LSI process technologies. Dr. Yamasaki is a member of the Japan Society of Applied Physics and the Institute of Electronics and Communication Engineers of Japan. *(Chapter 3)*

Foreword

For many years, GaAs MESFETs were regarded as devices of the future, but in spite of earlier reservations they are now accepted as just 'another' semiconductor component, widely used in systems and listed as standard products by semiconductor manufacturers. History is now poised to repeat for GaAs integrated circuits what has previously happened to the GaAs MESFET: once again, despite early reservations, many companies are taking GaAs ICs into manufacture and many countries are launching national programmes to ensure that the early research on this technology is capitalized upon. This book addresses a wide range of integrated circuits that can be fabricated on GaAs, discussing their theory, design, fabrication and application. The more detailed properties of GaAs, from material to MESFET, are well covered by many excellent books on the market and so these properties are simply included in this book in a condensed form for convenient reference.

Integrated circuits on GaAs can be broadly divided into two areas: analogue circuits and digital circuits, each warranting separate consideration. The focus on analogue has been primarily on monolithic microwave integrated circuits (MMICs). Recently, their frequency spectrum has been extending down to DC in circuits such as operational amplifiers, and up to millimeter-wave frequencies. Turner and Pengelly were amongst the first few pioneers who stimulated the early development of the MMIC. Since then, many intriguing technologies have been developed to improve the performance of the MMIC, such as air bridges and via-hole grounding. These structures in micro-engineering are most impressive when examined under the SEM. Many circuit ideas have been implemented in monolithic form on GaAs, such as the travelling wave amplifier which gives a real meaning to broadband design. Improvements in material and device structures such as the HEMT indicate that analogue integrated circuits operating up to 100 GHz can be expected in the near future.

Whilst the progress of analogue ICs has been steady, perhaps even uneventful, the role of GaAs digital ICs has created some of the most passionate debates in the semiconductor industry and, to a great extent, these debates are still raging today. The reason is that GaAs digital ICs are challenging head-on a captive market entirely dominated by silicon. Imaginations were fired by Van Tuyl and Liechti in the mid-1970s, who reported logic gates with propagation delays as low as 100 ps and counters operating at over 4 GHz. It took several years to overcome the early GaAs material limitations until Lee reported the first 1000–gate circuit in 1980. However, larger scale integration on GaAs only became a practical reality with the introduction of the low power direct coupled FET logic, DCFL, and with the introduction of various self-aligned technologies. The combination of these two approaches has led to gate propagation delay times of under 10 ps and circuit complexities capable of producing 16 kilobit SRAMs.

Although the rapid progress of GaAs digital ICs was spectacular, the improvements made on Si were equally impressive. It is generally recognised that such progress in Si was, in fact, prompted by the challenge from GaAs. The capabilities of GaAs and Si for digital applications are much better understood today, with better understanding of CAD and IC design techniques. It is now accepted that the performance margin of GaAs over Si is not as great as was optimistically speculated. The smaller performance margin (2 to 3 times smaller) is leading to the inevitable concluson that in the forseeable future GaAs digital ICs will probably remain as a specialist market.

Research in GaAs ICs is now moving into heterojunction-based devices. Some of these, like HEMTs (high electron mobility transistors) and HBTs (heterojunction bipolar transistors), are making rapid progress towards integration with optical devices such as lasers, LEDs and photodetectors leading to fully-integrated opto-electronic ICs. However, they are currently regarded as devices of the future . . .

I would like to thank most sincerely all the authors for their contributions. I would also like to thank the management of STC Technology Ltd for giving me the opportunity and environment to work in this exciting field of research which has made it possible for me to collate this book. Many colleagues have assisted me in many ways on this book and I would especially like to acknowledge the contributions of Dr. S. W. Bland, Miss S. A. Kitching, Dr. W. S. Lee and Mr. J. A. Phillips.

J. Mun, Harlow, June 1987

Chapter 1

Material for GaAs Integrated Circuits

DONALD T. J. HURLE

1.1 INTRODUCTION

The extent to which the physical properties of a 'new' material can be exploited to produce marketable devices which have novel or improved properties is dictated by the extent to which control can be obtained over the material's growth and device fabrication technologies. To illustrate: the dominant position held by silicon in microelectronics owes more to the unique physicochemical properties of its oxide than to its band structure.

This task of taming the material has been a long and arduous one for the gallium arsenide researcher; one which has been pursued over the last three decades since Welker first drew attention to the potential of the material and the other III–V compounds. Much of the added difficulty, over that experienced in the development of first germanium and then silicon, derives from the fact that GaAs is a compound having a finite, though small, phase extent. The control and maintenance of crystal stoichiometry is thus an added task for the materials scientist and one which is made the more difficult by the dissociative nature of the compound.

MESFET integrated circuit technology is heavily dependent on substrate quality and became possible in the early 1980s only after long-awaited improvements and cost reduction in bulk crystal growth occurred which yielded 2″ and 3″ diameter wafers of adequate properties such as purity, uniformity, reproducibility and yield [1]. The uniformity of threshold voltage across a wafer in an FET integrated circuit fabricated using an ion implantation technology can be a complex function of substrate defects and processing induced effects.

Early attempts to make substrates focused on chromium doping to achieve semi-insulating (hereafter SI) behaviour but the high solid state diffusivity and anomalous surface aggregation of chromium led to problems of conversion of the surface layer following thermal annealing. If this occurred during the implant-activation anneal then it adversely affected the implant profile and with it the control of device properties such as peak channel current and threshold voltage. The prospects for GaAs ion implanted ICs were transformed however by the development of large area wafers of undoped SI GaAs.

Historically Cr-doped SI GaAs was first grown by either a modified horizontal Bridgman technique or by gradient-freeze. In both methods the melt composition was controlled by controlling the temperature of a resevoir of solid arsenic.

Early attempts to apply the crystal pulling (Czochralski) technique, so successfully and widely used in the silicon industry, were beset with difficulties in preventing dissociation of the melt. The magnetically coupled syringe puller, in which the whole puller chamber was kept at a temperature above the condensation temperature of arsenic, provided a partial answer but was not readily adaptable to commercial production. The breakthrough came with the development of the liquid encapsulation Czochralski (LEC) technique [2] which adopted the elegantly simple solution of covering the GaAs melt with a layer of an inert molten glass – boric oxide.

In parallel with developments in techniques for bulk growth to yield wafers, a range of epitaxial techniques have appeared on the scene. Of these, liquid phase epitaxy (LPE) has been extensively exploited for the production of lasers and light-emitting diodes. Although the first FET digital ICs were made in material prepared in this way, the technique is not well suited to SI layer formation, because inherently it produces material to the Ga-rich side of stoichiometry and, as we shall see, the SI property of undoped material derives from a native point defect which is present in significant quantities only in material which is either nearly stoichiometric or which is As-rich.

Several vapour phase epitaxy (VPE) techniques, capable of yielding material of very high purity exist. The earliest of these utilizes chloride transport by flowing arsenic trichloride over molten gallium in a hydrogen carrier gas. In the related hydride technique, the gallium is transported directly as a chloride and the arsenic as its hydride, arsine. Both of these techniques operate close to thermodynamic equilibrium which means that it is possible to induce a reverse reaction at the substrate causing the wafer surface to be etched. This can lead to significant improvement in the structural and electrical quality of the layers but is not conducive to obtaining very electrically-abrupt junctions. Further, the efficiency of the process has a thermodynamic limit which generally means that there is an inefficient use of the reagents. However an exciting development of the chloride technique has just appeared which promises to overcome some of these problems [3].

The recent trend in VPE however has been strongly toward the use of organo-metallic reagents. Pioneered by Manasevets in the 1960s, the technique has enjoyed a dramatic revival and extension following demonstration that high purity layers could be obtained. The technique involves pyrolysis of a gallium alkyl (such as tri-methyl gallium – TMGa) and arsine on a heated substrate. It is therefore not a reversible reaction and takes place well away from equilibrium with high reagent efficiency. Whilst it does make more demands on the state of the semiconductor surface to achieve good epitaxy, it is well suited to the growth of abrupt junctions and quantum well structures. In this respect it is bettered only by molecular beam epitaxy (MBE) where, by directing beams of the elements from evaporation sources at a substrate surface in ultra-high vacuum (UHV), deposition control down to the atomic level can be obtained.

Ion implantation as a technique for producing a semiconducting active surface layer scores over all the above epitiaxial techniques on grounds of cost and of large area uniformity but makes considerably greater demands on substrate perfection and uniformity. However it is relatively inflexible in regard to optimization of the dopant profile. Nonetheless it is the route which has been chosen for the first generation of integrated circuits in gallium arsenide.

Before considering the above technologies in detail we first list the physical, electrical and crystallographic properties of GaAs which are relevant to its use in IC technology.

1.2 RELEVANT PROPERTIES OF GaAs

Current (1986) world consumption of gallium is around 50 tons, almost all of which is used by the semiconductor industry. It is obtained mainly as a by-product of the refining of bauxite for aluminium. At present it is extracted from only a small fraction of all the bauxite which is smelted so that the short-and mid-term future supply will be dictated by the economics of introducing new refining plant rather than by limited natural abundance. By contrast the annual production of arsenic is around five times greater, only 20% of which is used in the semiconductor industry. Obtained as a by-product of the smelting of zinc and lead ores, it costs only a small fraction of the price of gallium and its natural abundance does not impose any limitations.

Combined in equi-atomic ratios, these elements form a cubic zinc-blende (sphalerite) lattice having a high degree of covalent bonding which gives the material its semiconducting character. The phase diagram of the material is shown in Fig. 1.1. The partial pressures of gallium monomers and arsenic monomers, dimers and tetramers in equilibrium with the GaAs liquidus are shown in Fig. 1.2. The sphalerite structure consists of two interpenetrating face-centred cubic lattices one composed of the cation and the other the anion. Certain low index surfaces – {111} and {311} – terminate either in all Ga or all As atoms giving these planes a polar character. By convention {111} planes terminating in Ga atoms are designated 'A' and those terminating in As atoms 'B'.

The near equality of the atomic weights of the constituent atoms makes it difficult to determine the polar directions (e.g. (111) A as against (111) B) using X-ray techniques. However chemical etching is a cheap and ready method by which the non-equivalent {111} faces can be distinguished; the (111) A Ga-terminating face exhibits dislocation etch pits when etched in commonly used etchants whereas the (111) B does not [6]. Whereas the

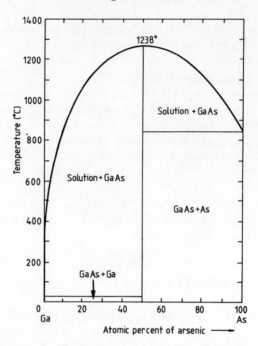

Fig. 1.1 Phase diagram of GaAs (after Koster and Thoma [4]).

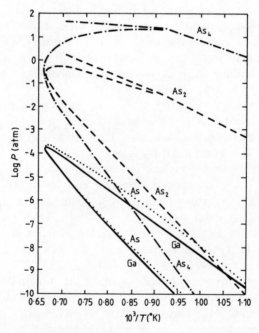

Fig. 1.2 Equilibrium partial pressures of vapour species along the GaAs liquidus (after Arthur [5]).

Group IV diamond cubic materials tend to cleave on {111} planes, in GaAs the dominant cleavage is along {110} planes. The intersections of two orthogonal sets of such {110} planes with the (100) surface of the wafer offers an easy method for dicing ICs fabricated on (100) substrates.

All crystal lattices above 0°K contain, at equilibrium, finite numbers of native point defects and in GaAs we have, in addition to vacancies on each sublattice and interstitials of both chemical types, the so called anti-site defects. The latter point defects consist of an atom on the wrong sublattice i.e. an As atom on a Ga site (denoted by As_{Ga}) or a Ga atom on an As site (Ga_{As}). The former is a much discussed entity as it is implicated in the SI nature of some undoped material (see below). The relative concentrations of the various native point defects depends on both the temperature of their formation and the arsenic activity in the melt or vapour from which the crystal is being grown. Crystalline GaAs has a finite phase extent (i.e. the vertical line drawn to represent the solidus in Fig. 1.1 has in fact a small but finite width which is strongly temperature dependent).

The net concentrations of point defects determine, at any temperature, the degree of non-stoichiometry of the crystal at that temperature. The absolute concentrations of these defects are not known but inferences can be drawn from measurements of a range of properties which, when inserted into a chemical thermodynamic model, yield estimates [7] which suggest that at the melting point of the material native point defects are present in concentrations in excess of 10^{18} cm^{-3}. One of the mysteries of the material (and of other III–V materials) is how, and in what form, these point defects agglomerate and precipitate as the crystal cools from the growth temperature to room temperature.

Dislocations, twins and other extended defects are not equilibrium defects but rather are introduced by deficiencies in the growth process or by thermal stress during cooling or during device processing.. It is the task of the crystal grower and the device processing engineer to avoid or minimize their occurrence for those applications where they produce known deleterious effects.

Physical properties of GaAs (measured at 300 K unless otherwise stated), which are relevant to its use for ICs are given in Table 1.1 (A comprehensive compilation of the properties of GaAs has recently been published [8].)

From the crystal growth viewpoint the dissociative nature of the compound is most significant. The dissociation pressure of arsenic (as As_2 and As_4 predominantly) from a stoichiometric melt at the melting point of GaAs (1511 K) is about one atmosphere as shown in Fig. 1.2 and to prevent surface loss of arsenic, the surface must be protected (e.g. by use of a boric oxide encapsulant) or equilibrated with an ambient arsenic vapour at this pressure.

Of significance to device fabrication is the congruent evaporation temperature (T_D) at which the equilibrium dissociation pressures of Ga and As are equal ($T_D \sim 630°C$). Marked dissociation occurs only at temperatures above this. The thermal conductivity of GaAs is only about one third of that

Table 1.1. Some physical properties of GaAs

Molecular weight			144.6
Melting point			1511K
Density	@300K	(solid)	$5.3165 \pm .0015$
	@1511K	(solid)	5.2
	@1511K	(liquid)	5.7
Lattice constant			5.654 A
Adiabatic bulk modulus			7.55×10^{11} dyne cm^{-2}
Thermal expansion	@300 K		6.05×10^{-6} K^{-1}
	@1511 K		7.97×10^{-6} K^{-1}
Specific heat	@300 K		0.325 J gm^{-1} K^{-1}
	@1511 K		0.42
Thermal diffusivity	@300 K		0.27 cm^2 s^{-1}
Latent heat			3290 J cm^{-3}
Band gap			1.44 eV
Refractive index @ 10 micron			3.309
Dielectric constant	(static)		12.85
	(infra red)		10.88
Electron mobility	@77 K		205 000 cm^2 V^{-1} s^{-1}
	@300 K		8500
Hole mobility	@300 K		400
Intrinsic electron concentration	@300 K		1.4×10^6 cm^{-3}
Intrinsic resistivity	@300 K		3.7×10^8 ohm cm
Electron effective mass			0.08
Lightest hole effective mass			0.082
Heavy hole effective mass			0.45
EL2 level	(by DLTS)		0.82 eV
	(by temp. dependent Hall)		0.75

of silicon and so power dissipation problems ın GaAs ICs are potentially much worse than with silicon. The low conductivity also gives rise to high thermal gradients in the crystal as it cools from the melt during crystal growth which, together with its relatively low critical resolved shear stress [9], readily leads to dislocation generation. In consequence it is proving to be much more difficult to grow dislocation free (so-called 'zero-d') GaAs than Si. Of course the other inherent deficiency compared to Si is the unsatisfactory nature of the native oxides (of Ga and As) as dialectrics or passivating layers.

The electrical properties of the material which make it highly attractive are its high electron mobility (8500 cm^2 V^{-1} s^{-1} at room temperature), its high electron drift velocity (2×10^7 cm s^{-1}) and the possibility, not possessed by silicon, of preparing it in SI form. Further, its light-emitting and photo-carrier generation properties offer the promise of an opto-electronics technology with on-chip lasers and photodetectors providing optical inter-connection.

1.3 BASIC MATERIALS REQUIREMENTS FOR ICs

The specification of wafers for integrated circuit production is a complex and evolving task closely coupled to developments in both crystal growth and fabrication technologies. Undoped SI material, having a resistivity $> 10^7$ ohm cm, is commonly specified but it is on the question of the uniformity of properties across the wafer that there is presently most interest. Current debate centres on the effects of dislocation structures on the scatter in threshold voltage of closely spaced FETs. A high electron mobility is required with values > 4000 or > 5000 cm^2 V^{-1} s^{-1} commonly specified. Required wafer orientation is $< 100 >$.

To understand the factors causing spatial variations in the electrical properties a knowledge of the mechanisms giving rise to SI behaviour is required. Chromium forms deep acceptor-like levels near mid-gap and is capable of compensating material which has a net shallow donor concentration, providing that the chromium is in sufficient excess. Material synthesized using silica boats or crucibles to contain the melt and then grown *in situ* into a single crystal tends to be very high in silicon impurity content ($> 10^{16}$ cm^{-3}) (located predominantly on the donor site Si_{Ga}^+) requiring such high levels of chromium to compensate it (to make it SI) that some precipitation of the chromium can occur. This fact together with the high solid-state diffusivity of chromium makes such material unsatisfactory for IC fabrication and the important breakthrough was the marked reduction in the silicon concentration obtained by direct synthesis of the GaAs in pyrolytic boron nitride crucibles followed by growth in the same crucible [10]. Undoped SI material has the following properties:

(a) a shallow acceptor level concentration (N_A) in excess of the shallow donor level concentration (N_D).

(b) a deep donor-like level known as EL2 present in a concentration (N_{DD}) which exceeds the net acceptor concentration *viz*:

$$N_{DD} > N_A - N_D$$

The EL2 level then pins the Fermi level close to mid-gap yielding resistivities which can exceed 10^8 ohm cm (see Fig. 1.3).

The dominant residual shallow acceptor in melt-grown material is carbon (C_{As}^-); the residual shallow donor is silicon (Si_{Ga}^+) arising from contamination from silica ware and other sources. The source of the carbon is more obscure but may well come from the graphite furniture in the puller, transported as CO.

The EL2 defect is believed to be related to the native arsenic anti-site defect (As_{Ga}) because its concentration in the crystal is dependent on the stoichiometry of the melt from which the crystal was grown. (For a review see [11].) Significant concentrations occur only for crystals grown from either As-rich or near-stoichiometric melts. Crystals grown from markedly Ga-rich melts are invariably p-type and of much lower resistivity. Carbon levels are typically in the region of 10^{15} to 10^{16} cm^{-3}.

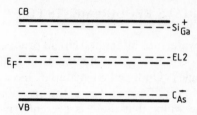

Fig. 1.3 Energy levels in unintentionally doped SI GaAs. Carbon acceptors are compensated by Si donors and the native point defect EL2 which pins the Fermi level near mid gap.

The neutral state of the EL2 defect is related to a broad absorption band at around 1 micron and studies have suggested that the distribution of EL2 around polygonized dislocation arrays is spatially non-uniform. This may contribute to non-uniformities in threshold voltage (said to be of the order of 200 mV between centre and boundary of the cell) (see Section 1.10.2).

Wafering and polishing are both exacting technologies. Firstly GaAs is much more brittle than Si and thicker wafers have to be used to reduce breakage loss to an acceptable level. Edge rounding is now common and flatness to better than a few microns across a 3 in. wafer is a firm requirement.

Dislocation density in the wafers is usually specified to be less than 10^5 cm^{-2} but this figure is related more to what can be achieved than by rational consideration of what is needed. As explained below, reduction of dislocation density is a subject of intense activity at the present time.

Ideally the device manufacturer would like to achieve the same (high) degree of activation of his implant in each and every wafer which he processes. In reality however, there is some variation from crystal to crystal and even along the length of each crystal. One possible causative factor is the variation in the residual net acceptor (carbon) concentration. The higher the carbon level, the more shallow acceptors there are to be compensated by the implanted donors before n-type semiconduction is obtained and hence the lower the apparent activation.

1.4 BRIDGMAN AND RELATED GROWTH TECHNIQUES

The basic configuration of a horizontal Bridgman equipment for the growth of GaAs is shown in Fig. 1.4. A charge of GaAs is contained in a boat, which in early work was made of silica, but which is now more commonly of pyrolytic boron nitride. A second boat holds a charge of solid arsenic. Both boats are contained within a sealed silica envelope which is placed within a multi-zoned furnace which maintains the whole ampoule at a background temperature of around 613°C thereby maintaining an ambient pressure of arsenic of around one atmosphere (composed predominantly of As$_4$ molecules). Crystallization of the charge can now be carried out either by melting a zone (as depicted in Fig. 1.4) or by employing a gradient furnace which permits melting of the whole charge with progressive solidification from one end obtained by steady lowering of the heater power. By placing a <111> oriented seed crystal at the appropriate end of the boat, single crystals can be obtained although achieving a high yield is not easy. <100> oriented seeds are not employed because only very low yields of single crystal can be obtained with this orientation. Commercial equipments tend to employ resistive heating which offers lower capital cost and maintenance than rf induction heating.

A crucial feature of the apparatus is the porous plug (E) which prevents convective transport of heat from the high temperature crystallization region to the arsenic reservoir. Without this plug, melting of a zone in the charge can induce a rise in arsenic reservoir temperature producing a catastrophic rise in the pressure in the ampoule.

Silicon take-up by contamination from the silica ware tends to make undoped material n-type and SI behaviour can be attained only if Cr doping is employed. (In some instances oxygen in the form of Ga_2O_3 is also added as a dopant.) The limitations of Cr doping for the manufacture of ICs by ion implantation has already been mentioned but use of chromium-doped substrates allied to an epitaxial technology for incorporating a buffer layer is still a common route to discrete FET production. Recently pyrolytic boron nitride boats have been used [12] eliminating the need for Cr doping. Reduced dislocation density (2–18×10^3 cm^{-2}) and higher chemical purity ($[C] < 10^{14}$ cm^{-3} and $[B] < 6 \times 10^{14}$ cm^{-3}) are claimed.

One production problem associated with the horizontal Bridgman technique is that the resulting wafers are not circular. To optimize yield,

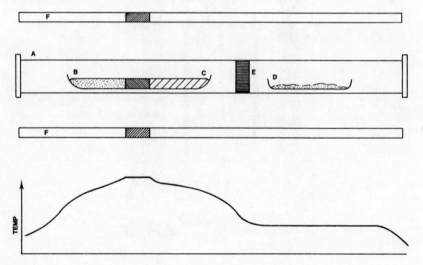

Fig. 1.4 Schematic representation of the horizontal Bridgman growth of GaAs. A, silica envelope. B, charge. C, crystallized material. D, arsenic source boat. E, porous plug. F, furnace.

crystals are grown on a [111] axis and the ingot is then sliced on an inclined (100) plane to increase the wafer area. The wafers are then restacked and waxed together and then ground to circularity at 2 in. and now also at 3 in. diameter.

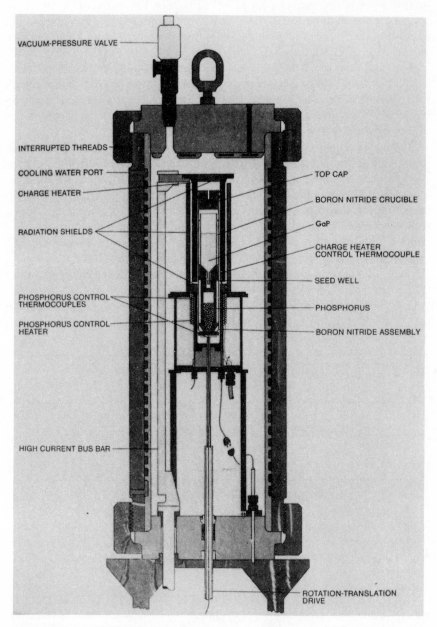

Fig. 1.5 Vertical gradient-freeze technique for the growth of III–V compounds. Shown configured for GaP but also used for GaAs [13].

The key limitations of the techniques are:

(a) Melt contamination by the quartz boats or poor crystallinity when pyrolytic BN boats are used due to wetting of the BN by the melt.
(b) Non-uniform properties across a wafer due to the inclined angle of slicing and gravity segregation and other mechanisms which cause non-uniform distribution of native point defects and chemical contaminants.
(c) The large wastage involved in rendering the wafers circular.

Its principal advantage is that size for size it yields material of somewhat lower dislocation density than does LEC growth (but see Section 1.5).

Very recently a vertical gradient freeze technique for the growth of seeded high quality single crystal GaAs (and GaP and InP) has been reported for 2 in. diameter crystals [13]. The apparatus consists of a vertical two-zone heater assembly mounted within a stainless steel-lined pressure vessel. A rather complicated growth vessel (shown in Fig. 1.5) is mounted axially within the furnace assembly. An arsenic reservoir is positioned at the base of the vessel and is heated by the lower furnace which provides control of the melt stoichiometry. The upper furnace heats a pyrolytic BN crucible having a tapered base with an extension to contain a seed crystal. A pyrolytic BN top cap is fitted to the growth vessel to minimise the rate of loss of arsenic vapour.

Much attention has been paid to the empirical optimization of the temperature field in the cooling crystal so as to minimize thermal stress and the concomitant dislocation generation by use of a combination of insulators, radiation shields and by control of heat input distribution. The growth vessel is also rotated to even out any radial thermal assymetries in the assembly. Growth is achieved by translating the growth vessel within the furnace assembly. Significant reductions in dislocation density are reported in undoped material which has a resistivity in excess of 10^7 ohm cm.

1.5   LEC GROWTH

1.5.1   Principles

If a pre-synthesized charge of GaAs is placed in a conventional crystal puller and melted, then virtually all of the arsenic will rapidly evaporate from the melt and condense on the cold walls of the puller chamber leaving, in the crucible, a pool of molten gallium.

To overcome this critical problem the LEC technique was developed by Mullin and co-workers at RSRE Malvern [2] from a principle used by Metz, Miller and Mazelsky [14] for the growth of lead chalcogenide crystals. The technique consists of floating a layer of molten boric oxide (B_2O_3) on the surface of the melt, dipping a seed crystal through the B_2O_3 and equilibrating it with the melt to achieve growth at the oxide/melt interface. Now provided that there exists an inert gas pressure in the chamber which exceeds the dissociation pressure of arsenic over molten GaAs (which is about one atmosphere at the melting point of a stoichiometric melt), then the formation of gaseous arsenic is avoided and arsenic can escape from the melt only by diffusion through the viscous boric oxide in which it is virtually insoluble. The central advantage of the technique is that, unlike Bridgman growth, the containing walls can be cooled and held at room temperature rather than at around 613°C, the temperature needed to prevent the condensation of solid arsenic in the Bridgman technique. (This advantage is most apparent when growing the group III phosphides [15], [16] where the dissociation pressures of the phosphorus are some tens of atmospheres and the design of pressure vessels operating at these pressures at temperatures in excess of 600°C are impracticable.)

It is no exaggeration to say that this simple concept of liquid encapsulation has revolutionized the production of bulk crystals of the III–V compounds including GaAs. Initially it was not favoured as a method for the production of Cr-doped SI substrates but its breakthrough came with the discovery that *undoped* SI material could be produced once the silicon contamination level was sufficiently reduced and that this could be achieved by the direct synthesis of the GaAs charge in the puller (preferably in a pyrolytic boron nitride crucible) [62] from its constituent elements. Molten gallium reacts exothermically with arsenic at around 800°C to form GaAs. However, at this temperature, the vapour pressure over solid arsenic is some tens of atmospheres. Pressure pullers, first developed by Metals Research Ltd – now merged with Cambridge Instruments Ltd – in collaboration with RSRE, were used initially for the growth of GaP. Later they were adapted for the *in situ* synthesis and growth of GaAs and have become widely used for this purpose. They are the basis of much of the research and production of undoped SI wafers at this time.

Subsequently, a method for direct synthesis in a low pressure puller was developed. The method employs a separate source of arsenic contained in an injection cell having a delivery tube which can be lowered into a gallium melt contained under boric oxide. By raising the temperature of the cell steadily, arsenic can be sublimed into the gallium melt until it has the required near equi-atomic composition. Initial enthusiasm for this low pressure technique, which avoided costly high pressure technology and permitted the use of redundant silicon crystal pullers, has been tempered by the experience of several manufacturers which has indicated that the length of boule which can be grown before the onset of polycrystallinity occurs is much smaller than can be achieved with high pressure pullers. Reasons for this are either not known or are shrouded in commercial secrecy.

In what follows therefore we concentrate on the high pressure LEC technology.

1.5.2 Technology

A photograph of the Cambridge Instruments Ltd CI358 LEC pressure puller is shown in Fig. 1.6. It is capable of routinely growing 3 in. diameter single crystals from 8 Kg melt charges. It can be operated at pressures up to 100

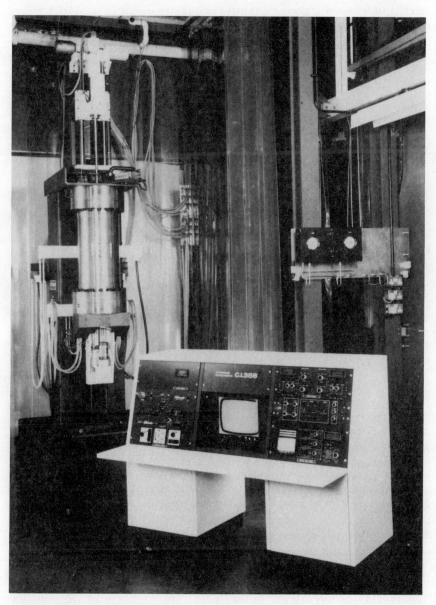

Fig. 1.6 C1358 High pressure LEC puller for the growth of 2 in. and 3 in. GaAs crystals from an 8 Kg melt. (Courtesy Cambridge Instruments Ltd.)

atmospheres and is fitted with a system for automatic control of crystal diameter developed at RSRE [17], which utilizes a crystal weight sensor. An active after-heater assembly and a superconducting magnet are optional additions.

A schematic of the process is shown in Fig. 1.7. The pyrolytic BN crucible is charged first with solid arsenic and then with rods of gallium (both typically of 99.9999% nominal purity) and the charge is capped with a pre-formed disc of boric oxide which has been purified and then dried to a controlled water content typically within the range 200 to 2000 ppm. The puller is then pressurised with dry argon.

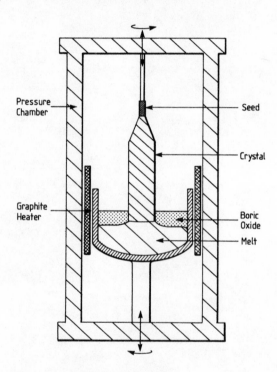

Fig. 1.7 Schematic representation of LEC growth.

A graphite 'picket fence' resistance heater is used to raise the crucible temperature first to around 450°C at which point the B_2O_3 glass softens and flows over the charge of arsenic and the, now molten, gallium. The ambient argon (or nitrogen) pressure is then raised to around 60 atmospheres and the temperature raised to about 800°C at which point the gallium reacts exothermically with the arsenic raising the temperature further. Time and care must be taken in this synthesis stage to control the rate of heat evolution. The charge temperature is then progressively raised until the whole charge is molten (the melting point of GaAs is 1238°C). The chamber pressure may now be lowered for the growth phase (typically to around 20 atmospheres). Some small loss of arsenic inevitably occurs during synthesis and there is also some evidence for the uptake of gallium by the boric oxide to a degree which is dependent on the water content of the latter [18]. The composition of the melt is dependent on these two factors and on the ratio of gallium to arsenic [Ga]/[As] in the initial charge and can vary by perhaps ± 1% or so. This lack of precise melt composition control and hence control of stoichiometry of the crystal is a present limitation of the LEC technique.

A <100> oriented seed crystal is attached to the pull rod by a chuck which is slowly lowered through the boric oxide to the melt surface and the melt temperature is carefully adjusted until a meniscus is supported by the seed crystal. A simple electrical circuit can be used to indicate when seed contact with the melt has been established. Seed crystal and crucible are both rotated at rates in the range 0 to ± 20 rpm determined empirically by each user. (Counter rotation of seed and crucible is common.) The crystal is pulled from the melt at a rate of usually less than 1 cm hr^{-1}. Drainage of the boric oxide from the crystal as it emerges from the melt is slow so that the crystal can be encapsulated in the molten glass. This reduces surface degradation of the crystal due to evaporative loss of arsenic.

Pyrolytic BN crucibles are very expensive (typically of the order of $5000 for a crucible suitable for the growth of a 3 in. crystal), but with care in cleaning and in handling it can be used ten or more times. Quartz crucibles are a factor of ten cheaper but can in general be used for a single run only. If quartz is used there is uptake of silicon by the melt but this can be suppressed (maintaining Si levels at below 10^{15} cm^{-3}) by the use of boric oxide with a relatively high water content (see Section 1.10.1). However this tends to increase the incidence of twinning in the crystal and a compromise has to be reached.

The diameter of the growing crystal can be controlled either by manual adjustment of heater power or performed by an automatic control system which measures the time rate of change of weight of the crystal (or the melt) and compares it with a reference signal, the difference signal being used as the error signal for a servo-controller. Unfortunately, because of 'anomalous' surface tension and buoyancy effects exhibited by GaAs (and by other III–V and IV semiconductors) weight is an ill-conditioned measure of crystal diameter and sophisticated signal processing must be performed on the error signal in order to obtain stable control [19]. Typical ADC-grown crystals are shown in Fig. 1.8. The use of alternative sensors such as optical imaging or meniscus angle sensing are impeded by the presence of the encapsultant. The automation of the pulling technique is currently being extended to full computer control of the process from seed-on to cool-down.

Recently both transverse and longitudinal steady magnetic fields have been employed in research to suppress growth rate fluctuations caused by turbulent convection in the melt. Interest in this work has been raised by attempts to harden the crystal lattice by doping with significant quantities of indium (0.1–1%) [20]. The magnetic field serves to remove the micro-inhomogeneities caused by melt turbulence. Superconducting solenoids are

Fig. 1.8 Single crystals of SI GaAs: <100> orientation 2 in., 3 in. and 4 in. diameter. (Courtesy ICI Wafer Technology Ltd.)

favoured because of the large bore size required. Useful field strengths appear to be in the range 0.1 to 0.5 tesla.

1.5.3 Ingot annealing

As-grown ingots have macroscopic and microscopic inhomogeneities in point defect concentrations and in electrical properties (see Section 1.10). It has been found that significant improvements in the uniformity of the electrical properties (resistivity and carrier mobility) across a wafer can be obtained by annealing the whole ingot in a sealed quartz ampoule (to prevent arsenic loss) at temperatures broadly around 1000°C and for times of some hours. (Actual values differ from supplier to supplier and tend not to be disclosed.) The macroscopic and microscopic distribution of EL2 in its neutral state is made much more uniform by this treatment.

1.6 WAFER FABRICATION AND SPECIFICATION

Of equal importance to the growth of the ingot is the wafering and polishing of that ingot. Indeed the costs associated with the latter generally exceed those of the former. Gallium arsenide wafer suppliers are fortunate in being able to draw on the vast experience of the silicon industry but GaAs brings some specific problems of its own. Most obvious amongst these is the greater brittleness of GaAs. This necessitates use of thicker wafers, typical values being 400 or 500 micron for 2 in. diameter wafers and 500 or 625 micron for 3 in. ones. The ingots are wafered using ID saws having diamond-edged cutting blades.

Integrated circuit applications are the most demanding in terms of structural and dimensional tolerances. Total thickness variations across whole wafers are generally required to be only a few microns. Precision edge rounding of the wafers is also now standard. Wafer surface orientations are specified to an accuracy of a small fraction of one degree and are either oriented to coincide with a (100) plane or are deliberately misoriented along a <011> direction by several degrees in order to accommodate the requirements of an epitaxy process. When the wafer is to be used as a substrate for the growth of an epitaxial layer, the required wafer orientation is dependent on the nature of the epitaxy (i.e. whether homo- or hetero-epitaxy) and on the epitaxial technique. Quite large misorientations are sometimes used for hetero-epitaxial structures as it has been found that these minimize the dislocation density arising from the lattice mis-fit and ensure the avoidance of anti-phase domains in hetero-epitaxy (see Section 1.7.6.).

Superficial (i.e. in-plane) orientation of the wafers is denoted by grinding a pair of flats on the edge of the ingot after it has been centreless ground to the required diameter and before wafering. These flats are of differing length (denoted 'primary' and 'secondary' facets) and are parallel to < 110 > directions.

Polishing of the sawn wafers is largely an undisclosed art. Whether to etch the wafer before polishing, or after polishing, or not at all; what polishing pad pressures to use and what kinds and grades of abrasive to use are all questions discussed only behind the closed doors of each supplier. Difficult trade-offs have to be established. Thus heavy pad pressure ensures flatness but leaves a higher level of residual damage which may subsequently reveal itself as a low activation of the ions implanted into it, especially if the energy and/or the dose of the ions is low. Conversely, a low pad pressure increases polishing time and can give inadequate overall flatness.

Whether to polish a single or both sides of the wafer is a further issue.

Initially single sided polishing was the norm but the demand for double sided polishing has grown steadily. Sophisticated – and costly – machines exist which polish both sides of the wafer simultaneously by imparting a planetary motion to a set of wafers held in a ring between a pair of differentially rotating pads. The practice employed in the silicon industry of deliberately introducing damage into the back surface of the wafer to 'getter' heavy metal impurities has, to date, received only little application to GaAs.

Finally, wafer cleaning, inspection, testing and packaging are operations requiring sophisticated equipment and at least Class 100 clean room conditions.

International standard wafer specifications are becoming established which are helping to bring rationalisation to the industry. However these specifications are not in mutual accord as to the location of the 'edge' facets. A typical manufacturer's specification for 3 in. SI materials is given in Table 1.2. State of the art 2 in. and 3 in. wafers are shown in Fig. 1.9.

Fig. 1.9 2 in. and 3 in. diameter polished wafers of SI GaAs. (Courtesy ICI Wafer Technology Ltd.)

1.7 EPITAXIAL TECHNIQUES

1.7.1 Introduction

Close control of both the doping and the thickness of the active region of any semiconducting device structure is vital to the specification of device

Table 1.2. Example of Gallium Arsenide Wafer Specification – 3 in. Polished Wafers

Single crystal gallium arsenide free from twins, lineage or cracks	
Diameter	76.2 ± 0.4 mm
Orientation	$(100) \pm 0.25°$
	or
	(100), $2° \pm 0.25°$ off towards nearest (110)
Thickness	500 or 625 micron $\pm$ 25 micron
Orientation flats	Major flat – parallel to (110) direction $\pm 0.5°$
	Minor flat – 90° in clockwise* direction from major flat when facing polished surface
Flat lengths	Major 20 mm
	Minor 10 mm
Edge rounding	All wafers to be edge rounded to (SEMI standard for silicon wafers)
Surface finish	Polish surface – free from scratches, haze and 'orange peel'
Backside	Etched

Flatness
TIR maximum 5 micron (slice held on vacuum chuck)
Warp
Maximum 30 micron (without vacuum)
Etch pit density
Maximum 10^5 cm^{-2} (5–point average at r/8, r/2 and centre on a (110) diagonal)
Slice numbering
All slices to be numbered in order from seed end of crystal

Gallium arsenide wafer specification – Electrical properties
(i)   *Undoped*
      Resistivity greater than 10^7 ohm cm
      Mobility greater than 5000 cm^2 V^{-1} s^{-1}
(ii) *Lightly chromium doped*
      [Cr] 5×10^{15} to 1×10^{16} cm^{-3}
      Resistivity greater than 10^7 ohm cm
(iii) *Heavily chromium doped*
      [Cr] 5×10^{16} to 1×10^{17} cm^{-3}
      Resistivity greater than 10^8 ohm cm
      Resistivity measured at 300°K $\pm 2°$
      Contacting – alloyed contents to be prepared at 400°C in forming gas atmosphere containing HCl vapour using tin dots
      Thermal conversion test
      Uncapped proximity anneal in As vapour or AsH$_3$
      Temperate 850°C
      Time 30 mins
      Resistivity of wafer to exceed 10^7 ohm per square after test

---

* Orientation flats: Major flat to be parallel to intercept of (111) A plane and (110) plane i.e. to small (Ga) facet on crystal boule.

performance. This is especially true of digital ICs where the characteristics of thousands of discrete devices must lie within rather close limits (e.g. variations in threshold voltage must lie within a few tens of millivolts or less). The low diffusivites of substitutional dopants and the propensity of the surface to dissociate at high temperatures makes solid state diffusion unreliable and epitaxial growth techniques and ion implantation are used instead.

The most utilized epitaxial techniques are liquid phase epitaxy (LPE) and vapour phase epitaxy (VPE) by chloride and hydride routes and these are the basis of the production of many opto-electronic devices (lasers and light-emitting diodes for example) and discrete microwave devices respectively. However both lack the capability to achieve extremely abrupt interfaces or controlled doping profiles over nanometric dimensions and the newer techniques of metal organic vapour phase epitaxy (MOVPE) and molecular beam epitaxy (MBE) are being developed intensively to meet the challenges posed by high electron mobility transistors (HEMTs), heterojunction bipolars, modulation doped-FETs and various quantum well structures. However a recent development of the chloride process by workers at AT&T Bell Laboratories promises an extended life for chloride epitaxy [3].

1.7.2 Liquid phase epitaxy

This technique is of historical interest in that the first GaAs digital ICs were fabricated by Hewlett Packard using this technology [21], [22]. The method consists of an arrangement whereby gallium, slightly supersaturated with arsenic is placed over a substrate surface on to which the layer is to be grown and growth is promoted by a careful and steady reduction in temperature. It is terminated by removing the substrate from the solution. Methods for achieving this are colloquially known as 'tipper', 'dipper' and 'slider' and are illustrated in Fig. 1.10. The mode of operation is hopefully self-evident; the 'slider' system, which is the most widely used, permits growth of multi-layer structures by passing different solutions successively over the substrate. Knowledge of the solid–liquid phase equilibria involved is necessary in order to control the process.

Used with careful attention to pre-baking treatments [23], [60], the technique is capable of producing n-type, but closely compensated, buffer layers with carrier concentrations below 1×10^{14} cm^{-3}. The process of 'wipe off' – i.e. the removal of the solution from the substrate at the end of growth – is crucial for IC fabrication since any surface imperfections can damage contact lithographic masks. Non-planarity due to the formation of meniscus lines, terraces and pits can occur; accuracy in orienting the substrate surface is important in minimizing these defects.

The introduction of computer control of the slider motion has made superlattice fabrication possible but not with the ultimate fine control obtainable with MBE. The technique also suffers from the disadvantage that

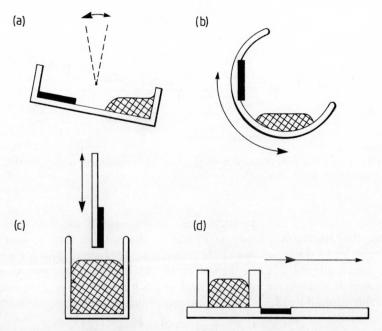

Fig. 1.10 Various configurations used in LPE. (a) 'Tipper': substrate on left is immersed in solution by tipping boat. (b) as (a) but with modified boat geometry. (c) 'Dipper': substrate attached to quartz or pyrolytic BN rod is dipped into solution. (d) 'Slider': (most commonly used) solution is translated over substrate which is embedded in a well.

it does not readily lend itself to the high throughputs needed for commercial production.

A useful and recent review of the technique has been given by Greene [24].

1.7.3 Chloride and hydride VPE

High purity buffer layers can also be grown by chloride epitaxy and this forms the basis of much discrete FET manufacture. The technique exists in two variants – liquid and solid source. With the liquid source technique the arsenic is obtained from arsenic trichloride and the gallium from a boat of liquid gallium. With the solid source technique the liquid gallium is replaced by a solid gallium arsenide source so that, in this case, the arsenic is derived from both the $AsCl_3$ and the GaAs source.

The apparatus is shown schematically in Fig. 1.11. Arsenic trichloride is contained in a thermostatically controlled bubbler and hydrogen passed through the bubbler to give 10^{-3} to 10^{-2} mole fraction $AsCl_3$. This gaseous mixture is flowed over the source containing either liquid gallium or gallium

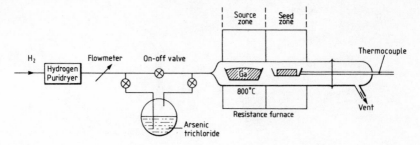

Fig. 1.11 Schematic representation of the chloride epitaxy of GaAs showing Ga source boat and AsCl₃ bubbler.

arsenide to give gallium monochloride. In the case of the liquid gallium source the arsenic is initially taken up by the gallium until it becomes saturated and a thin crust of gallium arsenide forms upon it. This is referred to as source saturation. Once the source is saturated both GaCl and As_2 pass downstream and over a substrate held at a somewhat lower temperature (typically around 750°C) where deposition occurs according to the reaction

$$2GaCl + As_2 + H_2 = 2GaAs + 2HCl$$

Thus the process utilizes the reversibility of the reaction which proceeds to the left at high temperatures and to the right at the lower (deposition) temperature. (The complete reaction sequence is actually somewhat more complicated involving significant concentrations of gallium trichloride and arsenic tetramer.)

The electrical purity of the layers is found to be strongly dependent on the molar fraction of arsenic trichloride flowing into the reactor. At low molar fractions (10^{-3}) carrier levels in the 10^{16} cm^{-3} range are obtained, but increasing the molar fraction up towards 10^{-2} can, in a well-designed system, reduce the carrier level to 10^{13} cm^{-3} or below. This 'molar fraction' effect (Fig. 1.12) is believed to be due to attack of the silica reaction tube by the HCl yielding volatile chloro-silanes which are reduced by hydrogen in the vicinity of the substrate to produce doping of the crystal by silicon donors [25]. It can be shown that the expected uptake of silicon should vary as the $-n$th power of the arsenic trichloride input pressure where $n = 2$ to 3 in agreement with observation. Thus by using a high molar fraction, very pure layers can be obtained. Controlled doping (e.g. with sulphur) is possible using doping lines.

A related process, known as the hydride process, utilizes arsine instead of arsenic trichloride as a source of arsenic and transport of gallium from a gallium-containing boat is obtained using HCl gas in hydrogen. The AsH_3 is added to the flow downstream of the gallium source boat. The process has the advantage of greater flexibility but the purity of available HCl can be a problem.

Both techniques are limited by the fact that they are not suited to the growth of GaAlAs because of the extremely high reactivity of aluminium trichloride which attacks the silica reactor walls.

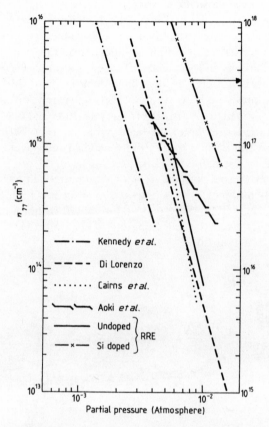

Fig. 1.12 Carrier concentration vs. arsenic trichloride partial pressure – the 'molar fraction' effect in undoped GaAs layers grown by chloride epitaxy. Several laboratories have all reported marked reduction in carrier concentration with increasing partial pressure of AsCl₃. Note that Si-doped material behaves similarly.

1.7.4 Metal organic VPE

The last few years have seen an almost meteoric rise in interest in metal organic vapour phase epitaxy (sometimes referred to as organo-metallic vapour phase epitaxy – OMVPE or as metal-organic chemical vapour deposition – MOCVD). Problems of reagent purity, which hampered the earlier development of the technique, have been largely overcome at least for the basic reagents such as tri-methyl gallium (TMGa) and arsine (AsH_3). GaAs with 77 K electron mobilities of 137 000 cm^2 V^{-1} s^{-1} have been obtained [26].

The technique offers a spectrum of advantages of which its flexibility and ready scaling-up to the production environment are perhaps the most important. The pyrophoricity of some of the metal alkyls and the severe toxicity of the Group V hydrides are however counter factors. The technique is being widely researched for the production of heterostructure and quantum well devices (including superlattices) as it is capable of use with all combinations of (Al, Ga, In): (As, P). Precise control of layer thickness and abrupt interfaces have been demonstrated. Most activity is in the opto-electronics field however and this brief review, directed toward IC fabrication, is confined to the growth of GaAs, GaInAs and GaAs/Ga_xAl_{1-x}As layers. Much broader reviews have been published (e.g. Stringfellow [27]). The first demonstration that MOVPE grown GaAs epi-layers could be used to fabricate FETs was made by Hirtz *et al.* [28].

A widely-used research scale reactor (commonly known as the Bass reactor after its designer – Bass [29]) is shown in Fig 1.13. Substrates are placed on a r.f. heated graphite pedestal mounted within a near-horizontal water-cooled quartz reactor cell. Separate inlets are provided for arsine/H_2 and TMGa/H_2 reactant flows.

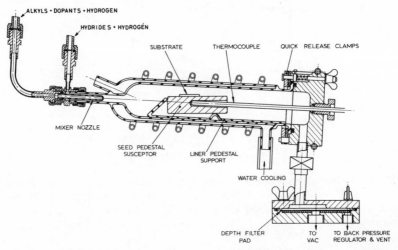

Fig. 1.13 'Bass' MOCVD reactor. The susceptor is rf inductively heated [29].

The deposition reaction is pyrolytic, the thermally unstable alkyls and hydrides being decomposed in the vicinity of the heated substrate. It is of course vital that the deposition reaction occurs heterogeneously on the substrate; gas phase reaction results in a dust of the precipitating phase.

The basic requirements of the organo-metallic reagents are that they have vapour pressures of at least 1 torr at temperatures near to room temperature and that they thermally decompose to yield up the required Group III or V element at the temperature of epitaxy (600–700°) for GaAs. Thus TMGa is a liquid which boils at 55.7°C. Its partial pressure in the reactor is controlled by carefully controlling the temperature of a bubbler through which hydrogen carrier gas is passed. Tri-methyl aluminium (TMAl) melts at 15.4°C and is also used in a bubbler.

Tri-methyl indium (TMIn) is a solid at room temperature and hence has either to be sublimed in a hydrogen stream or heated to above its melting point in a bubbler with downstream dilution with additional hydrogen. To prevent subsequent condensation of the TMIn, all the reactor components downstream of the In source have to be heated.

AsH_3 is usually supplied as a 1–10 vol % mixture in hydrogen. Tri-methyl arsenic (TMAs) is an alternative, less toxic, source of arsenic though it pyrolyses less readily than AsH_3 [27].

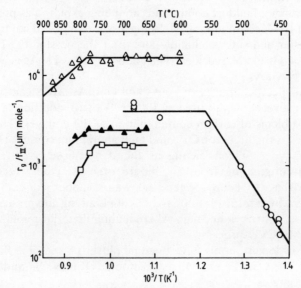

Fig. 1.14 Growth rate (normalized by the molar flow rate of TMGa) of GaAs MOCVD layers as a function of reciprocal substrate temperature; measurements from 4 different laboratories [27].

Growth rates of the order of 10^{-5} cm s^{-1} are achieved with total flow rates of several litres per min in reactors having cross sectional areas of a few sq. cm. The growth rate is relatively insensitive to the substrate temperature in the range where high quality epitaxy is obtained (Fig 1.14) and control of the latter to $\pm$ 5°C is generally adequate. Decomposition of the Group III alkyl is believed to take place in the gas phase adjacent to the substrate:

$$(CH_3)_3 Ga(g) + 3/2H_2 (g) = Ga(g) + 3CH_4(g) \qquad (1)$$

The cracking of the hydride by contrast is believed to be catalysed by the GaAs substrate, occurring heterogeneously on it:

$$AsH_3(g) = 1/4As_4 + 3/2H_2 \qquad (2)$$

The growth rate is controlled by the flux of gallium to the surface:

$$Ga + 1/4As_4 = GaAs \qquad (3)$$

Since reaction (1) goes to completion, the growth rate is controlled by the TMGa flow rate, AsH_3 being provided in large excess. The actual Group V excess is varied empirically in order to optimize the purity or electrical properties of the layer.

For mixed Group III sublattice crystals (e.g. $Ga_x Al_{1-x} As$) the composition is controlled by the ratio of the fluxes of the Group III precursors (TMGa and TMAl). Composition control (i.e. control of the value of x) requires very tight control of the temperature and the flow through the bubblers [30]. Typically, to control x to ± 0.01 requires bubbler temperature control to $\pm 0.2°C$.

A whole range of adduct compounds are being explored as precursors for MOCVD of III–V compounds [31]. Being less reactive (and not pyrophoric), they are easier and safer to handle and offer the possibility of dispensing with the toxic Group V hydrides. The adduct TMGa–TMAs is available for the growth of GaAs.

Whilst the electrical purity of GaAs and GaInAs layers can be very high (with carrier levels down to the low 10^{14} cm^{-3} level), addition of aluminium results in problems of carbon contamination of the epilayer. The carbon is incorporated as the acceptor C_{As}^- and high aluminium content ternaries are strongly 'p' type. Silicon, acting as the donor Si_{Ga}^+ is often a major contaminant originating from an impure organo-metallic reactant. It is vitally important to keep oxygen and water vapour contaminant levels extremely low by careful attention to reactor plumbing and pre-conditioning. Even when all this is done, high Al-containing material remains the weak link in the MOCVD process.

A range of dopant atoms have been employed: S, Se, Te, Si and Ge as donors and Zn, Cd, Be and Mg as acceptors. Of these Se and Zn are the most commonly used.

The widely used 'Bass' design of reactor is not well suited to scaling to a production size and several alternative reactor configurations are under development. These include barrel reactors (similar in concept to those used widely in silicon technology) and new designs of horizontal reactor where special attention has been paid to entrance geometry and thermal profile in order to achieve growth and dopant uniformity over large substrate areas.

A variety of gas-switching manifolds have been devised to achieve the rapid gas composition changes needed to produce abrupt junctions and control of graded compositions for High Electron Mobility Transistor (HEMT) and other structures. These involve reducing the 'dead' space in the reactor. An alternative approach is the use of reduced pressure in the reactor [32] which obviates the thermal convection effects present in atmospheric pressure reactors but does limit attainable growth rates.

The HEMT structures require highly 'n'-type doped layers of GaAlAs grown, with a very abrupt junction, on to a layer of high purity (and high mobility) GaAs. Aided by the near equality of the lattice parameters of GaAs and AlAs this has proved possible, but the reverse process of growing GaAs on to GaAlAs always results in material with poor mobility.

1.7.5 Molecular beam epitaxy

In essence molecular beam epitaxy (MBE) is the controlled evaporation of one or more neutral thermal atomic or molecular sources in the form of beams directed at a heated substrate which is maintained in ultra high vacuum. It utilizes a multiple chamber stainless steel assembly which is usually ion- and cryo-pumped and is fitted internally with large areas of liquid-nitrogen-cooled cryopanels. A state of the art research system is shown in Fig 1.15. Residual background pressures are typically 10^{-11} torr and this permits high quality growth to occur at quite low temperatures (550–680°C for GaAs and 630–680°C for GaAlAs). Relaxing the requirements on the vacuum is not possible since a reactive species having unity sticking coefficient will give monolayer coverage of a surface within one second at a vacuum of 10^{-6} torr.

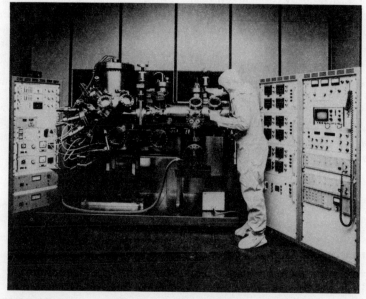

Fig. 1.15 V80H MBE reactor for the growth of GaAs and GaAlAs layers. (Courtesy VG Semicon Ltd.)

The atomic or molecular beams are generated in evaporation ovens. (The words Knudsen effusion sources are sometimes used but, in practice, the orifices have to be so large, in order to achieve the desired growth rates, that Knudsen conditions – i.e. 'thermodynamic' equilibrium within the oven – are not attained.) The ovens (which are commonly made of pyrolytic boron nitride) together with the resistance heaters used to heat them are the major source of contaminants. Traversable shutters are used to turn on and off the beams. The provision of a vacuum interlock – which obviates the need to take the growth chamber up to atmospheric pressure each time the sample cassette is loaded – is vital to the routine attainment of high purity layers.

Atom fluxes are monitored with an ion gauge and research machines are usually fitted with a range of diagnostic tools such as reflection high energy electron diffraction (RHEED) and Auger electron spectroscopy together with a quadrupole mass spectrometer for residual gas analysis.

Meticulous attention must be paid to substrate surface preparation. A common method is to use an oxidising etch which removes some deposited contaminants leaving the surface passivated with oxides which can be removed *in situ* in the MBE equipment by heating in an arsenic beam. Alternatively, the surface can be cleaned *in situ* using a uv/ozone process. The state of the surface so prepared is checked using RHEED.

The selected growth temperature for GaAs is commonly in the range 550–700°C and growth is initiated by first directing the arsenic beam at the surface followed by the gallium beam. Growth is carried out under As-rich conditions with an As flux typically 3–5 times greater than the Ga-flux. Under these conditions it is the latter which determines the growth rate. Growth rates are typically around 1 micron hr^{-1} (which is approximately one atom layer per second) and a high degree of spatial uniformity across the wafer is obtained by rotating the substrate.

For reasons to do with the atomic kinetics of the deposition process it was common to have the arsenic beam in the form of the dimer molecule As_2. If elemental crystalline arsenic is used as the source, then the beam is composed predominantly of the tetramer As_4. This can be 'cracked' to the dimer:

$$As_4(v) = 2As_2(v) \qquad (4)$$

by using a two-zone oven in which the tetramer beam generated in the first oven is passed into a second furnace at a much higher temperature. The temperature of the first oven is used to adjust the flux to the desired value; the second merely ensures cracking to the dimer molecule. An alternative method by which a dimer beam can be obtained is to use, as a source, a quantity of high purity GaAs as this will produce dimers directly when heated. Layers grown using a tetramer source were found to contain much higher concentrations of unwanted deep levels. However improvements in the purity of the source arsenic and in the background vacuum have changed this situation such that contamination from the tetramer 'cracker' oven, which has to be run at a very high temperature, is now the more serious problem and the trend has been to revert back to the use of tetramer sources.

Alloy layers (such as $Ga_xAl_{1-x}As$) involving a pair of Group III atoms can be grown as for GaAs described above if the kinetic factors which determine the ratio of the Group IIIs (the Ga/Al ratio) are known. At low temperatures the sticking coefficients will be unity in the presence of an excess arsenic flux (in the case of Ga, at temperatures below about 630°C) and the composition of the layer is determined simply by the ratios of the fluxes of the group III elements.

Alloys with mixed group V elements are more problematic in that one group V element will tend to be much more strongly absorbed than the other. In this situation one needs to grow with an excess of the more weakly absorbed group V setting the ratio of the Ga flux to that of the other group V at the value required in the layer.

A variation known as MOMBE – an acronym for metal organic molecular beam epitaxy – has recently been reported in which the usual 'Knudsen' sources are replaced by gaseous sources – for example arsine and a gallium alkyl such as trimethyl gallium $Ga(C_2H_5)_3$.. This avoids the problem of source depletion which requires breaking of the vacuum to reload the ovens and also the problems of contamination associated with those ovens. By careful nozzle design, improvements in growth rate uniformity across the wafer appear possible. The abruptness of the interface attainable is not degraded by changing to gaseous sources.

Dopant beams can be produced in similar ways to the matrix constituent beams – i.e. using Knudsen ovens or gaseous sources. Silicon and tin are the favoured donor dopants with beryllium for acceptor doping. Low sticking coefficients rule out the use of many of the commonly encountered donor and acceptor dopants.

The growth of high purity buffer layers has been demonstrated [33]. Carrier levels down to the 10^{14} cm^{-3} range with 77 K mobilities in excess of 160 000 cm^2 V^{-1} s^{-1} have been obtained.

These dopants can be used to make modulation-doped GaAs/$Ga_xAl_{1-x}As$ heterostructures in which narrow quantum wells (typically of the order of 100 Å) of undoped GaAs are formed in the undoped GaAs layer abutting the Si-doped $Ga_xAl_{1-x}As$ ($x > 0.2$) layers. Transfer of carriers from the donors in the doped GaAlAs to the GaAs quantum wells results in extremely high electron mobilities in the quantum wells because of the absence of ionised charge centres there. A wide range of devices have been demonstrated in MBE-grown structures including HEMTs operating at 60 GHz.

1.7.6 Epitaxial GaAs on Si

Very recently considerable advances in the quality of epitaxial layers of GaAs grown on Si substrates have been made and useful device operation has been demonstrated.

Two major problems posed by this technology are the large lattice misfit between Si and GaAs (around 4%) and the difference in the thermal expansion of the two materials. The larger expansion coefficient of GaAs produces tensile stresses in the film as it cools from the growth temperature to room temperature. The large lattice parameter difference leads to the formation of misfit dislocations and other defects and conditions must be found for which these do not propagate into the epitaxial layer.

The lower lattice symmetry of the zinc-blende lattice of GaAs compared to the diamond-cubic Si substrate allows the formation of anti-phase domains. Imagine, for illustration, an atomically flat (111) substrate surface; the first monolayer of GaAs can be formed either with As bonded to the substrate or with Ga. The problem arises for every growth direction for which the GaAs lattice is polar (including the commonly required <100>). Domains of each type tend to form and, in an attempt to avoid this, the practice is to absorb As over the whole surface before growth is initiated and then to grow slowly under As-rich conditions. A two-step growth sequence is favoured in which an initial buffer layer is produced by slow growth at a very low temperature. This is followed by an overlayer growth under conventional homo-epitaxial conditions. Success has been reported using both MOCVD and MBE techniques.

Akiyama *et al.* [34] have grown single domain layers onto a near <100> oriented Si substrate in a low pressure MOCVD reactor. To remove oxide from the substrate it was dipped in HF immediately prior to insertion in the reactor and was subsequently heated to above 900°C in a flow of H_2 and AsH_3. A buffer layer of about 200 Å thickness was grown at around 450°C followed by an overlayer at a conventional growth temperature of 700–750°C. An offset of the substrate wafer orientation from <100> was necessary to achieve single domain growth. This offset was only 0.2° in a <011> direction; much smaller than employed by most other workers. The layers were n type with carrier levels of around 10^{16} cm^{-2} with a room temperature mobility of 5200 cm^2 V^{-1} s^{-1}. One micron gate small signal and power MESFETs having good characteristics were fabricated on the layers. However the high tensile stress in the layers limited the epitaxial thickness that could be obtained without cracking. This problem has been addressed by Soga *et al.* [61] who employed a superlattice buffer layer composed of GaP and $GaAs_{0.5}P_{0.5}$ layers. They were able to grow crack-free layers 8 micron in thickness having mobilities around 4000 cm^2 V^{-1} s^{-1}.

MBE growth has been successfully employed by Koch *et al.* [35]. The substrate was misoriented from <100> by 3.7° along an in-plane <011> direction. An As pre-layer was first deposited followed by growth of a 100 nm buffer layer at a temperature of 405°C. The overlayer was then grown at an As_4/Ga ratio of 3.3 and a temperature of 575°C.

These developments in hetero-epitaxy are potentially extremely important since they offer the possibility of integrating GaAs and Si technologies on a single chip. Using MBE, Fischer *et al.* [36] have fabricated GaAs/GaAlAs MODFETs on a Si wafer containing NMOS devices without significantly degrading the performance of the latter. However the technological problems that have to be overcome in controlling dislocation density, grown-in stress and anti-phase domains on the route to a high yielding, high performance commercial process remain formidable.

1.8   ION IMPLANTATION

1.8.1   Introduction

Ion implantation directly into the substrate has emerged as the principal technology for the production of GaAs integrated circuits. There are a number of factors which have led to this situation of which the cost-saving obtained by the avoidance of the epitaxy stage is perhaps the most important. Other important advantages include:

(a) The total number of doping atoms (the fluence) can be precisely controlled by control of the beam exposure time.

(b) The depth of the doping profile can be controlled by varying the beam energy and, for the case of shallow layers, the dose.

(c) Spatial uniformity over the wafer. This is, to a degree, dependent on the scanning mechanism of the implanter. Non-linearities in electrostatically scanned systems can produce under or over-scanning gradations. Cosine-law variations are also possible but generally commercial machines are capble of uniformities of better than one or two per cent over 2 in. and 3 in. wafers.

(d) Selected areas can be implanted to define individual components.

(e) The process is compatible with a planar technology using fine line lithography.

1.8.2   Implantation

For ion species implanted into an amorphous target, the depth distribution of the species $N(x)$ is Gaussian:

$$N(x) = N° \exp\left[-(x-R_p)^2/2\,(\Delta R_p)^2\right]/(2\,\pi)^{2/3}\,(\Delta R_p) \tag{5}$$

$N°$ is the ion dose or fluence. The distribution is characterized by the projected range R_p and its standard deviation ΔR_p. Values for GaAs at implantation energies of 20 and 400 KeV are given in Table 1.3. The same distribution is obtained approximately for crystalline lattices also provided that the beam has been oriented with respect to the crystallographic axes of the target in such a way as to avoid the channelling of ions along low index, low atomic density channels (such as the $<110>$ direction in the zinc blende lattice) in the crystal. To achieve this, the ion beam is commonly misoriented by $7°$ from the normal to the $<100>$ wafer surface along a $<011>$ direction (Fig. 1.16). In addition the wafer can be rotated azimuthally by from $10°$ to $25°$. Avoidance of channelling of the incident ions can also be effected by implanting through a nitride cap which serves to partially randomize the beam before it enters the GaAs surface.

Implantation with $^{29}Si^+$ ions is used almost exclusively to produce the 'n' channel region of a GaAs FET. The energy and dose required depend on the application – monolithic microwave or digital IC. MMICs represent the

Table 1.3. Projected Range and Standard Deviation of Projected Range of Ions in GaAs. (Taken from Donnelly [58].)

Ion	20 keV		400 keV	
	R_p	ΔR_p	R_p	ΔR_p
p-type				
Be	0.062	0.041	1.092	0.205
Mg	0.022	0.014	0.453	0.127
Zn	0.011	0.006	0.157	0.064
Cd	0.009	0.004	0.098	0.040
n-type				
Si	0.018	0.012	0.351	0.121
S	0.017	0.011	0.307	0.110
Se	0.011	0.005	0.137	0.056
Sn	0.009	0.004	0.095	0.038
Te	0.009	0.004	0.092	0.036

Values are in microns.

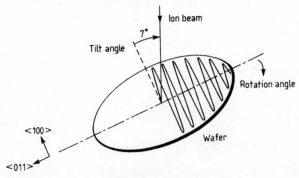

Fig. 1.16 Geometry of ion beam rastering configured to avoid ion channelling.

least demanding application since the large threshold voltages used go with a deep 'n' channel. By contrast, low energies giving shallow channels are required for digital applications and the activation of such implants is sensitive to residual damage and contamination of the wafer surface. The doping profile is fixed by the diffusive spread of the implant profile upon annealing. However, by the use of multiple implants of differing energies some tailoring of the profile is possible. For 'n' channel formation, fluences of 10^{12} to 10^{13} cm^{-2} are required. n$^+$ contacts can be formed by using fluences up to about 10^{15} cm^{-2}.

The profile of a dual-energy ^{29}Si implant into an undoped SI substrate is shown in Fig. 1.17. Shown dashed are the so-called LSS profiles (after Lindhart, Scharff and Schiott) for the two separate implants (which are plots of Equation (5) above for the respective values of R and ΔR_p). The full

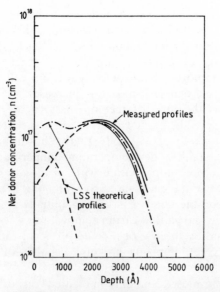

Fig. 1.17 Profile of a Si dual-implant through a Si₃N₄ cap (implant energies of 125keV and 325keV) into undoped Si GaAs [37].

lines are the actual distributions of carrier concentration measured by C–V profiling from which it can be seen that there is very little diffusion broadening. By comparison Fig. 1.18 shows a similar situation but with a chromium-doped SI GaAs substrate. The relatively poor reproducibility of the profiles is to be noted.

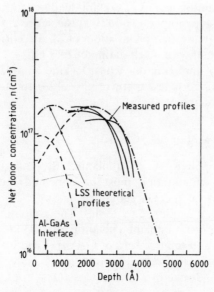

Fig. 1.18 As Fig. 1.17 but with implantation into a Cr-doped substrate [37].

A further application for ion implantation is to use proton bombardment to produce device isolation in a planar structure. This helps to reduce the unwanted device-to-device interaction known as 'backgating'. For applications requiring deep SI layers having good thermal stability, multiple implants must be employed. Unfortunately it is difficult to limit the lateral spread of the proton damage since this increases, as a fraction of the projected range, as the ratio of the ion mass to the target atom mass decreases. Lateral spreads of a few microns are associated with implants of MeV protons in GaAs. Speight *et al.* [38] have shown that isolated regions which are thermally stable at device operating temperatures of $\sim$ 250°C can be produced by multiple implantation with an initial energy of 0.9 MeV reducing by 0.1 MeV at each successive implant. The dose necessary depends of course on the number of carriers initially present in the GaAs. Recently it has been shown that carrier removal rates at least twenty times greater are obtained by the use of a deuteron beam rather than a proton one. Boron or oxygen implants are also capable of generating semi-insulating regions.

Whilst $^{29}Si^+$ is the most commonly implanted ion, other species can be used to produce both n and p type regions. Mg^+ and Be^+ are suitable acceptor implant ions. JFETs have been produced using 'p' type implanted layers. Buried 'p' type layers have been used to steepen the tail of the carrier profile produced by donor implantation. This gives a higher FET trans-conductance at low drain current, a condition essential for low noise operation. Another advantage of this technology is the improved threshold voltage uniformity [39]. The control of surface state density by use of a shallow 'p' region at the surface has been reported and a recent development is the fabrication of buried channel MESFETs. The device structures are fabricated by implanting a 'p' type dopant deep beneath the 'n' channel and using a shallow 'p' type implant to convert the surface. Claimed advantages of this technology are the elimination of drain current transients and oscillations associated with a self-gating effect [40].

In summary, ion implantation, whilst lacking complete control over the dopant profile, is nonetheless a powerful tool for the fabrication of a wide range of device structures.

1.8.3 Implant annealing

To obtain donor activity from implanted Si ions, the lattice damage produced by the implantation must be annealed out with the Si ions becoming localised on the Ga sublattice as Si_{Ga}^+. A variety of different techniques have been employed to do this. The temperatures which must be attained to get significant implant activation (800–950°C) are well above the congruent evaporation temperature of GaAs ($\sim$ 630°C) and so dissociation of the surface of the wafer is a major problem. The integrity of this surface is vital for subsequent device processing steps. This problem can be overcome in several ways:

(a) by carrying out the anneal in an ambient partial pressure of arsenic vapour sufficient to prevent dissociation;
(b) by coating the surface of the wafer with a protective cap (usually of Si_3N_4) either before or after implantation and before annealing;
(c) by attempting to carry out the process so rapidly that there is insufficient time for significant surface loss of arsenic to occur.

In the first method, known as capless annealing, the ambient arsenic pressure is generated either by passing arsine diluted in hydrogen over the wafer in an open-tube furnace or by having a closed ampoule in which a two-zone furnace is used to control the partial pressure over a source of either metallic arsenic or of GaAs. This partial pressure becomes an additional control parameter in the process. Typical anneal conditions are a temperature of 850–900°C held for 15–20 minutes.

In capped annealing, the surface of the water is coated with around 0.1 micron of Si_3N_4 which is deposited either by sputtering or by a plasma/CVD process. Typical conditions for plasma-nitride deposition are a substrate temperature of around 350°C with deposition occurring by pyrolysis of silane in the presence of nitrogen with a small hydrogen addition. Deposition rates of up to 100 Å/min are attainable at modest r.f. power. The variation in the thickness of the nitride is generally better than 5%. Prior reaction of the surface with the $N_2:H_2$ gas alone to remove native oxides from the wafer surface is necessary if reproducible results are to be obtained.

If the required implant is a shallow one, then this nitride cap can be deposited prior to implantation and implantation can take place *through* the cap by increasing the implant energy. The price to be paid is that there is a somewhat greater lateral spread of the implant, thus limiting device resolution. The advantage is that the wafer surface can be protected immediately after the polishing stage so that contamination is minimized. Ion channelling is also largely avoided by using this technique. Anneal temperatures and times are as for capless annealing.

Pulsed laser annealing of the implant using nanosecond to microsecond pulses which actually melt the wafer surface to a depth which is a little greater than the range of the implant has been attempted. However, both the degree and the reproducibility of the activation and the quality of the final wafer surface are inadequate and the method is not in general use. However, this research has spawned other 'transient annealing' approaches. Rapid thermal annealing using banks of incoherent lamps to raise the wafer temperature to the usual anneal temperature (800–950°C) and then lower it again all within a few seconds have been shown to yield reproducible results with little dissociation of the wafer surface. However, to ensure the retention of the integrity of the surface it is now common to cap the wafer before the anneal. The method is finding favour widely.

The degree of activation of the implant depends on a number of factors. These include the extent to which the lattice damage caused by the implant is annealed and this depends on the temperature and time of the anneal. The degree of lattice damage caused by the implant will depend on the temperature of implantation as well as on the dose and energy of the implant. Secondly, because silicon is amphoteric in GaAs, the differential apparent activation (η) expressed as the change in the number of carriers per implanted ion will necessarily be less than unity. One can write [37]:

$$\eta = \delta (N_D^+ - N_A^-) / \delta [Si] = \eta_{_A} \eta_{_{\bar{z}}} \tag{6}$$

$$\text{where } \eta_{_A} = \delta (N_D^+ - N_A^-) / \delta (N_D^+ + N_A^-)$$
$$\text{and } \eta_{_{\bar{z}}} = \delta (N_D^+ + N_A^-) / \delta [Si]$$

where δ [Si] is the total implanted number of Si ions. The net donor concentration is then:

$$n = \eta_{_A} \eta_{_{\bar{z}}} [Si] - (N_{Ao}^- - N_{Do}^+) \tag{7}$$

where N_{Do}^+ and N_{Ao}^- are the residual shallow donor and acceptor impurity concentrations respectively. This residual net acceptor concentration can increase the degree of compensation near to the channel–substrate interface and make the net donor profile more abrupt than the 'chemical' profile. We see from Equation (6) that to achieve reproducible carrier profiles in the channel we must have contol over $\eta_{_A}$, $\eta_{_{\bar{z}}}$, N_{Ao} and N_{Do} as well as over the fluence of the ion source. $\eta_{_{\bar{z}}}$ is approximately unity for sufficiently high anneal temperatures but high anneal temperatures serve to reduce $\eta_{_A}$ so that some optimization is needed. The optimum temperature lies in the range 850–900°C. Thomas *et al.* [37] have inferred the dependence of the amphotericity $[Si_{Ga}^+] / [Si_{As}^-]$ on implant anneal temperature from their results and this is shown in Fig. 1.19.

1.9 SURVEY OF MATERIALS CHARACTERIZATION TECHNIQUES

1.9.1 Introduction

The demanding nature of IC fabrication technology requires rigorous control of wafer and epi-layer properties. This is obtained only by careful and comprehensive materials characterization using a wide range of techniques to determine chemical, electrical, optical and structural properties and the degree of uniformity of these properties over whole wafer surfaces and from wafer to wafer. In an ideal world, this would be done by the wafer manufacturer and the IC manufacturer would buy against a specification confident in the knowledge that this specification would be met and that it would ensure a high yield of devices falling within specification. However, most manufacturers feel the need first to test process the first and last wafers from a batch taken from a single boule to check for adequate device properties and yield before consigning the rest of the batch to the processing line. They therefore tend to demand a large number of wafers per boule so that their pre-qualification test is proportionally only a small cost overhead.

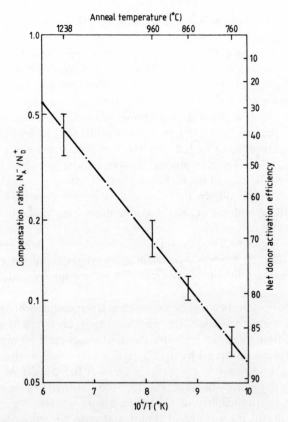

Fig. 1.19 Amphotericity and net donor activation of annealed Si-implanted layers as a function of the reciprocal anneal temperature [37].

As indicated in Section 1.5, this favours the high pressure LEC process over the low pressure one.

All this is not to say that the characterization of the material against an internationally agreed wafer specification is not vitally important and we next briefly review the techniques currently employed for such character-ization.

1.9.2 Chemical characterization

The requirements for chemical characterization can be classified into bulk and near-surface measurements. The first need is for analysis of starting materials to determine residual impurities at the sub-ppm level. Such measurements need also to be made at different longitudinal positions along as-grown crystals to ascertain the degree of purification by segregation and to identify contamination introduced by the environment of the growth process. The semi-insulating nature of the material makes the application of some of the ion- or electron-beam techniques difficult because of charging up of the specimen.

The workhorse tool for bulk impurity determination has been spark source mass spectrometry (SSMS). This has been complemented by the use of atomic absorption spectroscopy and atomic emission spectroscopy. In commercial mass spectrometers both electrostatic and magnetic deflection is employed to disperse the differing atomic masses in the ion beam onto a photographic plate. Sensitivities are typically 0.01 to 0.1 ppma and, because of the high voltages used to generate the ions, the sensitivities are not too strongly dependent on the atomic weight of the species. Nonetheless independent calibration of the technique is necessary if quantitatively useful information is to be obtained. Glow discharge mass spectrometers offer higher sensitivity and are replacing conventional machines but they pose new problems of calibration.

At the low impurity levels encountered, interference from residual gases in the spectrometer presents a problem and cryogenic pumping with liquid helium is necessary [41] to obtain reliable values for oxygen, silicon and sulphur.

Atomic absorption spectroscopy is an element-specific technique in which individual elements are analysed. The sample is dispersed in an oxygen-hydrocarbon flame and the impurity element is detected by measuring the optical absorption produced by it of a source consisting of the atomic line spectrum of the element in question. Since it is the density of the ground state atoms in the flame which is measured, the sensitivity of the technique is high but the reproducibility of specimen blanks can be a problem. The technique is useful for a range of donor and acceptor impurities in GaAs such as Si, S, Zn and Be.

Two distinct kinds of spatial information about the distribution of deliberately added dopants, and of the major residual contaminants such as C, Si, Zn and B, are required:

(a) Lateral – i.e. in the plane of the wafer surface. This is required to measure any radial non-uniformities introduced by the growth process or fluctuations due to dopant striations. The resolution required is not very high but where post-growth kinetic processes have produced segregation effects (such as at the cellular array of dislocation walls) there is a need for finer resolution down to a few microns.

(b) Depth profiling. This is used to obtain the dopant profile following implantation and annealing or through the thickness of an epitaxial layer. Ion sputtering is used to progressively erode through the layer to be profiled. Resolution on the nanometric scale is required for probing superlattice structures.

It is the latter – depth profiling – which is the most important and dynamic secondary ion mass spectrometry (SIMS) and sputter Auger electron spectroscopy (AES) are the most commonly used tools.

AES involves analysis of the secondary electrons ejected as the result of electron bombardment of the material with 1–10 KeV primary electrons.

The decay of an outer shell electron into a vacant inner shell state created by a primary electron can occur either with X-ray photon emission or by emission of a second (Auger) electron. The energy of the ejected Auger electron is characteristic of the emitting atom and this provides the method of detection. Auger electrons emitted at depths greater than about 20 Å are likely to have suffered inelastic collisions before emerging from the wafer surface and thereby to have lost their atomic identity; the technique therefore gives information only about the first few atomic layers. Lateral resolution is determined by the diameter of the primary electron beam so that there is a trade-off between resolution and beam intensity (sensitivity). Depth profiling can be performed by combining the technique with ion sputtering. UHV conditions are required to avoid spurious results arising from background contamination of the surface.

The Auger electrons are dispersed in energy with a cylindrical mirror analyser and detected with an electron multiplier. Signal differentiation is used to detect the small Auger peaks in the presence of the large continuum background resulting from the collection of the inelastically scattered electrons.

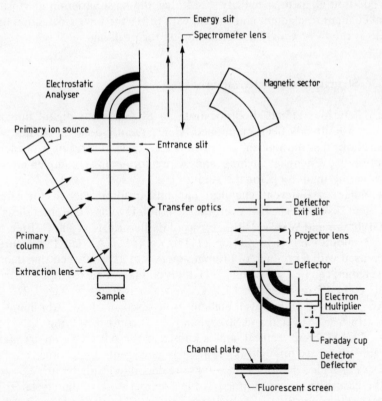

Fig. 1.20 Schematic representation of a dynamic SIMS equipment (Cameca IMS–3F) showing rastered primary source, electrostatic and magnetic separation and ion detection systems.

Secondary ion mass spectrometry (SIMS) has become the most widely used tool for profiling annealed implants and epilayers in GaAs (Fig. 1.20). Bombardment of GaAs with a beam of positive ions (commonly oxygen or caesium) of 1–30 KeV energy erodes the surface by ejecting neutral atoms but also produces, albeit with very low efficiency, secondary ions which can be collected and analysed in a quadrupole mass spectrometer. The depth resolution is around 50 Å. Since the incident ions erode the surface, it is a natural technique for dopant profiling. By scanning the incident ion beam the machine can be converted into an ion microscope. The basic technique can in fact be operated in several different modes with differing ion sources; for a comprehensive review see Blattner and Evans [42].

The sensitivity and wide dynamic range of the instrument (typically six or more orders of magnitude) make it a most valuable tool. Calibration of the technique is of great importance (requiring a wide range of standard samples) as is establishing the conditions of ion sputtering which give rise to a suitable erosion rate and eroded-pit profile for which the dopant profile can be unambiguously interpreted.

To summarize, SIMS, AES and SSMS can be compared as follows. All are good on element specificity. SSMS has the least variation in sensitivity from element to element and shares, with SIMS the lowest detection limits. SIMS is the most widely useful technique for profiling.

1.9.3 Structural characterization

The deleterious effects of dislocations in SI GaAs for digital integrated circuits has already been mentioned. Consequently detailed studies of the dislocation distribution in as-grown LEC crystals are of considerable significance. Chemical etching and X-ray topography complement each other as the main tools for the study.

A range of different chemical etchants produce pits at the sites of emergent dislocations. However, the currently favoured one, first described by Grabmier and Watson [43], consists of molten KOH at 300°C. Etch times of 2–3 minutes result in pits on {111} Ga, {111} As and {100} surfaces when used with a pre-etch of bromine-methanol to remove cutting damage. The etchant can be used with NaOH addition [44]. A micrograph of part of a KOH etched 2 in. diameter LEC GaAs wafer is shown in Fig. 1.21.

An intriguing and novel etchant was discovered by Abrahams and Buiocchi. (See Stirland and Straughan [6]). It consists of 8 mg $AgNO_3$: 1 g CrO_3: 2 ml H_2O and 1 ml HF and is known as the AB etch – an acronym of the initials of its originators. It reveals emergent dislocations on {111} surfaces provided that the dislocations are decorated with impurities – as will be the case if they were formed at high temperature in the crystal growth process. However, its real application is to {100} surfaces where it exhibits unique properties. Combined etching and X-ray topography has shown that raised linear features ('ridges') on etched surfaces are the projections in the surface of sections of dislocations which were in the bulk crystal which was removed by the etchant, i.e. there is an 'etch memory' of the dislocations in the immediately overlying material. Thus the etchant gives a measure of the volume density of dislocations rather than the superficial density obtained with conventional etchants.

X-ray topographic techniques have become near-routine with whole wafer imaging obtained using either conventional X-ray sources of high power (rotating anode tubes) or synchrotron sources. Important advantages of the technique are that it is non-destructive and can be used at successive stages of the device fabrication process to study the mechanisms by which process-induced defects occur. Double-crystal topography is a powerful technique for imaging very small amounts of both lattice parameter variations and lattice tilt variations over whole wafers.

Transmission electron microscopy (TEM) has been used to study precipitation on dislocations in GaAs. Dislocations in as-grown SI GaAs are observed in etching studies to be decorated with precipitates spaced every few microns along the length of the dislocation. They tend to be more abundant in the first part of the crystal to be grown. They have been identified as elemental arsenic in electron diffraction studies by Cullis *et al.* [45].

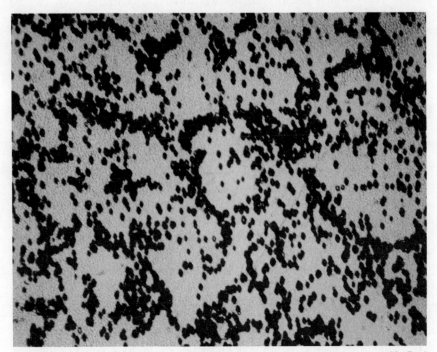

Fig. 1.21 Micrograph of part of a KOH etched wafer of LEC-grown SI GaAs showing the dislocation cell pattern. (Courtesy S. J. Barnett and G. T. Brown, RSRE, Malvern.)

1.9.4 Electrical characterization

The electrical properties of bulk material are commonly characterized by electrical resistivity and Hall mobility measured by the well-known van der Pauw technique. Measurement is usually made at room temperature, although, because of the problems of contacting to SI material, measurements at 400 K are sometimes given.

Annealed implants are characterized by their Hall mobility and sheet carrier concentration, again determined by the van der Pauw method. The depth profiles of the electrically activated donor implants are determined by capacitance-voltage profiling and Hall profiling. Clover-leaved samples are obtained by selective area implantation, by mesa isolation of uniformly implanted samples or by trepanning ultrasonically. Electrical contacts are made on the 'clover-leaf' pads by alloying tin dots in an HCl-forming gas mixture at around 400°C.

The donor concentration profile through the implanted and annealed layer can be obtained by depleting the region successively to greater depths by biassing a Schottky diode formed by a metal contact to the semiconductor surface. From a bridge measurement of the capacitance (C) of the depleted region one can obtain a measure of the donor concentration at the depletion edge. Automated equipment is available to plot $1/C^2$ versus the depletion voltage from which the profile of the donor concentration N_D can be obtained. Problems in determining the effective area of the Schottky contact limit the accuracy of the technique.

1.9.5 Optical characterization

Because SI GaAs is transparent in the near infra red, a number of i.r. optical techniques have been developed and utilized in its characterization. Amongst these local vibrational mode spectroscopy (LVM) is the most quantitative and the most useful. It is particularly suited to the study of boron, carbon and silicon in GaAs – all important impurities. Substitutional atoms whose atomic masses are less than that of the matrix have localized modes of vibration which lie in the i.r. region of the spectrum. The line frequency uniquely identifies the atom and the lattice site on which it is located. Thus, for example, Si_{Ga} can be differentiated from Si_{As}. Modes due to defect pairs such as $Si_{Ga}Si_{As}$ or $Si_{Ga}V_{Ga}$ can also be detected. From measurement of the area under the spectral absorption curve an estimate of the defect concentration can be obtained provided that prior calibration has been carried out. The measurements can be made at room temperature but there is sometimes a gain in sensitivity to be obtained by cooling to liquid nitrogen temperature.

It has been established that the neutral state of the EL2 defect produces a band of absorption extending from mid gap (0.7 eV) to the band edge (1.4 eV). A calibration of this effect from absorption measurements made at different wavelengths and temperatures has been made by Martin (see Makram-Ebeid *et al.* [11]).

A qualitative map of the distribution of neutral EL2 over a whole wafer can be obtained by imaging a suitably illuminated wafer with a silicon vidicon camera. A quartz-iodine projector is a suitable source when used with a Fresnel lens and a diffuser to give uniform diffuse illumination. Semi-quantitative line scans can also be obtained [46]. This involves the use of more than one wavelength in order to allow for absorption mechanisms due to causes other than the presence of neutral EL2. A whole-wafer image is shown in Fig. 1.22. A line scan is shown in Fig. 1.23. The fine scale fluctuations in the scan are related to the cellular dislocation array.

Other widely used techniques for the investigation of deep levels are photoluminescence and deep level transient spectroscopy (DLTS). However, the first of these cannot be made quantitative whilst the second does not permit spatial mapping with useful resolution.

The latest technique to be applied to the imaging of the distribution of deep levels across a wafer is SEM cathodoluminescence (CL) which gives a mapping with a resolution of the scanning electron microscope (SEM). The

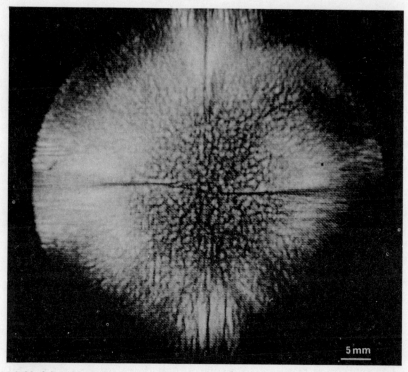

Fig. 1.22 2 in. SI GaAs wafer imaged in transmission using light of 1 micron wavelength to reveal the distribution of neutral EL2. Note the cellular array associated with the dislocation structure in the central region. (Courtesy G. T. Brown and M Skolnick, RSRE, Malvern.)

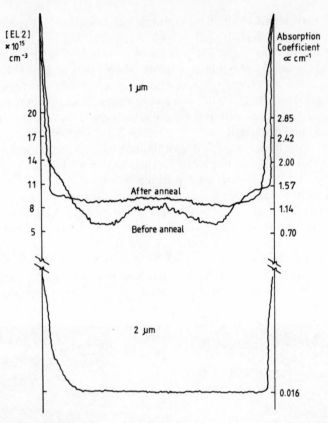

Fig. 1.23 Line scan of [EL2] (1 micron absorption curves) across a 2 in. SI GaAs wafer grown by LEC before and after annealing at 950°C for 5 hours. The 2 micron absorption curve is plotted to check for other spurious causes of absorption in the crystal [47].

use of this technique has been reviewed by Warwick [48]. With the technique one seeks to obtain an image of the specimen using the infra-red radiation emitted by the sample when it is irradiated with a rastered electron beam of 1–30 keV energy. The technique can be operated in one of two modes:

(1) wavelength dispersive CL where light of a particular wavelength, corresponding to emission by a specific centre in the lattice, is selected using a dispersive element such as a grating monochromator.
(2) so-called time dispersive mode in which the decay rate of the emission following excitation is measured.

The method is sensitive to levels present down to the part per billion range. It has the edge therefore on SIMS in sensitivity but it is much more difficult to make it a quantitative technique. Because the CL spectrum is thermally broadened at room temperature, it is necessary to cool the specimen, usually to liquid helium temperature, in order to resolve transitions due to near-band-gap states. The technique is therefore not truly non-destructive.

The intensity of the CL image is proportional to the fraction of the generated e^-h^+ pairs that recombine radiatively. Defects are non-radiative centres which reduce the CL image intensity in their vicinity. The technique is also of value in the study of surface damage which gives contrast due to regions where locally the concentrations of non-radiative centres resulting from residual polishing damage, is high.

1.10 STATE OF THE ART IN BULK GaAs WAFERS

1.10.1 Electrical properties

By avoiding any significant arsenic deficiency in the melt and by using pyrolytic boron nitride crucibles, SI single crystals of GaAs with resistivities of the order of 10^8 ohm cm can be grown routinely in a conventional high pressure LEC puller. Addition of a small amount of chromium increases the resistivity a little but produces some reduction in electron mobility.

However, the radial uniformity of the electrical conductivity across a wafer is generally poor exhibiting a 'W' shaped distribution when scanned along a <100> direction in the plane of the wafer. Quite early on this was shown to be correlated with the dislocation distribution in the wafer which also exhibits a 'W' profile. The distribution of neutral EL2 (as measured by 1 micron absorption) is similarly distributed. It has been found that the electrical resistivity (and with it the neutral EL2 concentration) can be made much more spatially uniform by annealing the whole boule after growth (see Section 1.5.3 and Fig. 1.23) and this practice appears now to be common amongst the world's commercial suppliers of GaAs. At least to superficial observation, the dislocation density and distribution appears to be unaltered by the annealing process. Marked improvements in threshold voltage uniformity are reported [49].

As outlined in Section 1.3 the carrier level is determined by the compensation of the residual net shallow acceptor by the native defect EL2. The electron mobility will depend on the concentration of ionized centres. Thus material with a high residual net acceptor concentration will require more EL2 to compensate it and will therefore be likely to have a degraded mobility.

The residual acceptor and donor concentrations depend not only on the purity of the starting charge materials (Ga and As) but aslo on the water content (and perhaps also on the purity) of the boric oxide. A number of impurities are gettered by increased [OH] content in the encapsulant. These include Si, the dominant residual donor, and carbon the dominant residual acceptor. The probable source of the carbon is the graphite 'furniture' in the puller and the method by which it is transported into the melt could be through reduction of Ga_2O_3 (formed by the action of the water in the boric oxide on the gallium in the melt) to Ga_2O and carbon monoxide. Thus a high water content favours crystal purity but unfortunately it also makes the crystal very prone to twinning during growth. There is therefore an optimum water concentration which gives adequate purity without reducing significantly the yield through the incidence of twinning. 'Dry' boric oxide also gives rise to high concentrations of boron in the crystal (as a result of reduction of the boric oxide by gallium). Boron concentrations in excess of 10^{17} cm are obtained with very dry oxide ([OH] < 200 ppm). However this boron does not produce any (significant) electrical effect nor appear to be deleterious in any other known respect.

As we have seen, the concentration of EL2 is dependent on the melt stoichiometry. However, again some optimization is necessary. As-rich material is characterized by a high concentration of microscopic elemental arsenic precipitates lying along the dislocations in the material, the effects on device performance of which are not known. Ga-rich material tends to be 'p' type with a dominant acceptor having an activation energy of around 77 meV which is thought to be related to a Ga-containing native point defect. Material which is near to the 'p' type transition, even if SI as grown, will tend to show thermal conversion upon annealing at temperatures commensurate with implant annealing or of epitaxial growth. The control of stoichiometry is therefore a key issue in the development of a reliable process. Two factors render this difficult in conventional LEC technology. Firstly, there is inevitably some arsenic loss during the synthesis stage. The water content of the boric oxide is again a factor; very dry boric oxide does not wet the boron nitride crucible readily and arsenic vapour loss tends to be greater. This loss is not quantifiable in advance and leads to uncertainty in the stoichiometry of the final charge. Secondly there is loss of gallium by uptake in the boric oxide as described below.

Thermal conversion in undoped SI LEC GaAs appears to be related to the diffusion kinetics of stoichiometric defects (and perhaps also due to the presence of carbon acceptors). In Ga-rich ingots where [EL2] is so low that the material is near to 'p' type conversion, annealing can produce conversion to large depths into the wafer, whereas in material grown from an As-rich melt, any thermal conversion is confined to the near-surface region. The behaviour of chromium-doped material is more complex because of the high diffusivity of chromium and its peculiar diffusion behaviour.

Recently it has been shown that water in the boric oxide changes the melt stoichiometry additionally by the uptake of a large amount of gallium from the melt as Ga_2O_3. The volatilization of some of this and its condensation on the graphite furniture where it is reduced, possibly contributes to the carbon uptake. This role of water has been demonstrated by Emori *et al.* [18] who showed that the degree of apparent Ga-richness of the melt (based on the weights of the starting charges) required to produce 'p' type material are much greater when wet boric oxide is used than with dry oxide (Fig. 1.24). Brozel *et al.* [50] have shown that segregation data can be used to map the

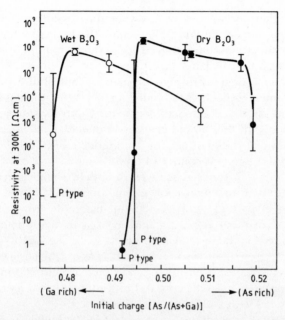

Fig. 1.24 Effect of water content of the boric oxide encapsulant on the resistivity of undoped GaAs grown from melts of different initial composition [18].

variation in stoichiometry down the crystal and that the results are consistent with the measurements of Emori *et al.* [18].

To achieve more As-rich conditions several groups have developed methods for injecting arsenic, using a bubbler, into the melt after the synthesis stage It is claimed that this leads to greater spatial uniformity of electrical properties and a reduced scatter in FET threshold voltage. Kirkpatrick *et al* [51] also report that As-rich conditions favour a lower dislocation density.

1.10.2 Dislocations

It is generally believed that the main cause of the relatively high density of dislocations in LEC crystals is plastic yielding in the presence of thermal stresses in the cooling crystal. This is much more prone to occur in GaAs than it is in silicon because:

(a) the thermal conductivity of GaAs is only about one third of that of silicon, so that the thermal gradients in the material are proportionately higher and
(b) the critical resolved shear stress for plastic flow is much lower than in silicon.

Attempts to reduce the dislocation density have therefore focused on reducing the thermal stresses by growing the crystal in a more nearly isothermal environment and/or by hardening the lattice by the addition of some electrically-inert solute atom. The isovalent indium atom is almost invariably chosen for this latter task [52].

Both of these strategies brings its problems. Control of crystal diameter becomes extremely difficult as the radial temperature gradient in the puller is reduced and improved automatic diameter control systems are urgently needed to cope with this. Rapid changes in diameter introduce additional stresses so that one does not benefit from reduction in radial temperature gradient unless one retains control of crystal shape.

The dislocation structure can be revealed both by chemical etching and by X-ray topography using either conventional or synchrotron X-ray sources. The dislocation patterns are similar in material grown in different laboratories throughout the world and consist of a pronounced lineage structure arranged in the form of a cross lying along the <110> directions in the plane of the growth surface between the arms of which is a cellular array of dislocations (Fig. 1.25) having a cell size of around 200 microns. This structure propagates down the crystal. Lattice tilts of 10 or more seconds of arc are associated with the lineage structure.

It is found that the concentration of neutral EL2 is higher in the vicinity of the cell walls and indeed there is an enhancement in the vicinity of single dislocations. This correlation of [EL2] with dislocation arrays can be seen by comparing Figs 1.21, 1.22 and 1.25. Using, at high resolution,

Fig. 1.25 X-ray topograph showing the cellular pattern of dislocations in an undoped SI GaAs wafer grown by LEC. (Courtesy S. J. Barnett and G. T. Brown, RSRE, Malvern.)

cathodoluminescence generated by a 0.68 eV emission known to be associated with EL2, Warwick and Brown [53] have shown a diminished emission adjacent to the cell walls and have concluded that the dislocation arrays getter the defect centres causing the emission. Hope *et al.* [54] have shown that EL2 is present, but at lower concentration, in small crystals which are dislocation-free and that its spatial distribution is much more uniform.

Reduction of thermal stress alone has proved inadequate for the removal of dislocations. After-heaters and thermal shields have ben employed to reduce the stress but numerical modelling and experimental measurement have indicated that the point of emergence of the crystal from the boric oxide is the region of highest stress. Accordingly attempts have been made to use very deep layers of boric oxide so that the crystal remains encapsulated throughout its growth. Its temperature is then allowed to fall slowly to a point at which it is strong enough for it to be safely withdrawn from the encapsulant. This technique seriously limits the length of crystal which can be grown in a pressure puller but is otherwise an effective method for reducing stress. It has been used in conjunction with indium-alloying to obtain nearly dislocation-free 2 in. diameter crystals [59].

Indium alloying is effective in dramatically reducing the dislocation density to the point that a large fraction of the cross sectional area of 2 in. crystals can be made dislocation-free. This results in very low standard deviation of FET threshold voltage and its improvements to LSI yield can be dramatic. However the indium concentrations necessary to achieve this ($1-2 \times 10^{20}$ cm^{-3}) produce a marked reduction in the efficiency of activation of implanted silicon donors and to unacceptable fluctuations in lattice parameter due to fluctuations in indium concentration caused by turbulent convection in the melt.

The relative high concentration of indium also renders the crystal prone to unwanted microsegregation phenoma due to the occurrence of constitutional supercooling. To avoid this a significant reduction in growth rate must be made. Even so, the rapid rise in In concentration in the tail of the crystal finally leads to cellular microsegregation and this portion of the crystal has to be discarded (Fig. 1.26). Not only is the density of dislocations reduced by In-doping but the pattern of them is markedly altered also. In particular, dislocations in indium-doped samples appear to remain on their slip planes, there being none of the polygonization into cellular arrays such as is seen in the undoped material. Lattice parameter mismatch at the crystal–seed interface can generate additional dislocations which propagate vertically down into the crystal (Fig. 1.27).

Dispute exists as to the mechanism by which indium alloying reduces the dislocation density. Recent measurements of the effect of indium alloying on the critical resolved shear stress of the material [9] casts doubt on the lattice hardening model of Ehrenreich and Hirth [55] and indicates a possible importance of the indium on the thermodynamics of point defect generation in the material.

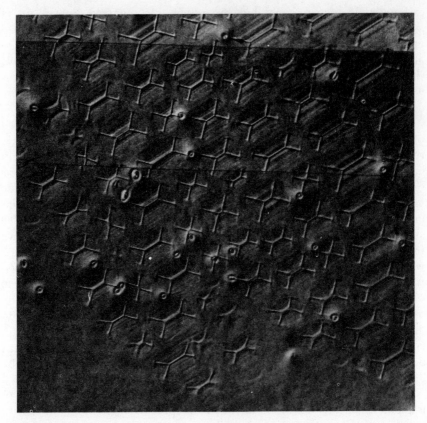

Fig. 1.26 Cellular structure caused by constitutional supercooling due to In doping of LEC-grown GaAs. Revealed by AB etching. (Courtesy D. J. Stirland, Plessey, Caswell.)

An additional problem posed by indium alloying is that the material is more brittle than undoped material with a consequently higher chance of wafer breakage. Nonetheless 16K static RAMs have been made in this material having a very low scatter in threshold voltage.

1.10.3 Use of a magnetic field

An applied steady magnetic field can be used to damp out turbulent convection in the melt and thereby obtain a much more uniform micro-distribution of indium (and other dopants) without the usual striation pattern. Both horizontally and vertically oriented fields have been tried but most interest centres on the use of a vertical (axial) field. The effect of such a field is to reduce the axial macrosegregation also since, by making the melt more nearly quiescent, the effective segregation coefficient of the dopant can be made to approach unity. The potential advantage of this is that each

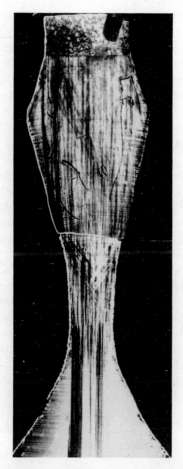

Fig. 1.27 Dislocations propagated from an undoped GaAs seed when growth is promoted from an In-doped melt. X-ray transmission topograph of (110) longitudinal section. (Courtesy D. J. Stirland, Plessey, Caswell.)

wafer taken from the boule would have the same indium concentration and hence the same lattice parameter and, one hopes, the same implantation characteristics.

Dislocation-free and striation-free indium alloyed 2 in. diameter GaAs crystals have been grown using a vertical magnetic field in conjunction with full encapsulation within the boric oxide [56]. However, by the same mechanism by which the axial segregation is avoided, any non-stoichiometry in the melt is much accentuated by the near-quiescent conditions. Marked changes in EL2 concentration – due to changes in melt stoichiometry at the interface – on applying a magnetic field have been observed [57]. In particular material which is only slightly Ga-rich when grown from a well-mixed melt can be so non-stoichiometric when grown from a magnetically damped melt that it actually becomes 'p' type.

The reader will therefore see that satisfactory methods for controlling dislocation density do not exist at present and the importance of indium alloying, the use of a magnetic field, full encapsulation, arsenic injection, improved automatic diameter control and other variants to the conventional high pressure LEC process have yet to be fully evaluated. It is therefore likely to remain a field of intense research activity for some time.

1.11 FUTURE MATERIALS NEEDS

The pressing immediate need, which is receiving much attention worldwide, is to remove the cellular dislocation arrays and lineage boundaries in 2 in. and 3 in. wafers because these give rise to variations in threshold voltage across a wafer which can be as high as 200 mV. Whilst this variation is tolerable in some (most?) MMIC applications, very much tighter tolerances (of the order of 20 mV) are needed for digital ICs.

Initial work on In-doped material suggests that individual dislocations lying in their slip planes do not give rise to a large scatter in threshold voltage. Rather, it is the climb of these dislocations into a cellular array which is most deleterious. This fact highlights the potential importance of In-doping but, as we have seen, the maximum acceptable doping levels are much smaller than those currently reported as being needed to achieve dislocation density reduction.

It has not been necessary to obtain 'zeroed' crystals in order to make the first generation of integrated circuits. However, for an integrated opto-electronic technology, with on-chip lasers, removal of dislocations will be vital since it has long been established that a single dislocation threading the active region of a GaAs laser causes a catastrophic reduction in the operational lifetime of that laser.

To achieve 'zeroed' GaAs in commercial-sized boules is likely to be a long, hard road extending the crystal grower's skills and ingenuity in pursuit of the control of thermal stresses and dopant uniformity. Another direction in which melt growth technology must develop is in the control of residual impurity levels and crystal stoichiometry since variations in these parameters cause a variability in the degree of apparent implant activation.

The scaling up of the LEC process, initially to 4 in. diameter and ultimately beyond, will follow the establishment of a firm market in GaAs ICs. Increased melt charge weight is perhaps a more important parameter than crystal diameter in this respect since a large number of wafers per boule is required for cost-efficient manufacture. The history of the development of silicon crystals suggests that there are no fundamental limitations to the scaling-up of the Czochralski process and that it will proceed steadily as markets develop. It could of course be that some new process will emerge to supplant the LEC process. However the extent of the investment in the LEC process to date makes this progressively less probable.

Developments in wafering, and in particular of preparing damage-free surfaces of adequate flatness, are at least as important as improvements in the growth technology. The ability to implant through a silicon nitride cap, albeit with some loss of lateral resolution, suggests that capping immediately after polishing to protect the prepared surface, is a desirable objective.

As the degree of sophistication and scale of integration evolve, epitaxy will become more affordable and MOVPE and MBE (or MOMBE) will be needed to grow $Ga_xAl_{1-x}As$ layers first for HEMT and heterojunction bipolar technologies and then for multi-layer devices based on low-dimensional effects. Hetero-epitaxy, and in particular that of GaAs on Si, is probably destined for a major role in the bid to marry the speed and opto-electronic capabilities of GaAs to the cheaper and more mature Si technology.

The inexorable trend toward ever higher frequencies makes the highest control of surfaces and interfaces imperative. This points down a road ultimately leading to device fabrication becoming an all-UHV technology where material preparation and device fabrication have finally become totally integrated.

REFERENCES

1. Nathanson, H.C., Thomas, R.N. and Oakes, J. G. (1980) 'Future of Systems Dependent on Wafer Availability' *Microwave System News* (June 1980) 37–56.
2. Mullin, J. B., Straughan, B. W. and Brickell, W. S. (1965) 'Liquid Encapsulation Techniques: The Use of an Inert Liquid in Suppressing Dissociation during Melt-growth of InAs and GaAs Crystals' *J. Phys. Chem. Solids* **26** 782–784.
3. Cox, H. M., Hummel, S. G., Keriamidas, V. G. (1986) 'Vapour Levitation Epitaxy: System Design and Performance' *J. Crystal Growth* **79** 900–908.
4. Koster, W. and Thoma, B. (1955) 'The Systems Ga–Sb, Ga–As and Al–As' *Z. Metallkunde* **46** 291–293.
5. Arthur, J. R. (1967) 'Vapour Pressures and Phase Equilibria in the Ga–As System' *J. Phys. Chem. Solids* **28** 2257–2267.
6. Stirland, D. J. and Straughan, B. W. (1976) 'A Review of Etching and Defect Characterisation of Gallium Arsenide Substrate Material' *Thin Solid Films* **31** 139–170.
7. Hurle, D. T. J. (1979) 'Revised Calculation of Point Defect Equilibria and Non-Stoichiometry in Gallium Arsenide' *J. Phys. Chem.* **40** 613–626.
8. EMIS (1986) 'Properties of Gallium Arsenide' *Data Reviews Series* No. 2 INSPEC. Institution of Electrical Engineers, London.
9. Tabache, M. G., Bourret, E. D. and Elliot, A. G. (1986) 'Measurements of the Critical Resolved Shear Stress for Indium-doped and Undoped GaAs Single Crystals' *Applied Physics Letters* **49** (5) 289–291.
10. AuCoin, T. R., Ross, R. L., Wade, M. J. and Savage, R. O. (1979) 'Liquid Encapsulated Compounding and Czochralski Growth of Semi-insulating GaAs' *Solid State Technology* (Jan. 1979) 59–62.
11. Makram-Ebeid, S., Langlade, P. and Martin, G. M. (1984) 'Nature of EL2: the Main Midgap Electron Trap in VPE and Bulk GaAs' *Semi-insulating III–V*

Materials (Ed. Look, D. C. and Blakemore, J. S.) Shiva Publishing Ltd, Nantwich, UK 184–203.

12. Burke, K. M., Leavitt, S. R., Khan, A. A. Riemer, E., Stoebe, T., Alterovitz, S. A. and Haugland, E. J. (1986) 'Characterisation of the Semi-insulating Properties of Undoped GaAs Grown by the Horizontal Bridgman Technique' *IEEE GaAs IC Symposium* 37–39.

13. Gault, W. A., Monbèrg, E. M. and Clemans, J. E. (1986) 'A Novel Application of the Vertical Gradient-Freeze Method to the Growth of High Quality III–V Crystals' *J Crystal Growth* **74** 491–506.

14. Metz, E. P. A., Miller, R. C. and Mazelsky, R. (1962) 'A Technique for Pulling Single Crystals of Volatile Materials' *J. Applied Physics* **33** 2016–2017.

15. Mullin, J. B., Heritage, R. J., Holliday C. H. and Straughan, B. W. (1968) 'Liquid Encapsulation Crystal Pulling at High Pressures' *J. Crystal Growth* **3/4** 281–285.

16. Bass, S. J. and Oliver, P. E. (1968) 'Pulling of Gallium Phosphide Crystals by Liquid Encapsulation' *J. Crystal Growth* **3/4** 280–290.

17. Bardsley, W., Hurle, D. T. J., Joyce, G. C. and Wilson, G. C. (1977) 'The Weighing Method of Automatic Czochralski Crystal Growth' *J. Crystal Growth* **40** 21–28.

18. Emori, H., Kikuta, T., Inada, T., Obokata, T. and Fukuda, T. (1985) 'Effect of Water Content of B_2O_3 Encapsulant on Semi-insulating LEC GaAs Crystal' *Japanese J. Applied Physics* **24** L291–L293.

19. Hurle, D. T. J. (1987) 'The Dynamics and Control of Czochralski Growth' *Advanced Crystal Growth: Proceedings of the 6th International Summer School on Crystal Growth, Edinburgh 1986.* (Ed. Dryburgh, P., Cockayne, B. and Barraclough, K. G.) Prentice Hall 97–123.

20. Jacob, G., Duseaux, M., Farges, J. P, van den Boom, M. M. B and Roksnoer, P. J. (1983) 'Dislocation-free GaAs and InP Crystals by Isoelectronic Doping' *J. Crystal Growth* **61** 417–424.

21. van Tuyl, R. L. and Liechti, C. A. (1974) 'High Speed Integrated Logic with GaAs MESFETs', *ISSCC Digest of Technical Papers* (Feb. 1974) 114–115.

22. van Tuyl, R. L. and Liechti, C. A. (1976) 'High Speed GaAs MSI' *ISSCC Digest of Technical Papers* 20–21.

23. Stolte, C. A. (1984) 'Ion Implantation and Materials for GaAs Integrated Circuits' *Semiconductors and Semimetals* **20** (Ed. Willardson, R. K. and Beer, A. C.) Academic Press 89–158.

24. Greene, P. D. (1987) 'Liquid Phase Epitaxy' *Advanced Crystal Growth: Proceedings of the 6th International Summer School on Crystal Growth, Edinburgh 1986* (Ed. Dryburgh, P. M., Cockayne, B. and Barraclough, K. G.) Prentice Hall 221–244.

25. Ashen, D. J., Dean, P. J., Hurle, D. T. J., Mullin, J. B., Royle, A. and White, A. M. (1975) 'The Incorporation of Residual Impurities in Vapour-Grown GaAs' *Inst. Phys. Conf. Series* No. 24 229–244.

26. Nakanisi, T., Udagawa, T., Tanaka, A. and Kamei, K. (1981) 'Growth of High Purity GaAs Epilayers by MOCVD and their Applications to Microwave MESFETs' *J. Crystal Growth* **55** 255–262.

27. Stringfellow, G. B. (1985) 'Organometallic Vapour Phase Epitaxial Growth of III–V Semiconductors' *Semiconductors and Semimetals* **22A** (Ed. Tsang, W. T.) Academic Press 209–259.

28. Hirtz, J. P., Larivain, J. P., Duchemin, J. P. and Pearsall, T. P. (1980) Growth

of GaInAs on InP by Low Pressure MOCVD' *Electronic Letters* **16** 415–416.

29. Bass., S. J. (1975) 'Device Quality Epitaxial Gallium Arsenide Grown by the Metal Alkyl-Hydride Technique' *J. Crystal Growth* **31** 172–178.

30. Betsch, R. J. (1986) 'Parametric Analysis of Control Parameters in MOCVD' *J. Crystal Growth* **77** 210–218.

31. Moss, R. H. (1984) 'Adducts in the Growth of III–V Compounds' *J. Crystal Growth* **68** 78–87.

32. Duchemin, J. P., Hirtz, J. P., Razeghi, M., Bonnet, M. and Hersee, S. D. (1981) 'GaInAs and GaInAsP Materials Grown by Low Pressure MOCVD for Microwave and Opto-electronic Applications' *J. Crystal Growth* **55** 64–73.

33. Morkoc, H. and Cho, A. Y. (1979) 'High Purity GaAs and Cr-doped GaAs Epitaxial Layers by MBE' *J. Applied Physics* **50** 6413–6418.

34. Akiyama, M., Kawarada, Y., Ueda, T., Nishi, S. and Kaminishi, K. (1986) 'Growth of High Quality GaAs Layers on Si Substrates by MOCVD' *J. Crystal Growth* **77** 490–497.

35. Koch, S. M., Rosner, S. J., Hull, R., Yoffe, G. W. and Harris J. S. Jr. (1987) 'The Growth of GaAs on Si by MBE, *J. Crystal Growth* **81** 205–213.

36. Fischer, R., Klem, J., Henderson, T., Peng, C. K. and Morkoc, H. (1985) 'Prospects for the Monolithic Integration of GaAs and Si' *IEEE GaAs IC Symposium* 71–73.

37. Thomas, R. N., Hopgood, H. M., Eldridge, G. W., Barrett, D. L., Braggins, T. T., Ta, B. and Wang, S. K. (1984) 'High Purity LEC Growth and Direct Implantation of GaAs for Monolithic Microwave Circuits' *Semiconductors and Semimetals* **20** (Ed. Willardson, R. K. and Beer, A. C.) Academic Press 1–87.

38. Speight, J. D., O'Sullivan, P., Leigh, P. A., McIntyre, N., Cooper, K. and O'Hara, S. (1977) 'The Isolation of GaAs Microwave Devices using Proton Bombardment' *Inst. Phys. Conf. Series* **33a** 275–286.

39. Umemoto, Y., Masuda, N. and Mitsusada, K. (1986) 'Effects of Buried p-Layer on Alpha Immunity of MESFETs Fabricated on Semi-insulating GaAs Substrates' *IEEE Electron Device Letters* **EDL–7** 396–397.

40. Canfield, P. and Forbes, L. (1986) 'Suppression of Drain Conductance Transients, Drain Current Oscillations and Low Frequency Generation-Recombination Noise in GaAs FETs Using Buried Channels' *IEEE Transactions on Electron Devices* **ED–33** 925–928.

41. Clegg, J. B., Grainger, F. and Gale, I. G. (1980) 'Quantitative Measurement of Impurities in Gallium Arsenide' *J. Materials Science* **15** 747–750.

42. Blattner, R. J. and Evans, C. A. (1979) 'Modern Ion Beam and Related Techniques for Material Characterisation' *Crystal Growth; a Tutorial Approach*, (Ed. Bardsley, W., Hurle, D. T. J. and Mullin, J. B.) North Holland Publishing Co., Amsterdam 269–306.

43. Grabmaier, J. G. and Watson, C. B. (1969) 'Dislocation Etch Pits in Single Crystal GaAs', *Phys. Stat. Sol.* **32** K13–K15.

44. Lessoff, H. (1985) 'Non-uniform Etching of Single Crystal GaAs' *Materials Letters* **3** 251–254.

45. Cullis, A. G., Augustus, P. D. and Stirland D. J. (1980) 'Arsenic Precipitation at Dislocations in GaAs Substrate Material' *J. Applied Physics* **51** (5) 2556–2560.

46. Brozel, M. R., Grant, I., Ware, R. M. and Stirland, D. J. (1983) 'Direct Observation of the Principal Deep Level (EL2) in Undoped Semi-insulating GaAs' *Applied Physics Letters* **42** (7) 610–612.

47. Rumsby, D., Grant, I., Brozel, M. R., Foulkes, E. J. and Ware, R. M. (1984)

'Electrial Behaviour of Annealed LEC GaAs' *Semi-insulating III–V Materials* (Ed. Look, D. S. and Blakemore, J. S.) Shiva Publishing, Nantwich, UK 165–170.

48. Warwick, C. A. (1987) 'Recent Advances in Scanning Electron Microscope Cathodoluminescence Assessment of GaAs and InP' *Scanning Microscopy* **1** 51–61.

49. Kasahara, J., Arai, M., and Watanabe, N. (1986) 'Threshold Voltage Uniformity of GaAs-FETs on Ingot-annealed Substrates' *Japanese J. Applied Physics* **25** L85–L86.

50. Brozel, M. R., Foulkes, E. L. Grant, I. R. and Hurle, D. T. J. (1987) 'Tin Segregation and Donor Compensation in Melt-grown Gallium Arsenide' *J. Crystal Growth* **80** 323–332.

51. Kirkpatrick, C. G., Chen, R. T., Holmes, D. E., Elliott, K. R. and Asbeck P. M. (1983) 'Substrate Materials for GaAs Integrated Circuits' *Extended Abstracts of the 15th Conference on Solid State Devices and Materials, Tokyo* 145–148.

52. McGuigan, S., Thomas, R. N., Barrett, D. L., Eldridge, G. W., Messham, R. L. and Swanson, B. W. (1986) 'Growth and Properties of Large Diameter Indium Lattice Hardened GaAs Crystals' *J. Crystal Growth* **76** 217–232.

53. Warwick, C. A. and Brown G. T. (1985) 'Spatial Distribution of 0.68eV Emission from Undoped Semi-insulating Gallium Arsenide Revealed by High Resolution Luminescence Imaging' *Applied Physics Letters* **46**(6) 574–576.

54. Hope, D. A. O., Skolnick, M. S., Cockayne, B., Woodhead, J. and Newman, R. C. (1985) 'A Comparison of the Deep Donor (EL2°) and Strain Distributions in Dislocated and Dislocation-free Semi-insulating Undoped GaAs' *J. Crystal Growth* **71** 795–798.

55. Ehrenreich, H. and Hirth, J. P. (1985) 'Mechanism for Dislocation Density Reduction in GaAs Crystals by Indium Addition' *Applied Physics Letter* **46** (7) 668–670.

56. Kohda, H., Yamada, K., Nakanishi, H., Kobayashi, T., Osaka, J. and Hoshikawa, K. (1985) 'Crystal Growth of Completely Dislocation-free and Striation-free GaAs' *J. Crystal Growth* **71** 813–816.

57. Terashima, K., Yahata, A. and Fukuda, T. (1986) 'Growth Condition Dependence of EL2 Concentrations in Magnetic Field Liquid-encapsulated Czochralski GaAs Crystals' *J. Applied Physics* **59** (3) 982–984.

58. Donnelly, J. P. (1977) 'Ion Implantation in GaAs' *Inst. Phys. Conf. Series* **33b** 166–90.

59. Nakanishi, H., Kohda, H., Yamada, K. and Hoshikawa, K. (1984) 'Nearly Dislocation-free, Semi-insulating GaAs Grown in B_2O_3 Encapsulant' *Extended Abstracts of the 15th International Conference on Solid State Devices and Materials, Kobe* 63–66.

60. Morkoc, H. and Eastman, L. F. (1976) 'Purity of GaAs Grown by LPE in a Graphite Boat' *J. Crystal Growth* **36** 109–114.

61. Soga, T., Hattori, S., Sakai, S. and Umeno, M. (1986) 'Epitaxial Growth and Material Properties of GaAs on Si Grown by MOCVD' *J. Crystal Growth* **77** 498–502.

62. Swiggard, E. M., Lee, S. H. and von Batchelder, F. W. (1977) 'GaAs Synthesised in Pyrolytic Boron Nitride (PBN)' *Inst. Phys. Conf. Series* **33b** 23–27.

Chapter 2

Digital Integrated Circuit Design

Marc ROCCHI, Bertrand GABILLARD, Etienne DELHAYE and Thierry DUCOURANT

2.1 ACTIVE DEVICES FOR GaAs DIGITAL ICs

2.1.1 Review of active devices of GaAs digital ICs

Ever since the advent of monolithic integrated circuits on silicon in the early 1960s, they have constantly moved towards higher speed and higher complexity. Nevertheless, in the early 1970s [1] and definitely after 1974 [2], it was shown that III–V materials and more particularly gallium arsenide (GaAs) would make it possible to produce ICs with even better performances than their silicon counterparts.

Among the various III–V compounds, GaAs turned out to be an excellent candidate for high-performance digital applications because of its direct wide energy band gap, high electron mobility and saturation velocity, and the availability of a semi-insulating GaAs substrate.

Semi-insulating GaAs substrates offer a natural electrical isolation between active devices. Moreover, they remain semi-insulating, even after being bombarded with fast neutrons or γ rays, which is of prime importance for military applications.

The high electron mobility (six times higher in n-type GaAs than in n-type silicon) and the poor hole mobility (about 15 times lower), along with the easy fabrication of Schottky barriers on n-type GaAs, have naturally led to the selection of n-channel MESFETs as the most appropriate active devices for the first generation of GaAs digital ICs.

In n-channel MESFETs, the thickness of the conducting path is controlled by the voltage on the Schottky gate which fixes the depth of the depletion region under the gate, much like in a junction field effect transistor (JFET). MESFETs can be made normally-on or normally-off by varying the active layer thickness and doping but normally-off MESFET's are more critical to fabricate. p channel MESFETs and n or p channel MISFETs have never been fabricated satisfactorily on GaAs [3] because of their low frequency instabilities. On the other hand, complementary JFETs have been demonstrated successfully [4]. However, they have so far been limited to radiation hard, medium speed digital applications.

From 1974 to the early 1980s, normally-on MESFETs have been widely

used in various logic gate arrangements (BFL, SDFL, . . .) to demonstrate
the high-speed capability of GaAs at the SSI/MSI level [5]. More recently,
NMOS-like logic gates (DCFL ⇒ direct coupled FET logic) based on
normally-off MESFETs have became possible thanks to better control of the
active layers.

LSI circuits and SRAMs (1 k, 4 k and even 16 kbits) have thus been
demonstrated with high speed and low power performances resulting in a
factor of merit (speed × power) about 10 times better than that of
equivalent silicon circuits [6].

Finally, after 10 years of active investigations, a first generation of digital
ICs based on 1 μm MESFETs is now getting out of the research labs [9].

It could be further improved using scaled-down FETs (reduced source to
drain spacing, reduced gate length, 0.5 μm, higher doping level in the
channel, $\geq 10^{18}$ cm^{-3}, self-aligned structures, . . .) to take full advantage of
the large overshoot of the saturation velocity of quasi-ballistic electrons in
submicron structures. MESFETs with 0.2 μm gate lengths, have thus
exhibited values of f_T as high as 60 GHz [7].

We will now briefly review the most promising types of transistors which
will certainly be used in the second generation of GaAs digital ICs.

The first type is usually referred to as 'the high electron mobility
transisotr' (HEMT), also called TEGFET, MODFET, etc . . . ([8], [9]). It is
basically a Schottky gate FET in which the conducting path is a 2-
dimensional electron gas (2 DEG) at the heterointerface between a highly
doped wide gap material (AlGaAs) and an undoped smaller gap material
(GaAs). Because the electrons are transferred from the wide gap material to
the smaller gap material, the electron density in the 2 DEG can be high
while the electron mobility is that of the undoped material. At 300 K
HEMTs exhibit equivalent high frequency performances to MESFETs with
0.25 μm shorter gates. The performance edge of HEMTs is much more
pronounced at 77 K [10] where the electron mobility in 2 DEG can reach
70,000 cm^2/V/s for sheet carrier concentration at 10^{12} cm^{-2}. Various digital
MSI and LSI DCFL circuits have been demonstrated with MBE HMETs. So
far, the overall performances are marginally better (20%) than those of
MESFET equivalents ([11], [12]).

A novel form of HEMT that could be called 'undoped HEMT', also called
'SISFET' for semiconductor insulator semiconductor FET, or 'HIGFET' for
hetero-insulating gate FET, seems more appropriate for digital applications.
Undoped HEMTs are based on undoped heterostructures, the electrons
being injected from the source n$^+$ region towards the drain n$^+$ region,
through the 2-dimensional potential well at the heterointerface when the
gate is forward-biased [13]. Such structures are potentially less sensitive to
process variations than MESFETs or even conventional HEMTs, since the
transistor threshold voltage is now independant of the distance between the
electrons and the gate. Complementary undoped HEMTs (n and p
channels) are also possible, which means that complementary logic circuits
can be envisaged, opening the way to GaAs VLSI [14].

The second type is a npn heterobipolar transistor (HBT) [15], with an emitter-base heterojunction implemented with a wide gap material (emitter = AlGaAs) and a smaller gap material (base = GaAs). The band gap discontinuity enables the over doping of the p^+-base ($\geq 10^{19}$ cm^{-3}) while keeping an emitter current injection coefficient higher than 0.95. Self-aligned 68 GHz HBTs with 2 µm emitters have been reported [16], demonstrating a performance edge over silicon BJTs.

HBT-based CML ICs have recently been reported [17] using a mesa process reminiscent of the early silicon bipolar processes. Planar lateral HBTs have also been developed for large GaAs I^{2}L gate arrays [18].

Beyond these candidates for the second generation of GaAs digital ICs in the early 1990s, speculative devices have been proposed and fabricated to reach ultimate performances with solid state devices. They are mainly based on quasi-ballistic hot electron transport and are therefore reminiscent of vacuum tubes. 30 GHz vertical field effect transistors like permeable transistors (PBT), [19], or the resonant hot electron transistors (RHETs) [20], are among the most exotic devices ever demonstrated.

Simultaneously, for future optoelectronic systems, including digital interfaces, similar types of transistors have been implemented on indium phosphide and related ternary and even quarternary compounds.

Schottky barriers are difficult to control on indium phosphide (InP) but good dielectric – InP interfaces have been demonstrated [21] allowing the fabrication of inversion mode MISFETs. Digital ICs based on InP MISFETs are still suffering from a residual low frequency drift of the drain current.

Using InP substrates, structures like JFETs on GaInAs have also been demonstrated [22]. More impressive are N-on and N-off nAlInAs/GaInAs HEMTs, as well as InGaAsP/InP HBTs ([23], [24]).

In conclusion (Table 2.1) selecting the right devices and the right process (active layers, planar process, self-aligned structure etc.) [25] has been historically determined by the ease of fabrication as much as by the achievable microwave performances (f_T, f_{max}, associated gain at $f_T/10$) (Fig. 2.1).

2.1.2 GaAs FET basics

Like Schockley's original silicon JFET (1952), [26], GaAs FETs consist of a conducting channel whose thickness is controlled by the depth of the depletion region under the gate electrode, and consequently by the gate voltage with respect to the channel. On n-type GaAs layers, the surface depletion region can be easily implemented with a Schottky gate (n-MESFET) or a p-type GaAs gate (n-JFET).

Conventional FETs can be split into an n-doped intrinsic part, directly under the gate and two n+ doped extrinsic parts often called the source and drain access regions.

The active channel spreads from source to drain and is squeezed between

Table 2.1. Active devices for III–V digital ICs (300 K)

Transistor Type	Subst	Active layer (examples)	Active layer preparation	Microwave performances (1987 state of the art)	Basic gate configurations	Threshold voltage uniformity/logic (%) swing
D-MESFETs (N-on)	GaAs	n-GaAs	I^2, VPE, MBE	f_T = 20 GHz (1.0 µm) gm = 200 mS mm^{-1} f_T = 60 GHz (0.2 µm) gm = 400 mS mm^{-1} f_T = 80 GHz (0.1 µm) gm = 600 mS mm^{-1}	Non-complementary buffered	Good 5–10%
E-MESFETs (N-off) 1 µm – 0.5 µm gate	GaAs	n-GaAs	I^2	f_T = 20 GHz (1.0 µm) gm = 250 mS mm^{-1}	Non-complementary buffered	Good 5 – 10%
HEMTs 1–0.25 µm gate	GaAs	AlGaAs/GaAs AlGaAs/GaInAs/GaAs AlInAs/GaInAs	MBE, OMVPE	f_T = 70 GHz (0.2 µm) gm = 430 mS mm^{-1} f_T = 70 GHz (0.2 µm) gm = 495 mS mm^{-1} f_T = 26 GHz (1.0 µm) f_T = 100 GHz (0.2 µm) gm = 700 mS mm^{-1}	Non-complementary buffered	Very good 2–5%
HBTs 3 – 2 µm emitter	GaAs InP	AlGaAs/GaAs	MBE OMVPE	f_T = 55 GHz (1.2 µm) gm = 6000 mS mm^{-1} f_T = 68 GHz (2.0 µm)	Common mode logic	Excellent 1%
UNDOPED HEMTs (MIS like)	GaAs	GaAlAs/GaAs n and p-channels	OMVPE	f_T = 20 GHz (1.0 µm) gm = 400 mS mm^{-1}	Complementary	10%
MISFETs	InP	n InP	I^2	f_T = 10 GHz (1.0 µm)	Complementary	10%
JFETs	GaAs InP	n, p-GaAs n, p-GaInAs	I^2, VPE OMVPE	f_T = 10 GHz (1.0 µm)	Complementary	Very good 5–10%
PBTs	nGaAs n-GaAs		VPE	f_T = 37 GHz		10%
Hot electron transistors	GaAs	heterostructures n+/n–/n+ InGaAs/InGaAlAs GaAlAs/GaAs	MBE, OMVPE	Optimum operation at 77K		10%

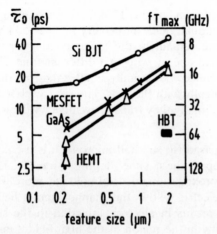

Fig. 2.1 Minimum switching time ($\overline{\tau}_o = 1/\pi \, f_{\text{Tmax}}$) versus feature size for silicon and GaAs transistors (for BJTs emitter width or emitter base spacing).

the surface depletion region (including that of the gate), and the interface depletion region at the interface between the semi-insulating substrate and the channel.

Mid bandgap traps, which are required to make GaAs substrates semi-insulating, control the low frequency response of these depletion regions. These parasitic effects will be analysed in terms of their effects on circuit performances in Section 2.1.3.

Electron transport in GaAs FETs normally assumes the gradual channel approximation with $L_g/d \geq 6$, where L_g is the channel length and d the thickness of the fully open channel [26].

Electrons injected from the source region are accelerated by the longitudinal electric field between drain and source.

For low drain to source voltage ($V_{DS} \leq 1$ V), electrons move with a high low field mobility which is usually close to the bulk mobility, except for electrons travelling near the boundary zones of the depletion regions because of a weaker coulombian screening.

For higher V_{DS}, the surface depletion region under the gate is deeper on the drain side (pinched channel), resulting in a non-uniform electric field with a transversal component. Locally the electric field can be very high (≥ 10 kV/cm) which can cause the electron velocity to saturate.

The maximum drift velocity reached in the channel of GaAs FETs depends strongly on the gate length.

(a) For greater than 1 micron channel FETs, the energy $\Delta\xi$ gained by electrons can be as high as 0.3 eV, which results in electron transfer into a higher conduction valley. A high field domain is thus formed under the gate. Current saturation is caused by the saturation of the electron velocity [27].

(b) For sub-half micron channel FETs, $\Delta\xi$ remains smaller than 0.3 eV, but the longitudinal field can be very high (≥ 30 kV cm^{-1}). Electrons under the gate remain in the lower conduction valley. They gain momentum and energy with two different time constants τ_m and τ_e respectively. Since τ_m is usually smaller than τ_e, the electrons do not relax energetically with the ions, and can reach a velocity as high as 3.10^7 cm s^{-1}. Finally they transfer to higher valleys between gate and drain.

When L_g is decreased (to sub-half micron), it is usually difficult to keep L_g/d constant (except if the channel doping is increased accordingly). The gradual channel approximation does not hold any more, i.e.: the transversel component of the electric field in the channel cannot be neglected. This 2-dimensional effect results in electron injection in the substrate or in the buffer layer primarily in the region of the high field domain [28].

Figure 2.2 shows the experimental variations of (f_T) max and (g_m) max of MESFETs with L_g ($0.2 \leq L_g \leq 1$ µm).

There is glaring evidence of the influence of the aspect ratio, L_g/d on f_T and $(g_m)_{max}$. With $L_g/d \geq 5$ transconductance as high as 250 mS mm^{-1} for $L_g = 0.7$ µm and 400 mS mm^{-1} for $L_g = 0.2$ µm have currently been achieved with MESFETs which is only about 20% smaller than the best HEMTs results. The same roughly holds for f_T. Optimizing a MESFET necessarily means increasing the channel doping level, in order to use the thinnest active layer compatible with the technological process.

However, at low temperature (77 K), HEMTs do exhibit a much better g_m and f_T than MESFETs. For instance, the transconductance of a 0.5 µm HEMT rises from 280 mS mm^{-1} at 300 K up to 450 mS mm^{-1}, at 77 K. This 60% improvement is also seen in f_T, while it is only 10% to 15% for MESFETs.

2.1.3 Parasitic effects influencing MESFET digital IC performances

Many parasitic phenomena have been observed on successive generations of MESFET devices. These effects have been widely held responsible for various limitations in the performances of both digital and analogue ICs.

However, a lot of work has been carried out to analyse their origins, understand their mechanisms and, if not completely eliminate them, at least minimize their undesirable effects [35].

2.1.3.1 *Physical origins*
Most of today's GaAs digital ICs are fabricated on silicon doped layers implanted directly into chromium-doped or undoped LEC semi-insulating (SI) substrate. Deep levels impurities are used to compensate for shallow donors or acceptors to make the substrate semi-insulating. These deep traps in the substrate, have low frequency responses ranging from microseconds

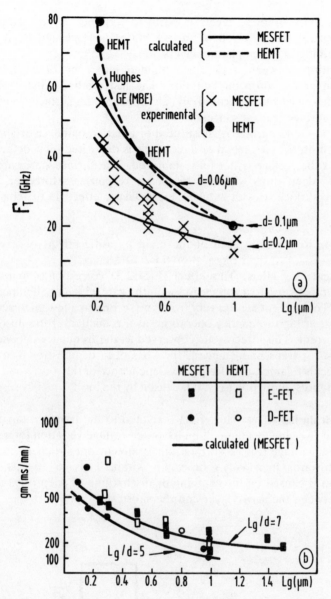

Fig. 2.2 F_{Tmax} (a) and gm (b) versus L_g and d for GaAs MESFETs and HEMTs.

up to seconds and have harmful electrical effects on MESFETs dynamic operation.

GaAs SI substrates also suffer from dislocation networks affecting the homogeneity of the ion-implanted active layers. However, slightly indium-doped GaAs substrates have enabled a reduction in the dislocation density from 10^5 cm^{-2} down to less than 500 cm^{-2} [36].

The uniformity of these layers is partly determined by the ability to carefully control the diffusion of surface states, surface and deep traps during annealing of the implanted layers.

The isolation between devices is obtained either by direct selective implantation or by boron implantation. The last method creates surface defects which have to be cured by an appropriate surface treatment minimizing surface leakage currents.

Finally, the GaAs surface is constituted of a native oxide layer and free ions buried in it. When a metal or a dielectric is deposited, new defects are induced and the surface is not fully stabilized. The surface states are also governed by deep traps with low frequency responses. Moreover, these layers create surface stresses inducing piezoelectric effects in the channel.

2.1.3.2 Electric description

It is possible to correlate each of the main parasitic effects observed in standard processes to their physical origins.

(a) **The gate lag effect.** This effect (Fig. 2.3) corresponds to a drain current lag in response to a voltage pulse on the gate with a small impedance drain load. This current lag can vary from one second to a few microseconds and appears, at high frequency operation, as a reduction of the maximum available current. This effect is also observed as the frequency dependence of the transistor transconductance [37]. It has been demonstrated that the gate lag effect is related to the frequency dependence of the source and drain access resistances; more precisely, it is caused by the low frequency response of surface states.

(b) **The drain lag effect.** This effect is related to the transient lag of the drain current I_{DS}, due to a large drain to source voltage variation for a given gate to source voltage (Fig. 2.4). First, the drain current switches abruptly, then lags towards its steady-state value within 1 μs to 100 μs. This phenomenon is caused by the charging or discharging of deep traps at the interface between the active layer and the substrate.

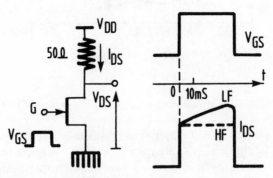

Fig. 2.3 Measurement of the gate lag effect.

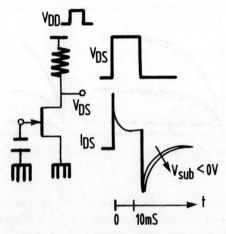

Fig. 2.4 Drain current lag associated with a pulse on the drain.

The time constants associated with these traps depend on the frequency and the amplitude of the electric field in the channel. Under high field conditions, electrons are injected from the channel to the interface depletion regions. They are trapped when V_{DS} is increased and are reemitted when V_{DS} is decreased as long as the time transients are not shorter than the trap time constants. The drain lag effect is therefore a combination of electron injection and electron trapping.

This results in various effects such as:

- the so-called looping effect,
- the frequency dependence of the real part of the output conductance (which increases by a factor of 2 to 3 from low frequency to high frequency),
- high low-frequency noise for MESFETs biased near V_T [38].

This parasitic phenomenon also acts as a dynamic offset current generator and has disastrous consequences for ICs requiring ultimate accuracy, such as fast comparators.

Note that the looping effect is drastically reduced when the drain to source voltage amplitude is limited to less than 1 V. This confirms the influence of the amplitude of the drain to source electric field variations and demonstrates the interest for low power supplies and low logic swing digital ICs (Fig. 2.5).

Recently, p-layers which form a high barrier interface with the n-channel have been shown not only to improve the output conductance (lower g_d), and the low frequency channel noise, but also to eliminate the drain lag effect completely [39]. Improved interfaces have also been demonstrated with undoped buffer layers or even AlGaAs buried layers. These solutions are however not compatible with a fully-implanted process.

(c) **The backgating or side gating effect.** This is related to imperfect isolation between two devices integrated on the same GaAs SI substrate. It

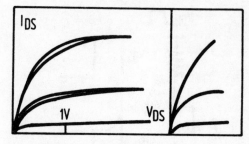

Fig. 2.5 Dependence of the looping effect on the amplitude of V_{DS}.

corresponds to the ability of a Schottky or an ohmic contact called a 'sidegate' to control the channel conductance of a near-by MESFET or resistor. A lot of work has been devoted all over the world to clarify the physical mechanisms which stand behind this parasitic effect ([40], [41]).

A few basic features, of prime importance to the GaAs IC designer can be pointed out:

- The sidegating effect only appears when the side gate is negatively biased with respect to the various electrodes of the influenced device.
- It only appears if the negative voltage difference is greater than a threshold voltage V_{SO} which varies approximately with the distance between the side gate and the controlled device (Fig. 2.6).
- However this threshold effect is reduced for ungated FETs (resistors), because of surface leakage currents.
- At high frequency (≥ 10 KHz), there is no marked threshold voltage.
- Low frequency oscillations are observed when the side gate voltage exceeds a certain threshold.

Measurements of I–V characteristics between two active regions have clearly demonstrated that there is a threshold voltage V_{TO} beyond which the current increases sharply as the 6th to 8th power of the applied voltage.

V_{TO} and V_{SO}, are process sensitive and for given surface conditions V_{TO} has also been found to be linearly dependent on the distance L between

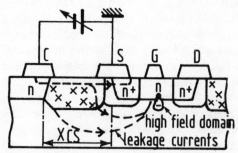

Fig. 2.6 Side gating test device.

active regions, which is different from the L^2 dependence in conventional bulk trap filled models.

It can be assumed that high density of surface states in the disordered region, form a surface conduction channel. V_{TO} represents a 'surface state filling voltage' that is proportional to L.

Different observations have been made as to the equivalence of V_{TO} and V_{SO}. V_{TO} values between $2V_{SO}$ and $\frac{1}{2}V_{SO}$ has been found. There is experimental evidence that V_{SO} and V_{TO} depend on the uniformity of the surface disordered region and on the existence of high electric fields. Such fields, result in accumulation of hot electrons and low V_{SO}.

In circuits, it may occur that a transistor drain is pulsed, while a side gate potential is also switched. The corresponding parasitic effect is a combination of the drain lag effect and the sidegating effect.

At very low or very high V_{DS} sweep frequencies, only the side gating effect is seen ($f \leqslant 1$ Hz and $f \geqslant 1$ KHz), since the interface traps responsible for the drain lag effect are frozen. For frequencies in between the effect is important and depends on the amplitude of the V_{DS} voltages and there is no threshold voltage. For high V_{DS} amplitude the quasi-static I_{DS} current is recovered. The kink effect observed here, may be caused by lowering of the interface barrier potential due to holes injection from the high field domain in the channel.

(d) **Influence of the various parasitic effects on the performances of the digital ICs.** Each of the parasitic effects described previously degrades the proper operation of the digital ICs. However their actual impacts are highly limited by process refinements and improvements. On the other hand, proper choice of the logic family can place each physical device (transistor, resistor) in such an electrical environment that the parasitic effects are minimized.

The first attempts to assemble GaAs devices to form a logic gate utilized the well-known Buffered FET Logic (BFL). This logic family requires two power supplies ($5\ V \geqslant V_{DD} \geqslant 1.5\ V$ and $0 \geqslant V_{SS} \geqslant -5V$). Since negative voltages are used within the circuit, backgating has to be seriously taken into account. The backgating threshold voltage imposes a minimum spacing between V_{SS} lines and active regions. This, of course, drastically limits the packing density. The same problem arises for SCFL (source coupled FET logic) which also requires a negative power supply.

On the other hand DCFL (direct coupled FET logic) using only one positive power supply ($1 \leqslant V_{DD} \leqslant 2\ V$) is almost completely free of backgating up to a very high packing density. (Only the load device in the logic gate, can be slightly affected.) Analysis of the drain current lag effect also concludes to the superiority of DCFL because most of the voltage transients inside a digital IC are limited by the logic swing. It is low enough (600 mV) in DCFL gates not to turn on the field induced parasitic effects, while this is not true in BFL or CML logic gates when logic swings as high as 2 V are used.

2.2 CAD TOOLS FOR GaAs DIGITAL ICs

2.2.1 MESFET CAD models

The purpose of developing an equivalent circuit model of a GaAs MESFET is to provide the designer with an accurate tool which will enable him to optimize carefully digital circuits as well as predict its ultimate performances. This requires an analysis of the physical phenomena involved in the device operation to exactly fix the limits of its use and to be aware of its limitations.

It can be pointed out that the equivalent circuit model does not need to be more accurate than the expected device variation associated with a given process. Moreover the trade-off between accuracy and CPU time (i.e. model simplicity) has to be optimized according to the number of elementary devices involved in a given design (especially for GaAs LSI design).

S-parameters of MESFETs can be measured to extract a precise equivalent quadripole. This method, well-adapted to small signal microwave analogue design can be applied to establish 'large signal' models for all possible gate and drain biasing conditions. Low frequency extraction of the equivalent circuit has also been widely used until recently, and is in fact complementary to the first method because g_m and g_d may be frequency dependent.

Two or three dimensional numerical models ([32], [46]) are used to solve in each bias point of the device, the general transport equations including non stationary effects. They are particularly suitable for ultra-short gate length devices but their extremely high CPU time requirements are making them practically unusable even for SSI digital IC simulation. However as long as they can include realistic boundary conditions (deep traps at the surface and interface) they can be used to extract very accurate CAD models.

Analytical models are used for long devices derived from Schockley's theory of JFETs [26]. For short channel devices, saturation velocity effect was introduced [47]. More recently, they have been derived from two dimensional models.

The principal features of this model are the gate-to-source and gate-to-drain diodes to represent the Schottky contact operation under normal bias conditions (for instance input impedance), as well as the drain current source controlled by gate-to-source (V_{GS}) and gate-to-drain (V_{DS}) voltages. External resistors and capacitors are added to account for the difference between the intrinsic and the extrinsic device.

If static I-V characteristics are well fitted in MESFET models, little attention is usually paid to the low frequency behaviour of the output conductance. A simple RC network can account for this effect in a first order approximation.

2.2.1.1 *GaAs MESFET analytical model: discussion*

(a) **Schottky contact modelling.** The Schottky contact of a MESFET should be ideally modelled by a distributed network since the channel is not equipotential under usual bias conditions. Actually, two separate diodes, one between gate and source and a second between gate and drain are sufficient to correctly describe MESFETs in digital circuits since the usual frequency of operation is always less than $f_T/2$.

The gate current I_G flowing through these diodes can approximately be expressed as:

$$I_G = Is \exp (qV/nkT - 1)$$

where Is is the diode saturation reverse current, V the voltage across the diode, T the absolute temperature, and where 'n' stands for the ideality factor of the contact. k is the Boltzmann's constant and q is the electronic charge.

The charge depletion region beneath the gate can be represented by two capacitances associated respectively with the gate-to-source and gate-to-drain diodes. Based on this simplification and assuming that the channel doping profile is flat and that the boundary of the depleted region is abrupt, these capacitances are given by the well-known law:

$$C (V) = Co/(1 - V/.\phi_D)^{1/2}$$

In fact, the channel has a finite thickness (d) and the capacitance decreases rapidly near the threshold voltage V_T.

When the diode is forward biased up to the flat band voltage ϕ_D, the modified Gummel-Poon expression has to be used:

$$C (V) = \frac{C_d}{(1-p)} \frac{[\delta + (1 - p) (1 - V/\phi_D)^2]^{c1 +p/2}}{[\delta + (1 - V/\phi_D)^2]^{(1+p/2)}}$$

where p and δ are fit coefficients.

Usually, the dependence of C_{GS} on V_{DS} is not considered. This roughly makes sense if C_{GS} and C_{GD} are evaluated for a mean value of V_{DS} (0.3 V for normally-off MESFETs).

Nevertheless, such an approximation can become invalid if the drain-to-source voltage reaches very high value as in N-on logic circuits with large logic swing. Variations of C_{GS} and C_{GD} are given in Fig. 2.7.

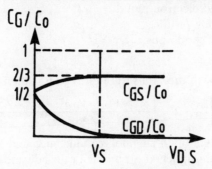

Fig. 2.7 C_{GS} and C_{GD} versus V_{DS} in GaAs MESFET (C_o = gate capacitance for $V_{DS} = 0$).

The capacitance threshold voltage is close to the value of the FET threshold voltage V_T only in the abrupt boundary approximation. Depending on this threshold voltage, the Gummel-Poon region can be the major part of the C–V characteristic (Fig. 2.8).

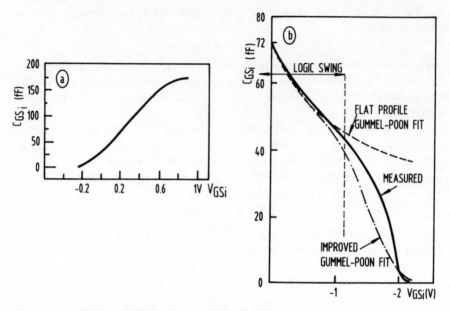

Fig. 2.8 C_{GSi} (V_{GSi}) curves for GaAs MESFET (a) 0.7×100 μm N-OFF MESFET ($V_{DS} = 0.7$ V) (b) 0.7×100 μm N-ON MESFET ($V_{DS} = 0$ V).

(b) **The voltage-controlled current source.** We will now review the most important models developed for the drain current law of MESFETs.

The Schockley approach of long-gate FETs is based on the following approximations:

- gradual channel ($L_g/d \geqslant 6$),
- constant low field electron mobility
- flat channel doping profile,
- abrupt boundary conditions.

The I_{DS} current law in the non-saturated region is then derived to be:

$$I_{DS} \text{ (Schockley)} = G_o \left\{ V_{DS} - 2/3 \left[\frac{(\phi_D - V_{GS} + V_{DS})^{3/2} - (\phi_D - V_{GS})^{3/2}}{(\phi_D - V_T)^{1/2}} \right] \right\} \qquad (1)$$

where G_o is the 'fully open channel' conductance and V_T is the threshold voltage, the channel is pinched-off on the drain side and I_{DS} saturates when $V_{DS} = V_{DSsat}, = V_{GS} - V_T$.

Beyond V_{DSsat}, I_{DS} is independent of V_{DS} in the Schockley's model.

A good approximation for (1) is given as:

$$I_{DS} = \frac{KW}{L_g}\left[V_{DS}\left(2\left(V_{GS} - V_T\right) - V_{DS}\right)\right] \tag{2--1}$$

for $V_{DS} \leqslant (V_{GS} - V_T)$

and

$$I_{DS} = \frac{KW}{L_g}(V_{GS} - V_T)^2 \tag{2.2}$$

for $V_{DS} \geqslant (V_{GS} - V_T)$ where $K = \dfrac{\varepsilon\mu}{2d}$

Channel length modulation and injection in the substrate beyond saturation can be taken into account by multiplying (2.1) and (2.2) by $(V_E + V_{DS}) / (V_E + V_{DSsat})$.

Nevertheless, this set of equations do not accurately model the actual MESFET I–V curves. In fact, the 'gradual' channel approximation is not realistic for current MESFETs with gate length smaller than 1 µm; moreover the electric field inside the device reaches its critical value which results in carrier velocity saturation. A modified version Shockley's expression including velocity saturation is [48]:

$$I_{DS} = I_{DS} \text{ (Shockley)} \cdot \frac{1}{\left(1 + \dfrac{V_{DS}}{E_c L_g}\right)}$$

where E_c is the critical electric field at which electron velocity saturates.

Using the modified Shockley's expression, it is still difficult to eliminate the discontinuity between the saturated and non-saturated regimes. To overcome this drawback, Curtice [49] has proposed the following fit expression:

$$I_{DS} = \text{ß} \left(V_{GS} - V_T\right)^2 \left(1 + \lambda V_{DS}\right) \tanh\left(\alpha V_{DS}\right)$$

which models the transition from linear to saturated regime where α, ß and λ are fit coefficients. More recently, Kacprzak [50] has proposed:

$$I_{DS} = I_{DSS}\left(1 - \frac{V_{GS}}{V_T}\right)^2 \tanh\left[\frac{\alpha V_{DS}}{V_{GS} - V_T}\right]$$

with:

$$V_T = V_{To} - \gamma V_{DS}$$

which takes into account the modulation of V_T by the drain to source voltage. Another approach consists of keeping the two regimes of operation and simplifying Shockley's modified expressions. Using a Taylor series expansion around $V_{GS} = 0$, this expression can be reduced to [51]:

$$I_{DS} = \frac{G_o}{1 + \theta V_{DS}}\left(V_{DS} - \frac{V_{DS}^2}{V_S\left(2 + \theta V_S\right)}\right)\left(\frac{V_{GS} - V_T}{\phi_D - V_T}\right)^m \tag{3.1}$$

GaAs Integrated Circuits

in the linear region, and beyond saturation:

$$I_{DS} = G_o \left(\frac{V_S}{2 + \theta V_S}\right) \left(\frac{V_{GS} - V_T}{\phi_D - V_T}\right)^m \tag{3.2}$$

where θ and m are fit coefficients and V_S the intrinsic saturation voltage, is considered to be V_{DS} independent.

The output conductance g_d is taken into account by a V_{GS} dependent impedance between drain and source. I–V characteristics (measured and modelled) of a 0.7 μm N–OFF MESFET are given in Fig. 2.9. It corresponds to a K factor of $1.9 \ 10^{-4}$ A/V^2. Simplified models for very long and very short gate FETs are summarized below:

	Shockley's model in long channel FETs	V peak approximation in short channel FETs
Assumptions	• $L_g/d \geqslant 6$ • $v = \mu E$ • $V_{DSsat} = V_{GS} - V_T$	• $L_g/d \leqslant 6$ • $v = v$ peak $= \mu \ V_s/L_g$ • $V_{DSat} = V_s$

MESFET parameters	• $I_{DSS} = KW (V_{GS} - V_T)^2/L_g$
	  with $K = \varepsilon \ \mu/2d$
	• $g_m = (2 \ K \ W/L_g) (V_{GS} - V_T)$
	• $C_{GS} = \varepsilon \ L_g \ W/x$
	  with $\dfrac{Nqx^2}{2 \varepsilon} = \phi_D - V_{GS}$
		• $I_{DSS} = v_{peak} \ W.Nq \ (d-x)$
		• $g_m = \varepsilon \ W. \ v_{peak}/x$
		• $C_{GS} = \varepsilon \ L_g \ W/x$
		  with $\dfrac{Nqx^2}{2 \varepsilon} = \phi_D - V_{GS}$

2.2.1.2 Extrinsic parasitic elements

Previous considerations only dealt with ideal instrinsic transistor. In fact, between the channel and the actual drain and source ohmic contacts, the I_{DS}

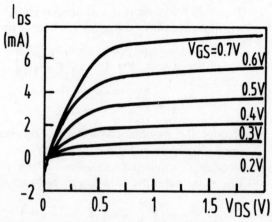

Fig. 2.9 I (V) characteristics of a 0.7 × 100 μm N-OFF MESFET ($V_T = 100$ mV).

current is flowing through resistive areas. Consequently, the intrinsic voltages V_{GSi} and V_{GDi} which control the drain current are somewhat different from those applied to the corresponding electrodes: V_{GS} and V_{DS}.

The parasitic capacitances of the Schottky and ohmic contact pads degrade the overall device performance. They include capacitances of these pads with respect to the ground plane (the substrate thickness is usually 400 μm) and the cross talk capacitances between pads. These capacitances can be deduced by using two dimensional simulation programme.

Channel edge capacitances have also to be accounted for. They can be approximated as constant capacitances $\left(C_{edge} = \dfrac{\varepsilon \pi W)}{2} \right.$ as long as pinch-off is not reached (especially on the drain side of the gate).

2.2.1.3 Extraction of a CAD model

As soon as the equivalent CAD model is determined, it is necessary to extract each of its parameter values from physical measurements. This can be achieved in a few steps: the following discussion applies, for example to Equations (3).

Since the I_{DS} current expression depends on the intrinsic drain to source and gate to source voltages, the first step is to determine the value of the source and drain access resistances as well as the low frequency output conductance. This last parameter is the slope of the I_{DS} versus V_{DS} curves which is generally also V_{GS} dependent. V_S, R_S and R_D are then derived from V_{DSsat} and I_{DSS} for various V_{GS} values.

In a second step, V_T (threshold voltage) and the fit coefficient m are extracted in the saturated region (Equation 3.2.). θ and G_o are then derived from the I–V curves in the linear regime (Equation 3.1.).

2.2.2 The interconnection issue and related modelling

The ability to calculate or to determine accurately the capacitances of interconnection wires on an integrated circuit chip becomes very crucial as the circuit packing density is increased and wire width and length decreased. Unwanted capacitive cross-talk noise limits the smallest spacing between neighbouring wires and thus the ultimate circuit density. One of the features of GaAs devices is that they are fabricated on a high resistivity substrate and as a result the fringing capacitances of electrodes play an increasingly important role in determining circuit performances.

(a) **Computation of interconnect capacitances.** 2D and 3D numerical methods have been developed using Maxwell's equation or the Laplace equation in order to determine the electric field distribution in the substrate ([52], [53]), and thus the capacitance and inductance matrix of a given set of conductors. Much work has been done to calculate these matrices analytically. The numerical methods are generally costly in CPU time and not very easy to use in CAD programs. Simple analytical formulae, deduced from either numerical simulation, theoretical approach, or experimental method have to be established for a given technology.

(b) **Modelling of interconnect lines.** Interconnect lines can be regarded as transmission lines. They are described by four parameters: line capacitance, line conductance, series resistance, and shunt conductances (Fig. 2.10).

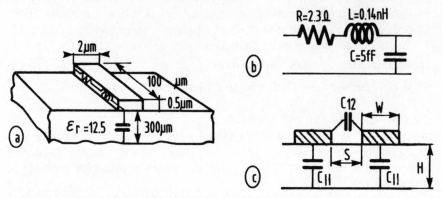

Fig. 2.10 Modelling of interconnection lines.

Shunt conductance, corresponds to losses in the semi-insulating substrate. It can be neglected for most digital circuit applications ($f \leqslant 5$ GHz). The series resistance (R_W) is process dependent. A sheet resistance value as low as 0.03 ohm/□ can be achieved. The contribution of the series resistance can be neglected as long as $R_W \leqslant 2.3\, R_{ON}$ [54], where R_{ON} is the on-resistance of the driver FET.

The series resistance only becomes a severe limitation for ultra high speed device which can deliver output signal with transition times less than 10 ps. The sheet resistance must then be lower than 0.03 ohm/□ to make sure that the 10–90% rise and fall times will not be degraded.

So in most applications today, the interconnection lines can be considered as purely capacitive. Using the microstrip transmission line formulae, the line capacitance of a 3 μm wide interconnection is 60 fF mm^{-1} on a 250 μm thick GaAs substrate and 155 fF mm^{-1} for a silicon substrate with 1 μm oxide on silicon. However these values became 52 fF mm^{-1} and 79 fF mm^{-1}, respectively for a 1 μm wide interconnection. This shows that the capacitances are scaled quasi-linearly and logarithmically versus line width for silicon and GaAs respectively (Fig. 2.11).

Cross-talk noise between two 2 μm wide lines 2 μm apart is less than 2% with silicon, but as high as 30% on GaAs susbtrate. This is accounted for by the dielectric constant of GaAs which is 3 times larger than that of SiO_2. Finally, the line-to-line capacitance C_{12}, in GaAs ICs is much larger than in Si, and also for the line to ground capacitance C_{11}.

Yoshira has shown that cross talk can be reduced from 30% down to 17% by using a thick low dielectric constant interlayer between the lines and the

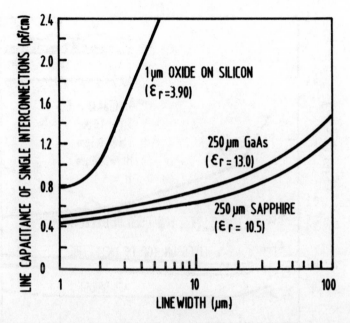

Fig. 2.11 Calculated interconnection capacitance on oxide passivated silicon, sapphire, and semi-insulating gallium arsenide substrates.

GaAs substrate [56]. Cross-talk can be further reduced down to 8% with an airbridge technology that consists in removing the dielectric interlayer and having the lines suspended in the air above the GaAs surface.

As shown in Fig. 2.12, the total capacitance $C_T = C_{11} + 2\,C_{12}$ on the middle line of a 3-line bus has been evaluated for three interconnect technologies with line spacing equal to line width (W) [57].

For $W = 1\,\mu m$, $C_T = 250$ fF mm^{-1} with lines on top of GaAs and embedded in dielectric ($\varepsilon_r = 4$), C_T is reduced down to 80 fF mm^{-1}, and even 50 fF mm^{-1} for lines on top of dielectric or air bridge lines respectively.

The same structure with a 1 μm thick SiO$_2$ layer on silicon, exhibits a total capacitance C_T of 150 fF mm^{-1}.

This clearly shows that it is possible to have much lower interconnect capacitances in GaAS ICs than in silicon ICs, as long as lines are not deposited directly on the GaAs substrate.

The ultimum interconnect technology is definitely a GaAs air bridge technology which results in an improvement of a factor of 3 over silicon.

(c) **Cross over capacitances and multi-level interconnects.** The packing density of GaAs digital circuits have long been limited by the number of interconnect levels (metal levels). However, it becomes now possible to have a half level directly on GaAs for very short interconnects ($\leqslant 100\,\mu m$) within logic gates, and two true levels for gate to gate interconnects. Via holes through dielectric isolation layers can be made as small as 1.5 × 1.5 μm.

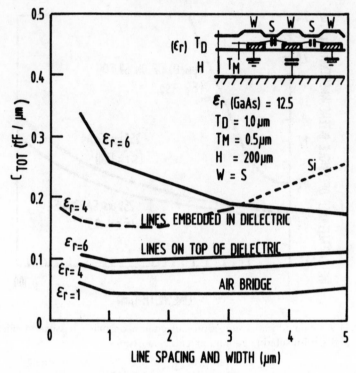

Fig. 2.12 Capacitances between tightly coupled lines.

If necessary a third metal level can be used, in order to reach the best design rules of silicon ICs today.

To minimize the high value of cross over capacitances, some laboratories have developed an air-bridge technology which gives the lowest capacitance possible [58] as discussed previously. The line capacitance to ground as well as cross over capacitances (0.15 fF μm^{-2}) can thus be reduced by 50%.

In conclusion, optimization of a GaAs interconnect technology is as important as that of the active device. We have seen that the GaAs substrate remains an advantage over silicon if an appropriate interconnect technology is developed to reduce cross talk between lines.

However, new CAD programs which can extract all the interconnect capacitances from the layout of an LSI GaAs IC are definitely badly needed.

2.2.3 CAD packages for GaAs digital design

2.2.3.1 Introduction

The impressive advance of microelectronics R&D and production raised, some ten years ago, a huge need in computer-aided design systems. Indeed, the design of an integrated circuit consists of several different steps, which can be summarized as follows:

- design and verification of the logic schematic (this first step also includes choices of an algorithm, an architecture, etc.),
- design of the required functional blocks (or linkage to a cell library),
- lay-out and verification,
- test pattern generation.

Most of these tasks were done by hand in the early years of integrated circuits. It is now taken for granted that the rising complexity of ICs, the reduced design and fabrication turnaround times and the need for ultimate performances with GaAs ICs require the use of high performance computers and dedicated software.

For these reasons, GaAs ICs designers take great advantage of the 'silicon experience' in computer-aided design. CAD packages are available today that can achieve many different tasks, such as logic simulations, timing verification, layout verification, etc. Those CAD packages are often included in so-called 'workstations', which are able to provide the user with an appropriate highly interactive graphic environment but which also allow easy communication with appropriate databases and libraries.

Most of the tools developed for silicon technologies can be used for GaAs digital ICs. However, specific, very refined, placement and routing software is necessary to optimize the wiring in a layout generated automatically.

2.2.3.2 Schematic capture and verification

At the very early step of a project, the design database must be fed in with the logic description of the circuit. This may be carried out either by a text description, in a given language, or graphically,. In this latter case, each logic cell is associated with a given schematic and connections between blocks are easily specified by lines, interactively drawn between inputs and outputs.

2.2.3.3 Simulations

Logic simulations enable the designer to check that the circuit or sub-block actually implements the proper logic function. Sequences of stimuli are applied to the inputs of the circuit and responses in terms of digital states can be analysed. Logic simulators often deal with four states and take into account the gate delays in nominal, best and worst cases.

These simulators enable the designer to perform an accurate simulation of the circuit, provided that the basic logic gates are accurately known (delay times, rise and fall times, power consumption etc.). If not, or if a few parts of the circuit has to be especially optimized, electrical simulations must be performed. The simulator (SPICE type) analyses the circuit at the component level in DC and transient modes.

These types of simulators obviously need models for the different electrical components. The accuracy of the simulations results being mainly dependent on the precision of the models. Also, electrical simulators are useful to simulate any external fluctuation away from the nominal operation of the circuit. For example, an accurate modelling of the temperature effects on the transistor will be required to predict any eventual degradation of the circuit performances under different temperature conditions (see Section 2.4.2.).

Most IC designers would like to be able to evaluate the fabrication yield that can be expected for their circuit. Worst case analysis could therefore be replaced by a pseudo-random statistical DC and transient analysis, taking into account the variations of the device electrical parameters over a short and long distances. This would accurately simulate the actual performance variations of circuits from chip to chip and the reproducibility from wafer to wafer. Statistical CAD models of active devices are required for such analysis. They can be developed with automatic extraction programmes as discussed in Section 2.4.2.

2.2.3.4 *Layout physical design*
This design step is usually considered to be the most important in the development of GaAs ICs. Not only the circuit performances but also the final commercial cost of the product can be greatly affected at this stage of the design for several reasons:

- Layout is usually time-consuming.
- It has to be error free, since the consequence of a mistake is always a complete layout/fabrication cycle.
- parasitic effects of pads and interconnects have drastic consequences on circuit performances.

This last point is of course especially important for very high speed GaAs ICs. Several approaches can be used to layout a circuit, each of them exhibiting its own advantages and drawbacks.

The first layout technique is the well-known full-custom approach. In that case, every device and wire between each block are drawn separately resulting in a highly optimized layout. Although this technique usually leads to a minimization of parasitic effects, it is always highly time consuming and may result in a large number of layout mistakes in the final implementation. These mistakes must obviously be detected and corrected, resulting in a still longer development time.

Various types of verification tools exist. These programs perform not only design rule and electrical rule checking but also verify the consistency of the layout by an extraction of elements and a comparison of the extracted electrical schematic and the initial schematic.

Other techniques, coming from the 'silicon world' can be used to layout GaAs circuits. These are the 'gate array' and 'standard cell' approaches usually referred to as the semi-custom approaches.

Software programs associated with these semi-custom solutions have been extensively studied, developed and sold for some years. The flexibility and algorithms of placement and routing programs often determine the performances of the circuit.

In terms of performance degradations, silicon manufacturers consider that any automation in the CAD line leads to an increase by 10% to 100% of the circuit area.

Therefore, the development of GaAs semi-custom circuits is now confronted with the following dilemma: Can GaAs circuit manufacturers afford to lose a little bit of their performance edge on other high speed technologies to reach lower development turn around time, or should the highest possible performances be kept as the overriding priority?

The answer to this question is surely very difficult but leads to the conclusion that progress on placement and routing software will be a determining factor in the future commercial success of GaAs ICs.

2.3 GaAs LOGIC GATES

2.3.1 Basic principles and design

Logic gate configurations using field effect transistors are basically technology-independent. GaAs logic gates can be very simply derived from silicon MOS logic gates. However a few specific features of GaAs MESFETs (or HEMTs) have to be taken into account. GaAs FETs do not have an insulating gate, and the high logic level has to be less than the gate barrier height namely 0.7 V to 1 V for MESFETs and 1.4 V for JFETs with respect to the source voltage [59].

Furthermore, GaAs FETs can be N-on or N-off transistors. N-off transistors are similar to enhancement n-channel MOSFETs, but a negative logic low level is necessary when N-on transistors are used, which can be achieved with appropriate level shifters in the logic gate.

Designing a GaAs logic gate involves the following steps:

- definition of the logic gate configuration,
- optimization of the static characteristics of the gate (transfer curve, noise margins etc) to ensure proper operation under worst case conditions,
- optimization of the dynamic performances of the gate (speed, power, fan-in and fan-out dependence, etc)
- optimization of the gate layout (minimum area, minimum pad capacitances, hot spots, etc).

The overall optimization of logic gate consists in determining the best trade-off between all of these design parameters.

Usually, GaAs gates are optimized for speed, and in some cases for low power consumption. The purpose of this section is to go through some basic considerations about these design steps and to serve as an introduction to Sections 2.3.2 and 2.3.3, where normally-on and normally-off gates will be more specifically discussed.

2.3.1.1 Logic gate configurations
A logic gate is always constituted of a logic block with n inputs (Fan in =

FI) connected to a positive supply voltage through a load. This logic block is essentially a voltage amplifier. It is usually followed by a current amplifier with a very low output impedance. This buffer stage has preferably two complementary outputs. The buffer stage should be capable of driving m gates (Fan out = FO) connected in parallel at the output.

In somes cases, the buffer stage can also perform some logic functions as well, or can even act as a level shifter in normally-on logic gates.

A normally-off buffered NOR-OR gate is shown in Fig. 2.13. The output buffer stage implements the wired-OR logic function.

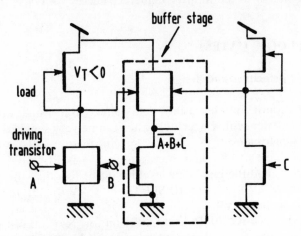

Fig. 2.13 Normally-off buffered NOR/OR logic gate.

As in silicon circuitry, logic gates can be static, but dynamic or transfer logic gates are also possible. Using GaAs FETs as pass transistors is however trickier than with silicon MOSFETs [60].

2.3.1.2 *Optimization of the static characteristics of the logic gate*

In most cases, we have to consider GaAs ratioed logic gates whether they use normally-off or normally-on transistors Ratioless logic gates have been developed with JFETs [4] and SISFETs [88] but they will not be discussed here.

Consequently, it is first necessary to ensure well defined low and high logic levels (V_L and V_H respectively) that are classically derived from the gate transfer curve as shown in Fig. 2.14 (the same holds for a non-inverting stage).

The gate threshold voltage V_{th} should be placed half way across the logic swing $\Delta V = V_H - V_L$, to maximize the worst case noise margins M and M'.

In ratioed logic, the load current has to be smaller than the maximum current through the driving transistor. This condition can be expressed by means of the linear gain G of the gate around V_{th}. G should be greater than 1. It can happen that in a sequence of logic elements a few successive gates

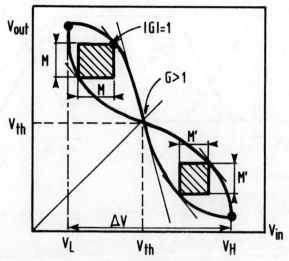

Fig. 2.14 Transfer curve of an inverting gate.

may have an overall gain slightly less than 1, in which case a strongly regenerative stage is required in order to restore the overall gain G to greater than unity. Assuming $G > 1$, $V_{th} = (V_H + V_L)/2$, we can relate G, ΔV, M and M' as follows:

$$M = M' = \frac{\Delta V}{2}(1 - 1/G)$$

As will be seen later in Section 2.3.1.3., high speed will require low gain, $(G = 2)$ which is only possible if all logic noises are well identified and smaller than the noise margins. Logic noise can be caused by cross-talk between independent gates, transient currents in supply voltages, and above all process non-uniformities which may result in improper operation and always results in performance variation.

2.3.1.3 Speed performance of logic gates

In practical MESFET circuit operation, logic signals have rise and fall times that are equivalent to the gate delay times. It is then possible, as shown in Fig. 2.15, to define the various time delays in a practical way, assuming linear transitions between V_H and V_L.

The dynamic gate threshold levels V_{th_+} and V_{th-} correspond to the voltage level reached by the input signal when the output signal starts to switch from V_L to V_H or V_H to V_L respectively.

t_{pdr} and t_{pdf} are the gate propagation delay times when the output reaches the dynamic threshold levels V_{th-} and V_{th+} after starting to increase from V_L or decrease from V_H respectively.

t_{pdr} and t_{pdr} are related to the 20% to 80% output rise and fall times (t_r and t_f respectively) as follows:

$$t_{pd_r} = (V_{th-} - V_L) \, t_r \, / \, (0.6 \, \Delta V) \approx 0.7 \, t_r$$

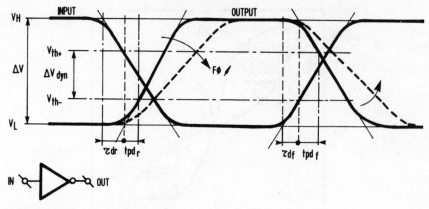

Fig. 2.15 Definition of logic gate delay times in the linear approximation.

and

$$t_{\text{pd}_f} = (V_H - V_{\text{th+}})\, t_f\, /\, (0.6\, \Delta\, V) \approx 0.7\, t_f$$

Finally two additional delay times τ_{dr} and τ_{df} have to be taken into account, they correspond to the time it takes for the output to change after the input has started to switch.

These delay times are physically caused by the Miller effect due to C_{gd} and the transistor transit time.

To evaluate t_{pdr} and t_{pdf}, we will first consider a logic gate without a buffer stage. The load capacitance C_L on the gate output node can be expressed as:

$$C_L = C_W + FI\, [C_{\text{GD}}\, (1 + 1/G) + C_D] + FO\, [C_{\text{GS}} + C_{\text{GD}}\, (1 + G)]$$

where:

C_W is the interwiring capacitance, C_D is the capacitance of the gate output node (drain pad), C_{GD} and C_{GS} are the mean values of the drain to gate and source to gate capacitances across the dynamic voltage swing ΔV_{dyn} (Fig. 2.16). G is the absolute value of the mean voltage gain across ΔV_{dyn} and finally FI and FO are respectively the gate fan-in and fan-out.

The propagation delay time of a gate can be approximately written as:

$$t_{\text{pd}_r} = C_L\, (V_{\text{th−}} - V_L)/\Delta\, I_r + \frac{R_W C_W}{3} + \tau_d$$

and,

$$t_{\text{pd}_f} = C_L\, (V_H - V_{\text{th+}})/\Delta\, I_f + \frac{R_W C_W}{3} + \tau_d$$

where $\Delta\, I_r$ and $\Delta\, I_f$ are the gate switching currents as defined in Fig. 2.17, R_W the resistance of the interwiring line, and τ_d is the delay time associated

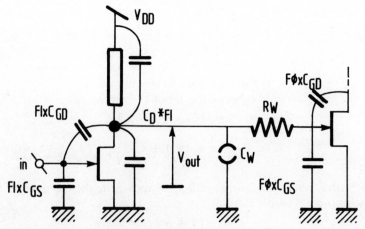

Fig. 2.16 Loading capacitances of logic gates.

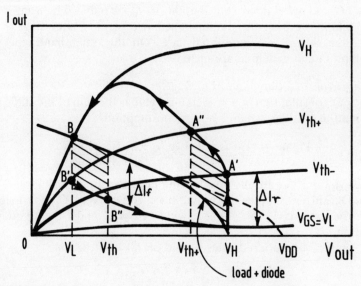

Fig. 2.17 Determination of switching currents in inverting gates.

with the feedback current $C_{GD} (dV_{in}/dt - dV_{out}/dt)$, that is not included in ΔI_r or ΔI_f.

It can be shown experimentally that:

$$\Delta I_r \approx I_{driver} (V_{GS} = V_{th-}, V_{DS} = V_H)$$

and

$$\Delta I_f \approx I_{driver} (V_{GS} = V_H, V_{DS} = V_L) - I_{driver} (V_{GS} = V_{th+}, V_{DS} = VL)$$

The $R_W C_W$ term can be neglected even in LSI circuits and memories as discussed in Section 2.2.2.

In order to get *tpd*r *tpd*f, it is then necessary to have

$$V_{th+} = V_{th-} = V_{th} \text{ (static)} = (V_L + V_H) / 2 \text{ and } \Delta I = \Delta I_r = \Delta I_f.$$

We then have:
$$t_{pd} = C_L/g_m + \tau_d$$

where $g_m = 2 \Delta I/\Delta V$ is the logic gate transconductance.

Let the mean transit time of the input transistor be

$$\tau_o = gm/(C_{GS} + ZC_{GD}).$$

We have:

$$t_{pd} = \tau_d + {}_{,o}\frac{C_{GD} (1 + 1/G) + C_D}{C_{GS} + C_{GD} (1 + G)} FI + \tau_o FO + \frac{C_W}{g_m}$$

If a buffer stage (current amplifier) is included, then a general linear expression for t_{pd} is still valid:

$$t_{pd} = \tau_d + a\, FI + b\, FO + C_W/g_m$$

where $1/g_m$ is the large signal output impedance of the buffer stage.

The *FO* and metal delay contributions to t_{pd} is reduced by using a buffer stage. Simply because the output impedance of the buffer stage is much lower than that of an unbuffered gate. On the other hand, the power consumption and gate area are increased.

2.3.1.4 *Power consumption of logic gates*
The power consumption of a ratioed logic gate is the sum of the static power consumption and the dynamic power consumption:

$$P/gate = \underbrace{V_{DD} I_{DD} + V_{SS} I_{SS}}_{\text{static}} + \underbrace{\Sigma_i (C_i \Delta V_i^2) f}_{\text{dynamic}}$$

where V_{DD} and V_{SS} are the positive and negative supply voltages, I_{DD} and I_{SS} the DC current supplied by V_{DD} and V_{SS} respectively, C_i any capacitance within the gate, ΔV_i the voltage swing across C_i and f the frequency of operation.

For simple non-buffered gates turned off 50% of the time

$$P/gate = V_{DD} \cdot \Delta I + C_L \Delta V^2 \cdot f$$

Adding a buffer stage will increase the static power consumption.

$$P/gate \text{ (buffered)} = \underbrace{V_{DD} \cdot \Delta I + C_L V^2 f}_{\text{logic stage}} + \underbrace{\frac{(V_{DD} - V_{SS}) I_{buffer}}{\text{buffer stage}}}$$

2.3.2 N-on logic gates – performance and limitations

In 1974, the first N-on (or depletion mode) GaAs logic gates were successfully demonstrated [2]. These gates used depletion mode MESFETs and an output level-shifting circuit was required for logic level compatibility between gates. However, about ten years before, the idea of a level-shifting buffer stage had been proposed in order to 'put more snap into silicon field effect logic gates' [62].

This clearly demonstrates that people do keep on reinventing the wheel, the basic logic gate configurations, are not technology-dependent, but mainly device-dependent. Other similarities between the historical development of GaAs and silicon FET logics have been observed.

As mentioned earlier in Section 2.1, the advent of a logic family is mainly dictated by the degree of maturity of technological processes. For instance, E/D NMOS processes were not controlled until the early 1970s, so prior to this E/E PMOS processes were developed instead. More recently, the more complex CMOS technology, has tended to replace NMOS for fast logic ICs when scaled down below 1 μm in minimum feature size.

In a similar fashion, it is well known that GaAs E-mode logic and complementary logic are badly needed, especially for large chips. However, everything started in 1974 with D-mode BFL gates (buffered FET logic), because at that time process and material development were not mature enough for E-mode logic. Some may argue that BFL was chosen for its high speed. This turned out to be irrelevant when five years later E-mode logic was finally shown to reach the same speed with a much lower power consumption [63].

2.3.2.1 Logic gate configurations

As detailed in Section 2.3.1, a logic gate consists of a logic block and, if necessary, an output buffer to improve the speed performance of the gate and its noise margins under heavy loading conditions. When D-MESFETs are used ($- 1.5$ V $\leqslant V_T \leqslant - 0.5$ V) the most appropriate logic levels are $V_{high} \simeq 0.5$ V and $V_{low} \simeq V_T$.

This results in large voltage swings (2 V $\geqslant \Delta V \geqslant 1$ V). The required V_T uniformity σ (V_T) is 50 to 100 mV over a wafer which is easily achieved by GaAs foundries today. This is the main reason why D-mode logics were developed early. Since it is now possible to achieve σ (V_T) as low as 20 mV, E-mode logics should soon completely replace D-mode logics.

The most straight forward and/or logic block as shown in Fig. 2.18 consists of transistors in series and/or in parallel with an active load, i.e. a D-MESFET with gate connected to source. In this way, 2-input NAND gates and up to 5-input NOR gates are possible. Multiplexing gates definitely demonstrate the high level of flexibility of such arrangements, which is, by the way, quite common in silicon MOS ICs. This logic block is biased with a positive supply voltage V_{DD}, and a level shifting stage biased between two supply voltages $V_{DD} \geqslant 0$ and $V_{SS} \leqslant 0$ is required for V_{in} to V_{out} compatibility.

The level shifter element is implemented with diodes in series biased by a D-MESFET current source. The level shifter circuit can be connected directly to the logic block output or through a common drain D-MESFET. In the latter configuration, the level shifter acts also as a buffer stage, and this type of logic has been called buffered FET logic (BFL), whereas the former unbuffered configuration is usually called FET logic (FL).

The number of diodes (n) in the level shifter circuit is related to the low logic level and consequently to the threshold voltage of the D-MESFETs.

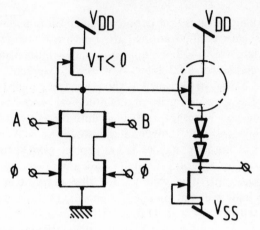

Fig. 2.18 D-mode multiplexing gate.

We have approximately:

$$V_T = -(0.3 + n\,0.8)\ \text{V}$$

To increase the speed of the FL or BFL gates, various capacitative feed-forward arrangements have been proposed such as feed forward FET logic (FFFL) [64], (Fig. 2.19).

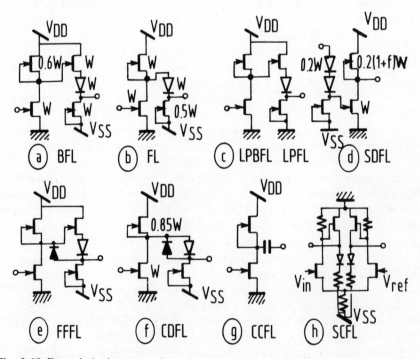

Fig. 2.19 D-mode logic approaches.

It has even been proposed to replace the level shifting stage by a capacitive coupling between gates, capacitor coupled FET logic (CCFL) [65]. It should be noted that CCFL gates require that the capacitor between gates be precharged and therefore cannot be operated down to DC which is not acceptable in most digital applications.

A different approach consists in using diodes in parallel for the logic block. This also has long been known from silicon. This type of logic has been called Schottky diode FET logic (SDFL) [66]. The SDFL inverter consists of a diode level shifter as the one used in the FL inverter followed by a D-mode inverting block. Two diodes in parallel implement an OR function. SDFL and BFL logics can be combined as shown in Fig. 2.20, to extend the logic flexibility.

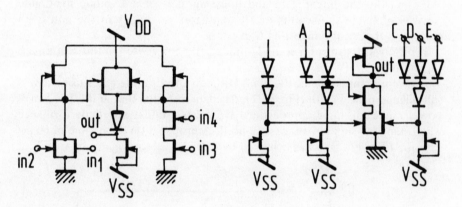

Fig. 2.20 Design flexibility of BFL and SDFL gates.

Recently it has also been proposed to use the common-mode logic approach with MESFETs. It is referred to as source coupled FET logic SCFL (Fig. 2.19 (h), and consists of a differential stage combined with two common drain, level shifting output buffers. SCFL is easier to use with N-off MESFETs and will consequently be discussed in Section 2.3.3.

Finally, a merge between CCFL and FL has been proposed [67] and called capacitor diode FET logic (CDFL) (Fig. 2.19(f)). The capacitive coupling is implemented with a reverse biased diode.

2.3.2.2. Power and speed performance

For many years, a lot of effort has been devoted to improving the speed-power performance of the original BFL gate.

Because of the level shifting stage, the early BFL gate had a high power consumption ($P = 10$ to 40 mW for $V_T = -1.5$ to -2.5 V).

Unbuffered gates seem then more appropriate for low power applications. This is not fully true since the DC power consumption in the level shifter element (SDFL or FL gates) has to be kept high to allow for large fan-out conditions.

No matter what type of D-mode gate is used, the power consumption P is roughly proportional to $V_{DD} - V_{SS}$ and V_T. The only way to really reduce P is therefore to use V_T as low as –1 V and even –0.5 V for BFL, SDFL or CDFL gates with $V_{DD} \approx 1.5$ V and $V_{SS} \approx -1.5$V.

At the boundary between D-mode and E-mode logics we have low power FET logic (LPFL) gates (Fig. 2.19(c)) using quasi normally-off MESFETs with $V_T \approx 0$ V and with $V_{DD} \approx 2$ V and $V_{SS} = 0$ V.

The power consumption of low power BFL, SDFL or CDFL gates range from 1 mW to 10 mW and from 1 mW to 3 mW for LPFL gates. This is 2 to 10 times higher than with unbuffered E-mode gates.

Optimizing the linear gain and noise margins of a D-mode logic gate consists of fixing the widths of all transistors or diodes in the gate with respect to that of the inverting transistors.

Typical width ratios for some D-mode gates considered in this section are shown in Fig. 2.19.

A recent comparison of the main D-mode and E-mode gates using a WN self-aligned process [70] (Fig. 2.21), has demonstrated that in the 1 to 5 mW power range, all logic approaches have a very similar speed performance when standard supply voltages are used. A propagation delay time of 60 ps/gate for $FI = FO = 2$ was achieved by LPFL at 2.9 mW/gate, by CDFL at

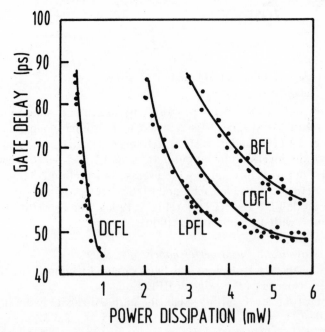

Fig. 2.21 Gate delay time versus power dissipation for low power GaAs logic approaches ($FI = FO = 2$).

3.7 mW/gate, by BFL at 5.2 mW/gate and by unbuffered E-mode gates (DCFL = direct coupled FET logic) at 0.5 mW/gate. The E-FETs had a V_T of 0.1 V and the D-FETs of –0.5V.

The main performance of D-mode logic gates are summarized in Table 2.2.

Because of the high power dissipation/gate, D-mode GaAs digital ICs are usually limited to 1000 gates (3 W). To prevent hot spots on the chip, it is important to place the output buffers (especially when ECL compatibility is required) as far apart as possible.

In laying out an MSI circuit with D-mode gates, it is also important to minimize backgating effect (Section 2.1.3.) by careful placement of the V_{SS} lines.

Table 2.2. Performances of GaAs logic gates (L_g = 1 μm)

	LPBFL/CDFL SDFL	DCFL	BDCFL	SCFL
V_T(V)	– 0.5/– 1	0.1 V/0.2	0.1 V/0.2	0.1 V/0.2
Supply voltages	$V_{DD} - V_{SS}$ = 3 V	V_{DD} = 1–2 V	V_{DD} = 2 V	$V_{DD} - V_{SS}$ = 3 V
P (mW)	2–10	0.2–0.5	1–2	2–10
t_{pd} (ps) (FI = FO = 2)	100–50	100–70	60–80	50–80
Logic swing (V)	0.8–1	0.5	0.6	0.4–0.8
Functions	OR NOR, NAND	NOR	WIRED OR, NOR	OR, NOR series gating
$\Delta t_{pd}/\Delta T$	0.2 ps/°C	< 0.2 ps/°C		
Packing density (gate mm^{-2})	100–200	500–1000	100–500	10–100

2.3.3   Normally-off gates – performance and limitations

2.3.3.1   *From D-mode to E-mode logics*
In 1977, GaAs E-mode logic gates were first reported [71]. The basic gate configuration used was directly derived from the Wallmark and Marcus unipolar logic proposed in 1959 [72]. The input transistor is an E-mode MESFET associated with a resistive or active load. A single supply voltage was used.

A level shifter was no longer required, since the low level obtained at the gate output (see Fig. 2.22) was sufficient to actually pinch off the input transistor of another gate. This type of logic is called 'direct coupled FET logic' (DCFL).

The total power consumption of the gate was significantly reduced ($\leqslant$ 1 mW/gate) as well as the device count per gate. The 1977 results from ring oscillators were very encouraging although the speed performance was not as good as that of BFL gates (t_{pd} = 300 ps, P = 0.2 mW/gate).

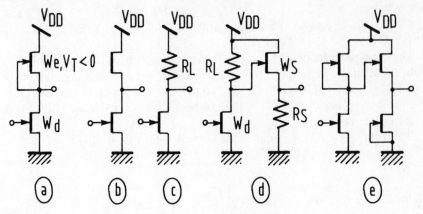

Fig. 2.22 Main E-mode logic approaches (a) E/D DCFL (b) E/saturated load DCFL (c) E/R DCFL (d) and (e) BDCFL.

During the following years, E-mode technological processes were rapidly improved. Better control of the MESFET threshold voltage and improved speed performances were achieved. Today's E-mode MESFET based gates usually exhibit t_{pd} in the 100ps range, with an associated power consumption ranging from 0.2 mW to 3 mW ([61], [73]) better results having been furthermore obtained by using HEMTs [74].

The load of the input transistor in the DCFL gate may be either a resistor or a D-mode MESFET with gate connected to source. Steady states transfer curves of the gate for both cases are shown in Fig. 2.23. The high level is clamped to 0.7 V, which is the built-in potential of the Schottky diode at the input of the next DCFL gate. For this reason, correct operation of this gate can be obtained with supply voltage as low as 0.8 V. However, typical supply voltage for DCFL gates ranges between 1 V and 1.5 V.

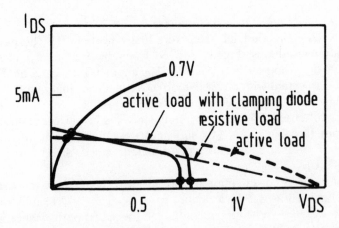

Fig. 2.23 Logic states of DCFL gates ($R = 35$ k Ω /W_d (μm), $W_e = W_d/3$.

The value of the load resistor, or the width of the normally-on FET must be adjusted for sufficient logic swing and noise margins, as discussed in Section 2.3.1. These parameters also have an influence on the speed performance of the gate (propagation delay time and fall times).

The use of either a resistive (E/R DCFL) or active (E/D DCFL) load for DCFL gates has raised a lot of controversy. As detailed in Figs. 2.24 and 2.25 E/D gates exhibit a better noise margin due to a better maximum static gain and a slightly larger logic swing. However, the power consumption of an E/D inverter is somewhat greater than that of an E/R inverter, if equivalent logic levels are achieved. The dynamic noise margins of E/D gates are affected by the frequency-dependent output conductance of MESFETs, which is not true of linear resistive loads.

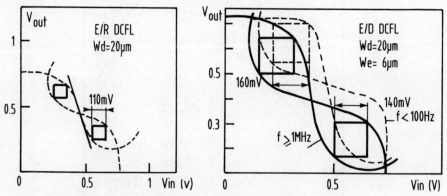

Fig. 2.24 Noise margins of DCFL gates ($P = 0.5$ mW/gate).

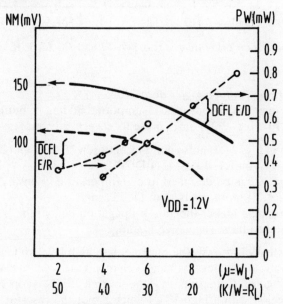

Fig. 2.25 Noise margins and power dissipation/gate versus load size in DCFL gates.

E/R and E/D DCFL propagation delay times are presented for equal power consumption and for equal noise margins in Fig. 2.26. For given speed and noise margins, the power consumption of E/D gates is higher by 60% than that of E/R gates, and the longer propagation delay time of E/D gates at a given power consumption is compensated by a better noise margin.

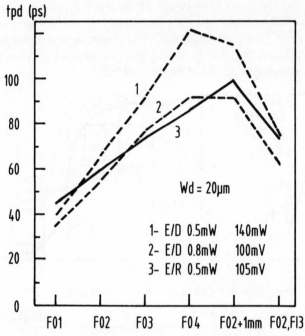

Fig. 2.26 Propagation delay time versus *FO*, *FI* and C_W for E/R and E/D DCFL gates.

2.3.3.2 *Various E-mode logics derived from DCFL*
Despite an increase in power consumption, adding a buffer stage to a classical DCFL (Fig. 2.22) gate, presents several advantages:

- As shown in Fig. 2.27, noise margins between 150 mV and 200 mV can be achieved with buffered DCFL (BDCFL) gates, at $V_{DD} \geq 2V$.
- BDCFL gates can be designed to be temperature insensitive up to 125°C (Section 2.4.2.) more easily than DCFL gates.
- The buffer stage makes the gate propagation delay time less sensitive to Fan-out and to large capacitive loading.

Indeed, with $V_{DD} = 2$ V, the intermediate high level voltage reaches 1.6 V while the output high level voltage is clamped to 0.7 V – 0.8 V by the input of the next gate. Consequently, the output FET is operating in the saturation region with a high V_{GS} (which is not the case for E/D DCFL), resulting in a high current flow in the output stage.

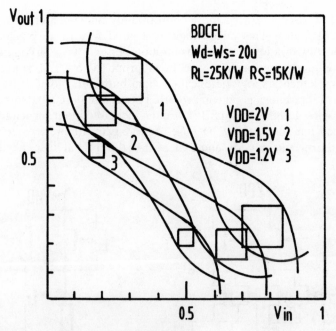

Fig. 2.27 Noise margins of E/R BDCFL gates.

Simulated and experimental performances of BDCFL gates are presented in Fig. 2.28. Compared with E/R DCFL. BDCFL gates are 50 to 100% faster than DCFL gates at the expense of a 6 times higher power consumption (3.4 mW versus 0.5 mW).

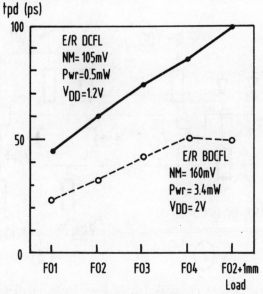

Fig. 2.28 E/R BDCFL delay time versus *FO* and C_W.

GaAs LSI circuits can be most appropriately implemented with DCFL and BDCFL gates. This mixed logic is a good way to keep high speed performance with a low power consumption per gate (≤ 1 mW/gate).

Other E-mode approaches have been proposed for LSI applications. Quasi-complementary logic gates called super buffer FET logic (SBFL) [75] (Fig. 2.29) represents a good compromise between buffered and unbuffered E-mode logics. Limitations of SBFL are the increased input capacitance and the reduced Fan-in capability because of large gate complexity.

Derived from silicon CML and ECL, GaAs differential logic called SCFL

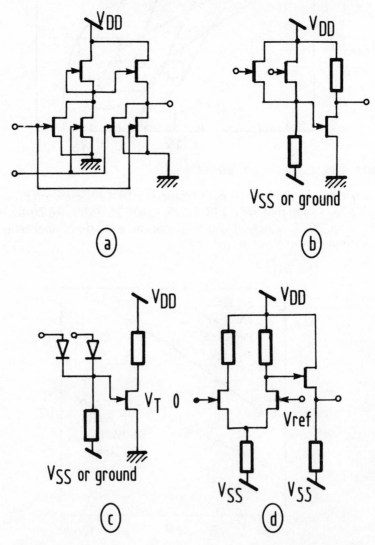

Fig. 2.29 E-mode logic approaches derived from DCFL or BDCFL (a) SBFL (b) and (c) FFL (d) SCFL.

(source coupled FET logic) can be implemented with D-MESFETs and preferably with E-MESFETs [76]. Full use of series gating enables very fast and very compact SSI functions to be achieved (1-bit full adder, master slave flip-flop . . .). Very large fan-in (up to 6) and fan-out (up to 4) are possible, as in silicon ECL. Excellent speed performances have been reported for SCFL: with 7.6 mW/gate, and a propagation delay given by: $t_{pd} = 40$ ps $+ 5$ (FI $- 1$) ps $+ 6$ (FO $- 1$) ps $+ 15$ ps $\times L$, where L is in mm.

All the main features E-mode gate configurations currently utilized are summarized in Tables 2.2 and 2.3. These E-mode gates are basic standard cells, that can be efficiently used together as long as they have been optimized for the same nominal supply voltage V_{DD} (1.5 V or 2 V).

There are several ways to increase E-mode logic swing. The Schottky gate metal can be replaced by amorphous silicon (MASFET) [59] which results in 1 V instead of 0.7 V barrier height. The same effect can be achieved with JFETs.

A Schottky diode can also be used in series with the transistor Schottky

Table 2.3. E-mode logic gates

Gate Configuration	Unbuffered gate (excellent for low power)		Buffered gate (excellent t_r and t_f matching and FØ sensitivity)		
	Direct coupled FET logic (DCFL)	Follower FET logic (FFL)	Super buffered FET logic (SFBL)	Buffered FET logic (BDCFL)	Source coupled FET logic (SCFL)
Comments on the main features	• *resistive load*: fast, optimum $V_{DD} = 1.2 - 1.5$ V • *active load* better noise margin high sensitivity to process variations • *complementary load* easy with JFET, difficult with MESFET • *compensating load* more complex two supply voltages	• derived from SDFL • larger logic swing, but much higher power consumption than DCFL	• excellent to drive large fan-out • fan-in capability limited by gate complexity	• E-mode version of BFL gates • good for large fan-out • a darlington arrangement is possible • wired-Or is also possible	• SCFL gates can be implemented with D-mode or E-mode gates • requires $|V_{SS}| \geqslant 3$ V to make full use of series gating • comparable noise margins to silicon CML gates

contact [77]. This theoretically results in doubling the logic swing, without degrading the speed performance of the gate. Sensitivity to process variations and to temperature are however increased. The extra Schottky diode can also be biased with a D-mode current source. This requires a negative supply voltage (Fig. 2.29 (d)).

In conclusion, it is clear that E-mode logics became possible a few years ago, because of improved material and processing quality. It is important to continue to improve the uniformity of material and processing, rather than making the E-mode gates as complex and as power consuming as D-mode gates.

Above all, simple E-mode gates are definitely better because they only require a single supply voltage. Comparisons between E and D mode logic gates (Fig. 2.21) definitely establish that there is no speed advantage in using D-mode logic gates and that LSI circuits are only possible with E-mode logic gates.

2.4 GaAs IC DESIGN

2.4.1 Design including performance variations

We have previously discussed the importance in chip design of some parasitic effects in GaAs ICs which are frequency dependent. Variations of process and material parameters must also be included when optimizing a circuit. Finally GaAs digital ICs have to be simulated over a wide range of temperature (at 77 K or room temperature for computer applications, from −55°C to 125°C for military applications).

Designing performance variations is partly related to fabrication yield evaluation. It consists of evaluating the distribution of the performances of functional chips on a wafer, the fabrication yield being given by the percentage of functional chips having their performances within the targeted specifications.

GaAs chips are designed with CAD models including input parameters. Variations of these parameters can be caused by temperature process or material variations. The resulting distribution law for each parameter can be extracted from statistical measurements.

The next step is statistical simulation of the circuit (DC or transient) to derive the distribution law of each of the performance parameters. The relationship between worst case simulations and statistical simulations can then be considered.

2.4.1.1 Sensitivity of CAD model parameters to process and material variations
Material variations affect CAD models because of fluctuations in doping profiles, mobility profiles and interface electrical responses. Process variations affect CAD models because of fluctuations of the surface

electrical response, geometrical dimensions of devices and active regions. It is not always that easy to separate the influence of material variations from that of process variations.

For example, since the drain conductance g_d is frequency dependent, it is also necessary to extract the model parameters from S-parameter measurements. g_d is usually found to vary by about a factor of 3 from dc to high frequency at $V_{GS} = 0.7$ V, and up to a factor of 10 close to $V_{GS} = V_T$. Statistical measurements on wafer can now be easily performed using the Cascade Microwave probe station (Fig. 2.30).

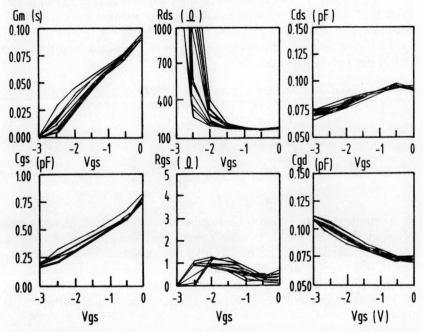

Fig. 2.30 Main electrical parameters of 17 (450 × 0.8 μm) MESFETs extracted from 2–26 GHz S-parameter measurements.

Finally it is important to determine the correlation coefficient between the model parameters. This means that eventually some fit parameters in CAD models will have to be revised to get a better agreement with the statistical data base on I (V) and C (V) curves, and even S-parameter data from which the parameters are extracted.

2.4.1.2 *Worst-case simulations versus statistical simulations*
Statistical CAD models are required to simulate the variation of the performances of all chips on a wafer [78].

Worst-case simulations have long been used in CAD design to build in lots of margin and make sure that the circuit will always meet the targeted specifications. For instance, it is still common to simulate at $2 f_{max}$, where f_{max} is the targeted maximum frequency of operation.

More recently, electrical simulators have been improved to the point where Monte Carlo transient analysis of circuit performances becomes possible. Yet statistical analysis remains CPU-time-consuming, and it is worthwhile comparing worst case analysis with Monte Carlo DC transient analysis.

Let us consider an E/R inverter and assume that the resistance value R, drain maximum current I_{DSS}, and threshold voltage V_T of the input MESFET, all follow Gaussian laws that are strongly correlated.

For a typical wafer, we have a mean threshold voltage V_T of say 100 mV with $\sigma(V_T) = 30$ mV. The drain maximum current is $I_{DSS} = f(V_T) \beta$ with $\sigma(\beta)/\beta = 5\%$. The load resistance is $R = g(V_T) \alpha$ with $\sigma(\alpha)/\alpha = 3\%$.

Mean $I_{DSS} = 75$ mA/mm and $R = 550\ \Omega$ for $V_T = 100$ mV.

In worst case analysis, the correlations are also included, and the worst case transfer curves correspond to:

$$V_{Tmax} = V_T + 3\ \sigma(V_T),$$
$$I_{DSSmin} = f(V_{Tmax})\ \beta_{min},$$
$$R_{min} = g(V_{Tmax})\ \alpha_{min}$$

and to:

$$V_{Tmin} = V_T - 3\ \sigma(V_T),$$
$$I_{DSSmax} = f(V_{Tmin})\ \beta_{max},$$
$$R_{max} = g(V_{Tmin})\ \alpha_{max}.$$

A comparison of DC Monte Carlo analysis (1000 trials) and worst case analysis in shown in Fig. 2.31. The nominal noise margins of the DCFL inverter considered here, are of 120 mV; the worst-case analysis and the Monte Carlo analysis yields 50 mV and 80 mV respectively.

It can be concluded, that even if worst-case analyses are pessimistic, they remain a safe design tool, which is far less time-consuming than the Monte Carlo analysis.

2.4.2 Temperature as a design parameter

Temperature is a fundamental parameter influencing the behaviour of GaAs devices. Its effects have to be carefully analysed to be able to predict the ultimate performances of various digital ICs over a wide range of temperature (from 77 K up to 400 K). Two approaches can be envisaged to integrate this parameter into CAD models.

The first approach consists of deriving analytical laws for the variation with temperature of the various parameters in the MESFET models.

The second approach consists of extracting all the different input parameters of the MESFET models for each given temperture. This method is of great interest especially when it turns out to be impossible to define simple analytical laws for some particular parameters (fit coefficients for

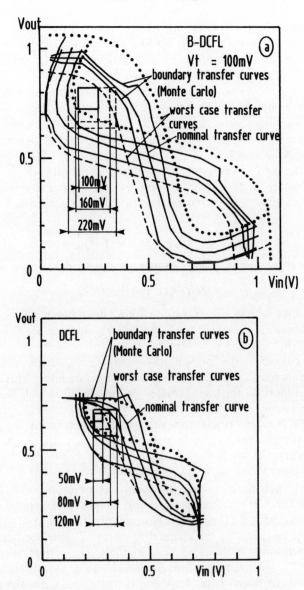

Fig. 2.31 Worst case transfer curves of DCFL and BDCFL gates.

example). This approach multiplies the number of models but has the advantage of better accuracy and reduced CPU time [79].

We will now go through the effects of temperature on the main GaAs devices (MESFET, resistors, etc.) used in digital ICs:

- ion-implanted resistors (n^+ GaAs layers)
- Schottky diodes
- MESFETs.

(1) **Resistors.** Generally speaking, most GaAs digital ICs processes utilize the same active layer for both MESFETs channels and resistors. The corresponding sheet resistances, typically from 150 Ω to 2 KΩ, are achieved with doping levels higher than 10^{17} cm^{-3}.

Consequently, the value of ion-implanted resistors increases approximately linearly with temperature and their temperature coefficient is very low because of the high doping levels used. Temperature coefficients ranging from 0.1%/deg C to 0.35%/deg C are observed for sheet resistances between 100 Ω and 1500 Ω at 300 K.

(2) **Schottky diodes.** The equivalent circuit of a Schottky diode consists of a perfect Schottky diode with an associated capacitance and leakage resistor in series with an access resistance. The current flowing through the device is given by:

$$I_G = I_s \exp(qV/nkT - 1) \tag{1}$$

The saturation current, I_s, is also temperature dependent and it is given by:

$$I_s \propto T^2 \exp(-q \, \phi_B/nKT) \tag{2}$$

It turns out that ϕ_B can be expressed with a linear law: $\phi_B = -aT + b$.

The coefficients $(-a)$ and (b) are highly process-dependent. The slope $(-a)$ ranges from -0.2 mV/deg C to -0.8 mV/deg C.

The I–V characteristics of Schottky diodes are shifted towards lower voltage with increasing temperature (effect of ϕ_B mainly). This characteristic is very important for logic families such as DCFL which have a diode limited logic swing.

(3) **MESFETs.** The drain current, which depends on V_{GS}, can be described with an analytical law including input parameters such as the threshold voltage (V_T).

We know that: $V_T = \phi_B - q \, N_D \, d^2/2\varepsilon$

The barrier height (ϕ_B) decreases with temperature. But the actual electrical thickness of the channel, d, is also temperature-dependent since it is modulated at the same time by the temperature behaviour of the Schottky barrier height (ϕ_B) and by the channel to substrate interface barrier which are both decreasing with rising temperature. This effect increases the variation of V_T with temperature and results in a negative temperature coefficient ranging from -1 to -2 mV/deg C (compared with -0.8 mV/deg C for ϕ_B).

The relative V_T variation is small for depletion mode transistors but can turn enhancement mode transistors ($0 \leqslant V_T \leqslant 0.6$ V) into depletion mode transistors when T is increased from 25°C to 125°C.

The open-channel conductance (G_o) decreases slightly with temperature. The combined effects of V_T and G_o result in a slight increase in saturation current (10%) from -55°C to $+125$°C for $V_T = +100$ mV at 25°C.

The injection of carriers via the substrate (leakage conductance) increases nearly linearly with temperature. This results in degraded slopes of the I–V characteristic (in the saturated region) as temperature increases (Fig. 2.32).

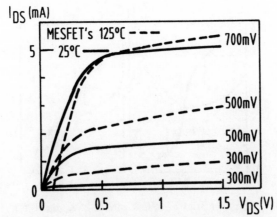

Fig. 2.32 I–V characteristics of N-off MESFET at 25°C and 125°C.

The effect of the temperature on the C–V curves is simply a shift towards negative voltages with increasing temperature. These variations are exactly correlated to the negative shift of the threshold voltage V_T. Under these conditions, these capacitances increase rapidly with temperature. As it is the same situation for the resistors, the overall performances of GaAs digital gates are better for low temperatures. If the target is to cover the military temperature range, design and worst case simulations should be made at 125°C.

The most significant parameter variations degrading GaAs devices versus GaAs ICs with increasing temperature can thus be summarized as:

> Increase in:
> – sheet resistance
> – capacitances (C_{GS}, C_{GD})
> Decrease in:
> – diode barrier height (ϕ_B)
> – MESFET threshold voltage (V_T)

These variations cause more or less pronounced degradations depending on the type of logic family used. In BFL gates, the variations of the diode level shifting result in degradation of the logic levels.

In DCFL gates, the high level is defined by the clamping of a Schottky diode while the low level is fixed by the transistor saturation current. At high temperature the threshold voltage becomes negative and the input transistor may be turned off. On the whole, DCFL is less temperature-dependent than BFL, CBFL and LPFL [70].

A smart way to minimize the degradation of the overall performances of DCFL gates is to use a buffer stage. Buffered direct FET logic (BDCFL) turns out to be less sensitive to temperature. Fig. 2.33 gives the transfer curve for both DCFL and BDCFL gates at 25°C and 125°C. It is clear that the noise margin of BDCFL gates at 125°C is nearly as good as that of DCFL gates at the 25°C.

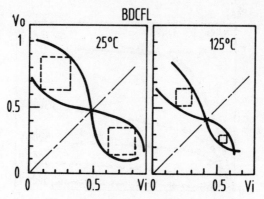

Fig. 2.33 Transfer curve of a BDCFL NOR gate at 25°C and 125°C.

Under these conditions BDCFL dividers by two have been reported to operate correctly at 4.5 GHz without any degradation from 25°C to 125°C. This capability of high temperature operation is one advantage for GaAs digital circuits over silicon equivalents [61].

Recently, excellent temperature behaviour of GaAs digital ICs has even been reported up to 200°C with a WN SAGFET process. The effect of temperature on the power delay product for various E or D mode logics is shown in Fig. 2.34.

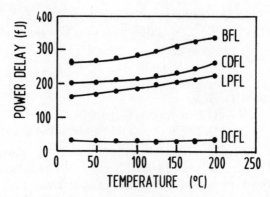

Fig. 2.34 Power delay product versus temperature for various E- or D-mode logic approaches.

2.4.3 Full custom versus standard cell IC design

In the early 1970s, with the advent of large and very large scale integration in silicon MOS technologies, one chip systems became possible and flourished.

System designers realised that they could place the equivalent of many of their PC boards on a single chip. Simultaneously, IC foundries investigated cheaper solutions for offering a quick turn around time (design + fabrication + testing) for application specific integrated circuits (ASICs). The emergence of this new way of thinking required the development of new fabrication and design techniques and tools.

The fundamental difference between standard products (memories, microprocessors) and specific digital circuits is obviously the number of dice to be fabricated. Consequently, whereas standard product fabrication can allow a long and optimized design, ASIC manufacturers' strategy will be to speed up as much as possible the overall turnaround time to avoid a drastic increase in chip cost.

The first of these techniques is the gate array approach, where wafers are processed with unconnected basic logic gates.

Gate arrays have pre-defined routing channels, free of gates and reserved for interconnection wires. As a result, the total die area is much larger than with a full custom layout, which influences directly the circuit speed. A 'real' gate array approach, with pre-positioned gates over the circuit surface will often result in LSI circuits with additional parasitic capacitances at critical nodes, which would seriously degrade the circuit performance unless each gate is buffered.

For example, the area penalty in Si technologies is considered to be as much as 1.8 to 2 times that of hand-crafted custom layouts.

On the other hand, standard cell design allows the use of buffers only on especially critical nodes (which may be determined after a first placement of cells). Moreover, total freedom (depending on the placement and routing software used) is achieved regarding the position of the basic components of the LSI, resulting in a more optimized layout.

When it comes to comparing the performances of a custom-designed and a semi-custom implementation of an LSI function, we are faced with many parameters. First of all the design target may have been for high speed or for low power consumption. Moreover, device intrinsic performances are still determining factors in reaching the ultimate speed–power product and they vary from laboratory to laboratory.

The best GaAs gate arrays ever reported are shown in Table 2.4 and Table 2.8 shows the recently reported results in fast full custom or semi-custom multipliers. The interpretation of these figures may be the following: at present, material uniformity, device performance and maximum expected yield still determine the final performance of any complex GaAs digital circuits. Eventually, improved CAD systems will make it possible to fabricate real semi-custom GaAs digital circuits with high performance and fabrication yield.

2.5 DESIGN AND PERFORMANCE OF GaAs DIGITAL ICs

We now review the main digital functions that have been implemented with GaAs technologies. It should be first pointed out that there are fundamen-

Table 2.4. Performances of GaAs and silicon gate arrays

	Logic	Gate count	Packing density (gate/mm²)	t_{pd} (ps) (FI = FO = 1)	FO delay (ps)	FI delay (ps)	Metal delay (ps/mm)	P (mW) gate
Toshiba	DCFL 1.5 µm	2 K	500	42	16	11	59	0.5
NTT	silicon ECL 0.5 µm	2.5 K (7 K)	250	78	19	14	64	2.6
Fujitsu	HEMT 1.2 µm	1.5 K	100	85 / 53 (77 K)	66 / 29 (77 K)		44 / 14 (77 K)	4 / 2.6
LEP	SCFL 0.7 µm			40	6	5	15	7.5
TI	HBT I²L	4 K		(F4)400				1
Triquint	BDCFL 1 µm	3 K		120	19.5	10	41	1
Toshiba		6 K		76	45	10	45	1.2

tally no digital functions that could not be implemented with GaAs MESFET, HEMT or HBTs technologies.

However, limitations do exist, such as the process maturity required for LSI fabrication, power dissipation per gate, etc. The development of GaAs MESFET technology started with SSI circuits such as prescalers, then moved to MSI parts such as interface or telecommunication ICs: multiplexers, demultiplexers, decision circuits, counters, A-to-D and D-to-A converters etc. Finally LSI circuits such as 1 to 16 Kbit SRAMs and arithmetic circuits, 8 × 8 and 16 × 16 parallel multipliers, were demonstrated. Recently the first 8-bit GaAs microprocessors have been opening the way to GaAs VLSI.

2.5.1 GaAs prescaler and telecommunication IC design

2.5.1.1 Introduction
Flip-flops have always been considered to be excellent test vehicles for evaluating various logic approaches and technological processes. Unlike ring oscillators, they consist of a small number of logic gates operating in a real environment (sensitivity to threshold voltage variation causing self oscillation, fan-out conditions, etc.).

Flip-flops are used in two large areas of applications:

• Static and dynamic flip-flops are basic sequential SSI blocks which are necessary to implement shift registers, stack counters, coders, decoders, etc. They will thus be part of large data paths for future GaAs LSI or VLSI chips.

- Flip-flops and more particularly T flip-flops are the building blocks of prescalers. Variable modulus prescalers are commonly used in various telecommunication systems and instrumentation equipment.

In receivers, they are an essential device for phase locked loops to synthesize local oscillator frequencies. They are also included in the timing or synchronization block of multiplexers, demultiplexers and switch matrixes for instance.

2.5.1.2 GaAs flip-flop design

Static and dynamic flip-flops and latches can be implemented with depletion or enhancement mode gates. 6-NOR gate edge triggered flip-flops and ECL master slave flip-flops have been adapted from silicon MOS or bipolar technologies. Yet a lot of effort has been made to get the best speed out of them. We will now discuss the most usual configurations.

The conventional 6-NOR gate D flip-flop is depicted in Fig. 2.35a.

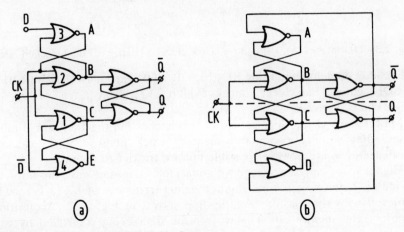

Fig. 2.35 6-NOR gate D and T flip-flops.

When used as a divide by 2 circuit, it is rather slow because of the maximum delay time across the flip-flop which results in a maximum toggling frequency $f_{max} = 1/(5t_{pd})$ (where t_{pd} is the mean propagation delay time/gate in the flip-flop). With a fully symmetrical arrangement f_{max} as high as $1/4t_{pd}$ can be achieved (Fig. 2.35b).

A conventional method for implementing a T flip-flop is to use a master-slave flip-flop.

A master-slave flip-flop is nothing but a one bit shift register. Dual clock phases are used to isolate input from output. In large chips there will be a timing block that supplies non overlapping clock signals on a clock bus.

Typical latches used as master or slave cells are shown in Fig. 2.36. They

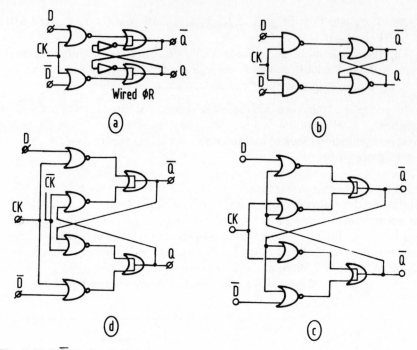

Fig. 2.36 D$\overline{\text{D}}$ latches (a), (b) and (c) single clocked latches (d) dual clocked latch.

are readily implemented using BDCFL OR/NOR gates, NAND-NOR BFL or SDFL gates. Dual clocked latches exhibit a 1 t_{pd} delay instead of 2 t_{pd} for single clocked latches.

Master-slave D flip-flops consist of two latches put together and in most cases, they require two complementary clock phases. Single clocked configurations, are however possible but are trickier to optimize.

Higher speed can be achieved with dual clocked master-slave flip-flops. A divide by 2-circuit with a maximum toggling frequency of 1/2 t_{pd} is readily achieved with the master-slave flip-flop shown in Fig. 2.37. Acquisition (master) and memorization (slave) of the data being controlled by two complementary clocks phases, each of the complementary output signals are

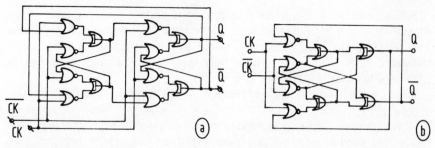

Fig. 2.37 ½t_{pd} master-slave T flip-flops (a) full master-slave configuration (b) reduced configuration.

generated simultaneously in the slave cell, and consequently the skew between the two outputs is minimum.

Although this kind of flip-flop is inherently fast, it is very sensitive to the skew time between the two complementary clock signals. Indeed, if it occurs that CK and $\overline{CK}$ are simultaneously high, both outputs are forced to zero. This phenomena can easily occur under low frequency operation because the wired-or gate threshold voltage is not exactly half way across the voltage swing. Consequently, care must be taken actually to minimize the rise and fall times of the clock signal, even when low frequency sinewave signals are applied to the clock generator.

Conversely, overlapping of CK and $\overline{CK}$, at low-logic level, can cause oscillation. In some cases self-oscillation can also be observed when the clock inputs are DC-biased.

The low cut-off frequency of T flip-flops for sinewave clock signals, can be improved by using a high gain clock generator [61].

More conventional clock signal generators can be implemented with differential stages [82]. Static and dynamic offset voltages are however a limiting factor resulting in overlapping clock signals at high frequency, so preventing circuits from reaching the f_{max} that can be expected from the dual clocked master-slave flip-flop.

Finally, the best way to get around the clock generator issue and really achieve an f_{max} of $1/2\ t_{pd}$ is to use (quasi) normally-off MESFETs in SCFL gates [76]. Conventional SCFL master-slave flip-flops are implemented with SCFL latches as shown in Fig. 2.38.

Dynamic operation of GaAs MESFETs, in a frequency division loop, seems to be interesting for achieving very high speed performances, associated with a very low power consumption.

However, this type of frequency divider is not appropriate for operation

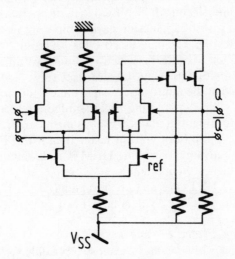

Fig. 2.38 SCFL D$\overline{\text{D}}$ latch.

at low frequency (say below $1/10\, f_{max}$) because of the parasitic off-resistance of MESFETs used as pass transistors. Indeed, when the pass transistor T_S in Fig. 2.39 is turned off, during a memorization phase, the sub-threshold current of T_S, degrades the stored value on node M with a time constant = Roff (T_S).C, leading to a minimum operation frequency of the divider at least a few MHz.

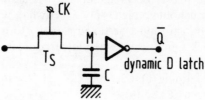

Fig. 2.39 Dynamic D latch.

Nevertheless, this kind of dynamic approach can be considered a good solution for narrow-band applications and excellent speed-power performances have been achieved with dynamic divide by 2-circuits [83] (1.9 GHz/ 0.25 mW).

The main difference with silicon MOS dynamic prescalers results from the non-insulating gate of MESFETs. Therefore, the logic levels of the clock signals have to be optimized independently [84].

2.5.1.3 Speed-power performances of GaAs T flip-flops

We will now review the performances of GaAs T flip-flops that have been reported since the early work from Rory Van Tuyl and Charles Liechti at HP [85]. Circuits dividing by 2, 4, 8, 32, and 256 have thus been reported ([86]–[90]).

The speed power performances of T flip-flops are shown in Table 2.5. Flip-flops using HEMTs, HBTs and silicon bipolar technologies have been included [91–94] and silicon MOS. Frequency of operation above 10 GHz has been achieved with $1/2\, t_{pd}$ master-slave T flip-flops using BFL, SCFL or dynamic logic gates. f_{max} as high as 20 GHz [95] and 26 GHz have been obtained with 0.3 μm MESFETs, which is about two times faster than the best silicon results with equivalent feature sizes (Figs 2.40, 2.41).

The performance of the interconnect capacitances already appears critical in small circuits as flip-flops. Using an air bridge interconnect technology [89], a speed improvement of about 50% has been demonstrated. It can be noted that HEMT T flip-flops are not faster than MESFET equivalents at 300 K but the real advantage in using HEMTs is definitely at 77 K, where f_{max} is two times higher than at 300 K.

Early results with HBTs are very encouraging (f_{max} = 13.7 GHz for 2 μm emitters). Recently, 22 GHz dividers have been demonstrated.

2.5.1.4 Frequency dividers in systems

High-speed frequency dividers have extensive possible system applications, especially in professional telecommunication equipment.

Table 2.5. Best speed-power performances of GaAs and silicon T flip-flops.

	Technology		Ref.	Feature size (μm)	Logic swing (mV)	Type	$N = 1/f_{max}\, t_{pd}$	f_{max} (GHz)	P (mW)
Silicon	Bipolar SiCOS		[91]	0.5 0.2	500 mV	T M/S	2 2	6.0 10.0	45 90
	SST		[92]	0.35	200 mV	T M/S	2	10.6	175
	NMOS		[93]	0.4		6 NOR	5	2.5	17.5
GaAs	MESFET	SCFL	[89]	(air gap) 0.5	400 mV	5 M/S	2	11.0 9.7	149 52
		CDFL	[7]	0.2			2	14.5	98
		BFL	[7]	0.2		T M/S	2	17.92	657
		DCFL		1.0	500 mV	T 6 NOR	4	5.15 3.5	78 2.9
		BDCFL		.8	500 mV	T M/S	2	6.6	20
		Dynamic logic	[90] [95]	0.5 0.3		T M/S	2 2	13.2 20.2(26)*	115 350
		Dynamic logic	[83]	1.0		T M/S	2	1.9	0.25
	HEMT	DCFL		0.35		6 NOR	4	8.0	25
		DCFL		0.7		6 NOR	4	5.5 13.0 (77 K)	10 25
	HBT	CML	[91]	3.5 2.0		M/S	2 2	11.0 13.7(22)*	150 160

* Recently published results

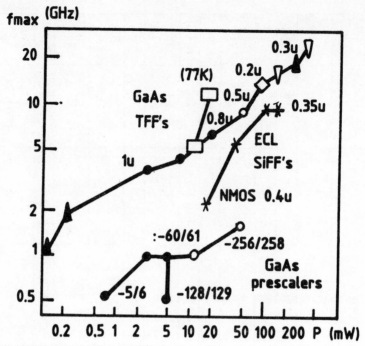

Fig. 2.40 Best speed power performance of GaAs and Si prescalers (TFFs, and variable modulus frequency dividers)

GaAs static FFs
- ▲ dynamic FFs
- ● DCFL, BDCFL
- ◇ BFL, CDFL
- ○ SCFL
- □ (HEMT) DCFL
- ▽ (HBT) CML

Si FFs
- * Si ECL and
- Si NMOS

Variable modulus prescalers and counters can be used in:

- stabilized frequency source,
- counters and heterodyne counters,
- period and time interval measurement,
- heterodyne RF receivers.

As an example, divide by 60/61 prescalers [96], based on a dynamic divide by 5/6 cell, shown in Fig. 2.42, were fabricated with a 1 μm E mode process and can be operated from 400 to 1.2 GHz with a power consumption of 7 mW. These excellent performances make this IC most suitable for land mobile radios, handheld pocket portables, modular radios, etc.

To evaluate fully the potential of this type of prescaler for use in frequency synthesis systems, phase noise measurements have been carried

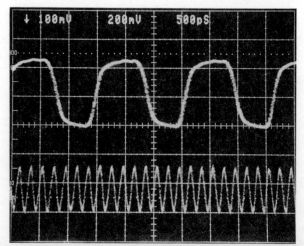

(a)

(b)

Fig. 2.41 5 GHz divide by 8 GaAs IC (commercially available) (a) input and output waveforms (b) microphotograph of the GaAs circuit.

out using an HF generator, a mixer, a phase-locked generator and a spectrum analyser.

The dynamic divide by 5/6-circuit and the whole divide by 60/61-circuit have exhibited a phase noise of –137 dBc/Hz and –124 dBc/Hz, respectively, at 25 KHz off the 1 GHz carrier. These figures are comparable and even better than silicon ECL prescalers.

2.5.1.5 *Circuits for digital telecommunication systems*
Telecommunication systems could be a most appropriate field of applications for GaAs digital ICs. Modern digital telecommunications are based on broad band switching systems and high bit rate links.

Using quasi-complementary E-mode gates, a 4 × 4 cross point switch was designed and fabricated with a 1 μm refractory gate N-off process [97].

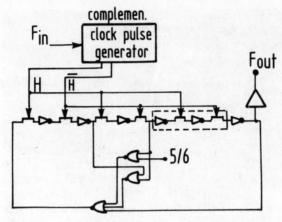

Fig. 2.42 Dynamic divide by 5/6 cell.

A four-channel digital time switch with a 2 Gb/s throughput and 500 Mb/s channel data rate, was developed with a self-aligned 0.5 μm process. The time switch was designed with low power source coupled FET logic gates (SCFL) [98].

However, high bit rate links over long or short distances are only a short-term application and market for III–V technologies.

Fibre optic transmission systems require light emitters (lasers) and light receivers (photo diodes, etc.) which can only be fabricated with GaAs or InP technologies. The input interface is a multiplexer (typically 4, 8 or 12 channels). The output interfaces after the receiver are a decision circuit and a demultiplexer with the same number of channels as the multiplexer. The whole system is synchronized by a fixed clock signal which is carried over the fibre optic link and extracted from the coded data by a clock recovery circuit (Fig. 2.43). The bit rate on the link is equal to the clock frequency.

These three circuits, multiplexers, decision circuits and demultiplexers, are three telecommunication ICs that have been most reported on. *Monolithic integration of a multiplexer with a laser has even been achieved* [99]. However separate chips seems still a better way to go in the foreseeable future, in order to achieve high fabrication yield.

Multiplexers consist of a multiplexing block which can be a multiplexing

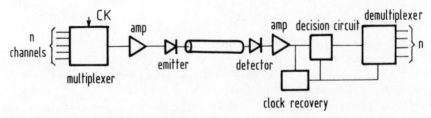

Fig. 2.43 Fibre optic link for multigigabit telecommunication systems.

gate or more conventionally a shift register and of a timing circuit which is a prescaler (Fig. 2.44).

The multiplexing gate principle consists in connecting electrically one input channel among n to the output during one period or half period of the system clock signal. For instance 4:1 multiplexers can be designed with a clock frequency f_{ck}, half the output bit rate, then a simple divide by 2-circuit is all that is necessary. In fibre optic links, since the clock signal has to drive the decision circuit, the clock signal frequency is normally equal to the bit rate and a divide by 4-circuit should be used in the multiplexer (Fig. 2.44).

The conventional shift register approach consists of loading an n-bit shift register every n periods of the clock signal and unloading the data serially on the output line at the system clock frequency. For a 4:1 multiplexer a divide by 4-circuit is necessary as in the multiplexing gate approach. The shift register approach utilizes $1.5\ n$ D flip-flops for n input channels.

For 4-channel multiplexers, the most compact solution is based on a SCFL multiplexing gate [100], shown in Fig. 2.45, which is derived from the best design in silicon ECL ([101, [102]). As already mentioned complementary inputs to SCFL gates are necessary to compensate for large offset voltages in differential stages.

The maximum data bit rate at the output of the multiplexers is given by the maximum frequency of operation of GaAs SCFL flip-flops ($1/2\ t_{pd}$). Consequentely, 5–10 Gb/s GaAs multiplexers could be fabricated today if speed is the design target. For instance, a 2 Gb/s time switch [98] previously described includes 6 GHz SCFL T flip-flops which could be used for a 5 Gb/s 4:1 multiplexer.

The best result reported so far with SCFL gates, are based on the shift register approach: a 4:1 multiplexer with a data bit rate of 1.6 Gb/s and $P = 380$ mW [100] and a 8:1 multiplexer with an output bit rate of 3 Gb/s [103]. In both cases the flip-flop speed was traded off against the power consumption, which resulted in medium speed performance.

Using simpler and more refined design, Si ECL multiplexers have achieved better speed performances [101]. A 4-channel multiplexer with a bit rate of 5.5 Gb/s, $P = 600$ mW ($f_{clock} = 2.75$ GHz) has been fabricated with an oxide wall isolation 2 μm silicon bipolar process. This circuit includes only the multiplexing gate and a divide by 2-circuit. It should also be noted that with a sub-half micron NMOS technology, researchers at Bell Labs have fabricated 2 Gbs/s 16:1 multiplexers.

Early GaAs multiplexers were implemented with low power normally-on gates [107]. A 5 Gb/s 8:1 multiplexer was demonstrated with BFL gates as part of a large word generator.

More recently various multiplexers (4:1, 8:1 and 12:1) have been reported using CDFL [104] DCFL [103], BFL ([105], [108]) and finally SBFL [106]. In most cases slow flip-flops ($1/4\ t_{pd}$) are used and the maximum bit rate achieved varies from 1.1 Gb/s to 3 Gb/s with a 1 μm D-mode process [105].

We can conclude that, as already discussed previously for prescalers, GaAs MSI circuits such as multiplexers can reach better performance than

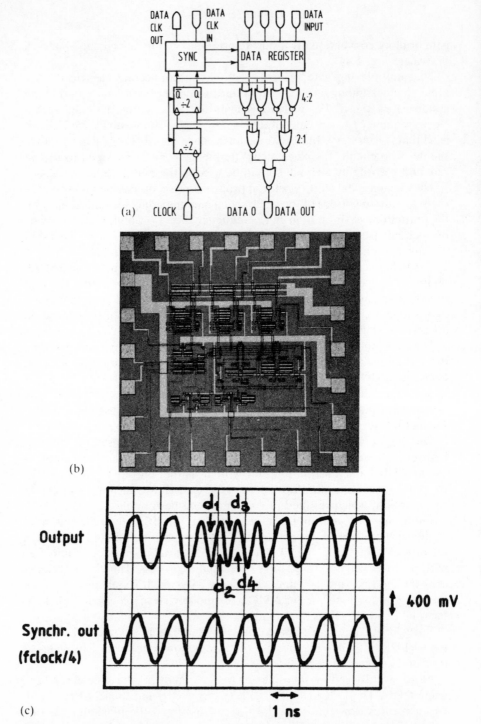

Fig. 2.44 multiplexer design example (a) block diagram (b) chip microphotograph (c) 3 Gb/s measured output waveforms. ($d_1 = d_2$ = synchr. out/4 = f clock/16, $d_3 = 0$, $d_4 = 1$).

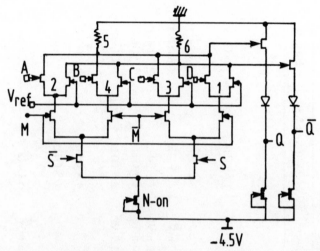

Fig. 2.45 SCFL multiplexing gate.

Si equivalents as long as the same design refinements are used. Table 2.4 summarizes the best silicon and GaAs multiplexer results.

Demultiplexers are normally based on an n-bit shift register serially loaded at the rate of the system clock signal (f_{clock}) and unloaded in parallel into n D flip-flops. The toggle frequency of the D flip-flops is f_{clock}/n, which is generated by a divide by n-circuit, usually a ring counter.

Direct demultiplexing is also possible as shown in Fig. 2.46. A demultiplexing gate consisting of a tree of D flip-flops switches the input towards the allocated output channel among n. Here again a divide by n timing circuit is required.

The best bit rate reported for GaAs demultiplexers is 2.1 Gb/s (Table 2.6)

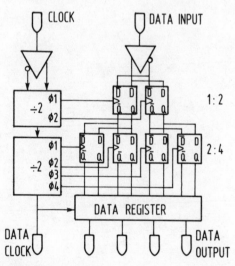

Fig. 2.46 Demultiplexer design example.

Table 2.6. Telecommunication ICs for fibre optic links

Function and logic approach [ref]	Channel number	Approach	Complexity	Power	Clock frequency	Bit rate
Multiplexers						
GaAs CDFL 1 μm [104]	8	Multiplexing gate	200 gates	1.49 W	1.6 GHz	1.6 Gb s^{-1}
GaAs SCFL 0.5 μm [103b]	4 and 8 — 1	Shift register			3 GHz	3 Gb s^{-1}
GaAs SCFL 1 μm [100]	4	Shift register	40 gates	380 mW	1.65 GHz	1.65 Gb s^{-1}
GaAs SCFL 1 μm [LEP]	4	Multiplexing gate	20 gates	350 mW 400 mW	1.2 GHz 1.5 GHz	2.4 Gb s^{-1} 3.0 Gb s^{-1}
GaAs BFL (word generator) [107]	8	Multiplexing gate	400 gates	1.9 W	2.5 GHz	5 Gb s^{-1}
GaAs DCFL 1 μm	4	Shift register			1.8 GHz	1.8 Gb s^{-1}
GaAs BFL 1 μm [105]	8	Multiplexing gate			1.5 GHz	3 Gb s^{-1}
Si ECL 0.5 μm [101]	4	Multiplexing gate		576 mW	5 GHz	5 Gb s^{-1}
Si ECL 2 μm [104]	4	Multiplexing gate	10 gates	600 mW	2.75 GHz	5.5 Gb s^{-1}

Demultiplexers

GaAs CDFL 1 μm [104]	1	8	Shift register	200 gates	1.5 W	1.6 GHz	1.6 Gb s⁻¹
GaAs SCFL 1 μm	1	8	Shift register			2.1 GHz	2.1 Gb s⁻¹
GaAs SCFL 1 μm [100]	1	4	Shift register	50 gates	500 mW	1.9 GHz	1.9 Gb s⁻¹
GaAs DCFL 1 μm [108]	1	12	Shift register		100 mW	1.3 GHz	1.3 Gb s⁻¹
GaAs BFL 1 μm	1	8	Shift register			1 GHz	2 Gb s⁻¹
Si ECL 0.5 μm	1	4	Shift register gate		1148 mW	1 GHz	4.8 Gb s⁻¹

Decision circuits

Si ECL [109]	1	(2 flip-flops)	11 gates	600 mW	1 GHz	2 Gb s⁻¹
GaAs DCFL [111]	1	(1 flip-flop)	7 gates	30 mW	2 GHz	2 Gb s⁻¹
GaAs BFL [110]	1			350 mW	3 GHz	3 Gb s⁻¹
GaAs SCFL	1			1.1 W	2 GHz	2 Gb s⁻¹

while it is 4.8 Gb/s with a 0.5 μm silicon bipolar process. The performance of GaAs D flip-flops makes it possible to reach bit rates above 3 Gb/s. However, the power consumption is almost as important as the bit rate for telecommunication systems, and GaAs n-off technologies definitely present a better speed-power trade-off than silicon.

A very important circuit in the fibre optic link is the **decision circuit**, which reshapes the data at the output of the receiver (photodiode). A decision circuit is essentially a one-level latched comparator. In some designs the comparator is not latched and reshaping is done with a D flip-flop.

The f_{max} of such decision circuits fixes the highest bit rate of the link. It is possible to improve the situation, by using two identical flip-flops operated in complementary mode and recombining the reshaped data at the output of the flip-flops in a 2:1 multiplexer. With such a design, the highest bit rate of the link is $2 f_{max}$ [109].

It is then possible to design the three interface circuits, multiplexer, decision circuit and demultiplexer, so that their f_{max} is half the maximum bit rate on the link. 2 Gb/s 600 mW decision circuits have thus been fabricated with a 2.5 μm silicon bipolar process. As usual, speed improvement results in a very high power consumption.

Conventional decision circuits have been reported with GaAs D-mode and E-mode technologies ([110], [111]). A 2 Gb/s bit rate has been achieved with a very low power consumption of 30 mW and 30 mV input sensitivity. Rise and fall times (20–80%) below 150 ps into 50 Ω were observed.

2.5.2   GaAs ADC and DAC design

2.5.2.1   Introduction

Analogue-to-digital and digital-to-analogue converters (ADCs and DACs) have become fundamental blocks in many signal processing systems. In principle, any analogue signal can be digitized without information loss; it can then be digitally processed, or simply transmitted and finally reconstructed in a highly reliable way.

The overall market today is estimated to be of about $500M worldwide; this market is mainly for low-frequency (low sampling rate), high-resolution circuits. However, since data processing systems are more and more demanding, the trend is inevitably toward higher sampling rates while keeping a sufficient resolution; a good example is the video bandwidth which is now widening up to 30 MHz and which requires to be sampled in the 100 Ms/s range with a resolution of 8 bits. In another field, instrumentation and more precisely in advanced oscilloscope systems, analogue or digital signals with bandwiths in the GHz range have now to be handled. Finally, military applications such as real-time radar signal processing or EW systems require medium/high resolution in the Gb/s range (from 500 Ms/s up to 6 Gs/s).

In the meantime, state-of-the-art silicon converters can hardly cope with these challenging requirements.

2.5.2.2 GaAs: three port devices for A to D and D to A conversions

(a) **MESFET.** While well adapted to digital circuit design, there has been some disappointment concerning the MESFETs ultimate accuracy in analogue functions.

The main building blocks in ADCs are differential amplifiers (for comparators) having simultaneously large bandwidth and low input offset voltage. The ultimate speed of a FET depends on the gate length: typical values are now 0.7 μm leading to (F_T) max of about 20 GHz.

However, the transconductance is rather poor compared with bipolar devices, say $g_m \approx 200$ mS/mm. This is not helped by the effects of process and material variation. Nevertheless, there have been drastic improvements in the last few years by using In-doped pre-annealed material and the standard deviation of offset voltages [112] has been reduced to 10 mV for V_{GS} close to V_T.

Unfortunately, MESFETs still suffer from parasitic effects which have been discussed in details in Section 2.1.3.

(b) **HEMTs.** The rapidly emerging heterostructure transistors offer new exciting possibilities to ADC designers. High electron mobility transistors (HEMTs) have been used for comparators and 5-bit converters [113]. Though the optimum performances are located in the low temperature range, good results are obtained at room temperature too. Stringent control of MBE layers as well as clean processes have demonstrated a standard deviation of threshold voltage of 4 mV over 1 mm². Moreover, it seems that the various epi-layers are somewhat playing a shielding role for the 2 DEG, thus limiting the parasitic effects occurring with MESFETs.

(c) **HBT.** Even more recently, Heterojunction bipolar transistors (HBTs) have been demonstrated satisfactorily on GaAs. 40 GHz f_T as well as high transconductance (5000 mS/mm) have been achieved with 1.6 μm emitter HBTs. This makes them excellent candidates for A to D conversion applications. The first comparators [114], exhibit offset voltages and hysteresis as low as a few mVs (one order of magnitude better than MESFETs).

However, unlike MESFETs, their speed-performance depends on their bias current which is high at maximum f_T for today's emitter dimensions (1–2 μm). Under these conditions, as with silicon bipolar transistors, very high speed will be only achieved at a very high power consumption level which will always be a limitation.

2.5.2.3 Architecture and design

(a) **Comparators.** These are the key elements for ADCs. Their design is derived from the existing ECL silicon biplar comparators, but it has to be adapted to the specific characteristics of GaAs devices: the clock-feedback structure is often used to compensate for the limited gain of the input differential amplifier (either FET or HBT); moreover, it has the advantage of retiming the outputs. Offset, backgating and $1/f$ nosie can be minimized (mainly in FET technology) by self-calibration technique, though it complicates and slows down the circuit [115]. As for hysteresis reduction in MESFET comparators, the use of enhancement-mode FETs has been found to be a major improvement with respect to N–on MESFET. In fact, since the logic part of the comparator can be operated with a reduced logic swing, it limits the feedback from the logic block to the analogue input pair [112].

   (b) **D to A converters.** D to A converters can be implemented using the well-known current mode logic circuitry (Fig. 2. 47). n-bit binary weighted currents sources (for n-bit resolution) are flowing or not into a load depending on the input digital word. These highly accurate current sources can be implemented 'off-chip' by silicon circuits or 'on-chip' using NiCr resistors. GaAs MESFETs are naturally adapted to form the differential switches and enable the circuit to operate at much higher frequencies than with silicon equivalents. Since the current sources are well defined and fixed, GaAs DACs are not as sensitive to parasitic effects as ADCs and high performance (1 Gb/s) circuits have already been reported up to 12-bit ([117], [116]) using conventional D-MESFET processes.

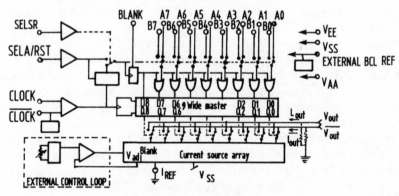

Fig. 2.47 Design example of 8–bit D/A converter.

   (c) **A to D converter.** The most utilized architecture for fast A to D converters is the well-known flash or parallel converter arrangement (Fig. 2.48). It requires 2^n-1 comparators (for n bit resolution) to operate at maximum bit rate with ultimate accuracy but produces one digital word each clock period. 4-bit at 3 Gs/s, 150 mW with a E-MESFET process [112] (Fig. 2.49) and 5-bit at 350 Ms/s with a HEMT process [113] have been demonstrated. It is likely that MESFET or HEMT converters will be limited by offset and hysteresis, to the 4–6 bit range. This may be the same for HBTs because of their higher power consumption. To reach higher resolution (8–12 bit), other architectures have to be considered; the series parallel feed forward structure [118] seems to offer an optimum speed/resolution trade-off.

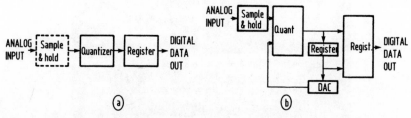

Fig. 2.48 A/D converter designs.

2.5.3 GaAs SRAM design

2.5.3.1 Introduction

Sophisticated electronic systems, particularly superfast supercomputers, will require a large quantity of very-high-speed memory circuits. For example, fast cache or scratch pad memories will be an essential part of high-speed signal processors for programmes or data storage.

Development of GaAs SRAMs has been based on different design approaches from 1982.

- SRAMs using D mode gates only, such as BFL, SDFL or CML gates [119]
- SRAMs using E/D gates only with MESFETs or HEMTs ([120], [121])
- SRAMs using E/D gates and BFL gates [122]
- SRAMs using complementary JFET gates [4]
- SRAMs using E/R gates only [123].

1 K bit SRAMs have been successfully fabricated with all of these approaches. However, fast access times (≤ 2 ns) have mainly been acheived with memories based on Normally-off MESFETs or HEMTs. This section will deal with the last design approach, and when necessary, specific features of the other approaches will be included and discussed.

2.5.3.2 SRAM design

(a) **Memory cell.** The performance of the memory cell strongly determines that of the whole memory (Fig. 2.50) and directly influences the main design aspects of the peripheral circuitry. Conventional memory cell design is based on two cross coupled inverters accessed by two pass transistors requiring four transistors and two resistors with the E/R approach.

As with bipolar memory cells, Schottky diodes can be used as access elements.

The size of FETs in the cell has to be optimized with respect to speed, power, read/write stability and area. Cell stability means achieving no change in the state of a cell when reading from the cell or when writing a complementary state into another cell.

The main factors affecting the cell stability are: the driver FET current, load current, read and write current through pass transistors, bit line current and capacitance.

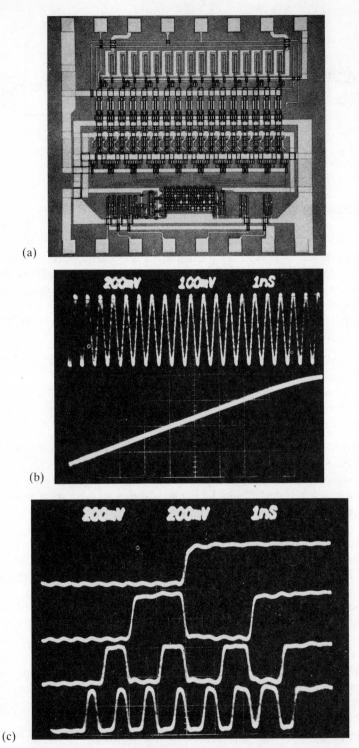

Fig. 2.49 2 Gb/s 4-bit GaAs A to D converter (a) microphotograph of the chip (1.4 × 1.4 mm²) (c) 2 GHz coding of a full scale saw tooth input signal, shown in (b).

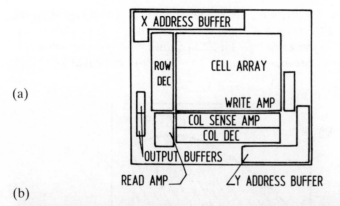

(a)

(b)

Fig. 2.50 GaAs 1K SRAM (a) block diagram (b) microphotograph of the chip (1.7 × 2.5 mm²).

Load current has to be very low in order to reduce power consumption, thus the speed is achieved by precharging of the bit lines which is also a prerequisite for the dynamic stability of the memory cell. A circuit including three cells in two columns is analyzed with transient simulation to optimize the dynamic stability of the cell. Furthermore due to variation of the technology the threshold voltages V_T of the four MESFETs in the cell have to be mismatched to simulate worst-case operations of the memory cell. In the read mode, the stability which is V_T-dependent, mainly depends on the threshold voltage difference ΔV_T, the driver to switch width ratio and the high voltage level on the word line. Fig. 2.51 shows the maximum allowable ΔV_T versus nominal threshold voltage V_T for a width ratio equal to 3 [124].

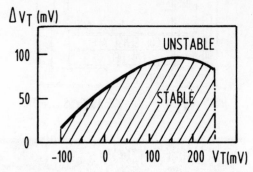

Fig. 2.51 Maximum allowable ΔV_T versus V_T for a cell stability factor of 3.

(b) **Precharging of the bit lines.** The simplest technique to precharge the bit lines is to use pull-up resistors (Fig. 2.52). They reduce both the signal swing ΔV_B and the rise time on the bit lines. Moreover a small logic swing ΔV_B results in a short access time T_{aa}. ΔV_B is proportional to the product

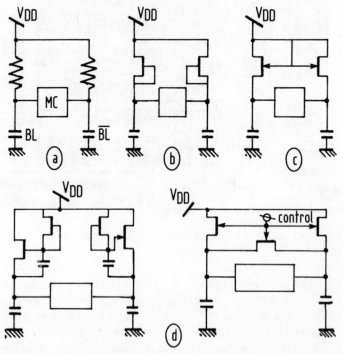

Fig. 2.52 Bit line precharging techniques (a) resistive load (b) and (c) MESFET load (on and off) (d) bootstrap load.

of the pull-up resistor R_B and the width of the pass transistor. Nevertheless, if ΔV_B is less than 100 to 150 mV, high speed sensing will be difficult, so 200 mV is a recommended minimum value. Furthermore, R_B must be high enough to ensure a low voltage level, less than 250 mV on one of the bit lines in the write mode. Most GaAs SRAMs utilize N-off or N-on pull-up FETs instead [125] (Fig. 2.52). The N-off approach where $V_{DG} = 0$ V implies a supply voltage equal to or less than 1 volt to ensure a low voltage on the bit line in the write mode and fast precharging of the bit lines. Too many different supply voltages will never be accepted by system designers. This problem does not occur with N-on pull-up FETs. However, due to non-uniformity of the Schottky diode (gate length, channel recess, build-in potential Vbi), variation of the access time will be higher than with a pull-up resistor.

(c) **Peripheral circuits.** A PLA architecture is used for the address decoder. The high capacitance of the word line is driven by a push-pull stage. For the read stability of the cells, the high voltage level on the word line is clamped to 0.7 Volt. With a fully N-off approach, careful design can lead to a propagation delay time of $1.5 \times$ gate t_{pd} instead of $3 \times t_{pd}$ across the selected stage decoder of a 1 K SRAM.

Some GaAs SRAMs make use of D-mode FET source follower circuits with level shifting diodes as word line driver [126]. D-mode pass transistors can then be used in the memory cell to speed up discharging of the bit line capacitance and thus reduce the access time. Furthermore, the pass transistor can be pinched off beyond V_T with a large negative V_{GS} voltage ($V_{GS} \leq V_T$) which reduces memory cell leakage currents and thus parasitic interference between adjacent cells. Temperature behaviour of the whole memory will also depend on the control of the D-FET switches in the cells. Since MESFETs tend to become more normally-on as the temperature is increased, a negative low voltage level on the word line, is necessary for proper operation at 125°C of fully normally-off SRAM's.

In order to increase the speed of the decoder, OR-type address pre-decoding circuits can be applied to reduce the load capacitance of the address stage. Power dissipation, die size and access time are thus decreased.

(d) **Access time variations.** The access time variation within a SRAM is mainly related to that of the data line signal amplitude and to the large RC time constant [125]. With the column differential sense amplifier approach, a cross-coupled translator stage can be inserted between the bit lines and the differential amplifier in each column (Fig. 2.53) [123]. With this circuit, the input voltages of the differential amplifiers are shifted by 500 mV resulting in a negative V_{GD} voltage and the Miller capacitances in the $(N-1)$ unselected differential amplifiers are drastically reduced. A faster rise and fall time signal along with a reduced variation of the delay time on the data bus can thus be achieved. At the 1 K-bit level, the read access time is thus shortened by a factor of 20% and output transition time improved by a complementary E/D push-pull write amplifier which must ensure a write operation in less than 1 ns. Final read amplifiers for SRAMs consist of a

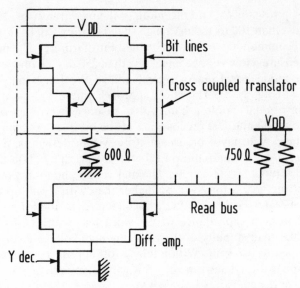

Fig. 2.53 Improved sensing circuit to minimize access time variation.

cross-coupled translator stage, a follower stage, two cascaded differential amplifiers and two complementary output buffers.

The cascaded differential amplifiers are needed to achieve output rise and fall times shorter than 200 ps (Fig. 2.54). The output buffers are designed to provide 800 mV into 50 Ω load. The address access time is 1.8 ± 0.2 ns.

(e) **Memory cell stability.** The most sensitive part in SRAMs is the

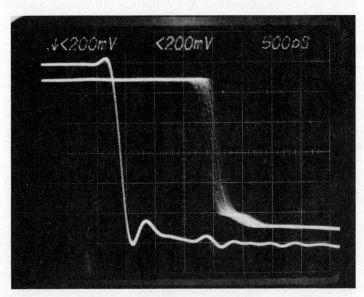

Fig. 2.54 Read access time variation in E/R 1K SRAM.

memory cell array and the read-mode stability of the memory cell has to be simulated accurately. A first analysis of all the parameters shows that the transistor threshold voltage V_T is the main parameter to be considered. Computer simulations show the cell stability in the read mode depends on three parameters (Fig. 2.5):

- The ratio α between the gate width of the driver transistor to access transistor. The greater the ratio, the higher the cell stability. However too high a stability in the read mode will prevent proper write operation and optimum trade-off has to be determined.
- The high level on the word-line voltage, which must be lower than 0.7 V. If the Schottky gate of MESFETs is forward biased ($\geqslant 0.7$ V), access current flows from gate to source and can destroy the stored information.
- The threshold voltage difference (ΔV_T) between the two cross-coupled driver FETs of the memory cell. This last parameter is V_T-dependent, V_T being a function of the electrical uniformity of the active layer and of the variation of the process parameters. In order that the memory cell does not lose its data during the read mode, the maximum allowable difference in threshold voltages ΔV_T (V_T between the driver transistors) is deduced from transient simulation for a worst case unbalanced memory cell. Considering a ratio α of 3, a maximum ΔV_T of 90 mV is found which corresponds to V_Ts between 150 mV and 250 mV (Fig. 2.51).

The short range distribution of ΔV_T is characterized by its standard deviation σ (ΔV_T). A 'design yield' can be derived from the stability condition of a memory cell, $\overline{V_T}$ and σ ($\overline{\Delta V_T}$) which are respectively the mean value of V_T and σ (ΔV_T) over the whole wafer. For 1 K SRAMs, 95% design yield can be expected with a mean σ ($\overline{\Delta V_T}$) equal to 15 mV. For σ ($\overline{\Delta V_T}$) of 20 mV and 30 mV, the design yield is only 75% and 40% respectively (Fig. 2.55).

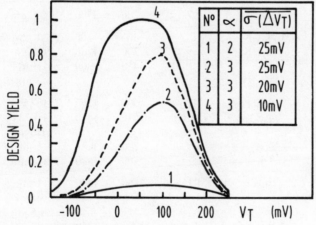

Fig. 2.55 SRAM design yield versus V_T.

Very high quality substrate material has to be used in order to expect reasonable fabrication yield with GaAs SRAM. The best uniformity at the microscopic level is definitely required.

2.5.3.3 State of the art

Although GaAs SRAMs started to be developed in 1981, they are still in their infancy compared with silicon equivalents.

If we are to take a realistic view of GaAs SRAMs performances, we can state that the 1 K bit level is now under control and that 4 K and 16 K bit SRAMs will require further development. It is above all required to establish advanced interconnect technology. 1 K, 4 K and 16 K bit SRAMs have been fabricated, but only 1 K bit and 4 K bit SRAMs have so far been reported to have passed marching tests. The bet access time reported for 1 K SRAMs is as low as 1 ns [127] with MESFETs and 0.6 ns with HEMTs [129]. Typical access times are in the 1.5–2 ns range [123] with a power consumption of 250 mW. 4 K SRAMs dissipate 700 mW to 2 W with best access time of 2.7 ns [121]. Two companies have reported 16 K bit SRAMs. NTT [127] with a memory cell area of $41 \times 32.5 \ \mu m^2$ has achieved 4.1 ns with 1.46 W power consumption. Fujitsu [129] with HEMTs and a memory cell area of $23 \times 30 \ \mu m^2$ has achieved 3 ns at 77 K with 1.3 W power consumption. Table 2.7 gives the performances of the best GaAs SRAMs reported to date.

The power consumption of GaAs SRAMs is already remarkably lower than Si bipolar SRAMs (0.1 to 0.2 mW/bit). However the speed perfomance

Table 2.7. Performance of GaAs SRAMs

	N (bits)	t_{access} (ns)	P (mW)	Cell area (μm^2)	Logic (DCFL)	Design rules (μm)
LEP/RTC	16 × 4	0.7	200	2500	0.8 µm	3
Fujitsu	1 × 1 K	1	300	1400	1 µm	2
LEP/RTC	1 × 1 K	1.5	200	1200	0.8 µm	3
Rockwell	1 K	0.8–(0.6)	450	2200	HEMT 1 µm	2
Hitachi	4 K	1	1600	1200	0.7 µm	2
Fujitsu	4 K	2 (77 K)	1540	1600	HEMT 1.2 µm	3
Mitsubishi	4 K	2.5	200	1044	1 µm	2/3
Fujitsu	4 K	2.7	700	1400	0.8 µm	1.5
NTT	16 K	4.1	1460	1300	1 µm	1.5
NEC	16 K	4	1600	700	SiECL 0.5 µm	
Mitsubishi	4 × 4 K	5	1000	750	0.8 µm	1.5

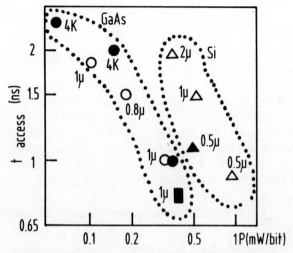

Fig. 2.56 GaAs SRAM performances. GaAs (○ 1K – ● 4K) ■ 1K HEMT, Si (Δ 4 × 256 ▲ 16 × 256).

is still greatly limited by the design rules. Improvement by a factor of at least 2, 1 K SRAMs with less than 0.5 ns access time, is expected in the near future (Fig. 2.56). A 4 × 1 K SRAM in HEMT technology has recently been reported with a t_{access} = 0.5 ns and P = 5.7 W.

2.5.3.4 Pipelined SRAMs

One way to enhance the performance of a system is to take advantage of pipelined operation. At 1 ns cycle time [128], no skew between the SRAM inputs could be tolerated and a synchronous SRAM will definitely be required. A solution is to include on-chip latches and an on-chip clock signal which would synchronize inputs and regulate outputs. In this case, timing complexities due to skew are reduced even at gigabit per second data rates. The pipelining of both input and output enables a read data flow or a read cycle time which is half of the read access time in the asynchronous mode to be obtained.

2.5.4 GaAs arithmetic IC design

2.5.4.1 Introduction

For a couple of decades, digital signal processors (DSPs) have considerably widened their field of application. Indeed, while digital techniques appeared to give many more possibilities in signal processing than analogue approaches, a number of application areas, such as radar detection, speech or image processing and automation, have raised a dramatic need for ever higher performance digital monolithic arithmetic circuits.

Meanwhile, trends in supercomputer design indicate that ultra high speed arithmetic circuits and memories will be strongly demanded for the next generation machines.

The main arithmetic functions to be found in large VLSI systems have been extensively demonstrated with GaAs:

- adders: such as 32-bit BFL adders [130] with a maximum add time of 2.9 ns and a power consumption of 1.2 W.
- Arithemic logic units (ALUs): such as 4-bit DCFL ALUs with 3.5 ns maximum delay time and 15 mW power consumption [131].
- multipliers: several methods are available to implement the multiplication of two binary words. In serial-parallel multipliers, one of the two words to be processed is synchronously fed in serially, and an addition is performed during each period. This approach requires only one adder synchronously addressed (Fig. 2.57). In a fully parallel approach, each word to be added is processed by a different adder, resulting in a concurrent operation which achieves better speed performance for large bit numbers (≥ 8).

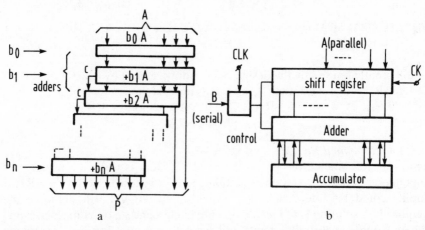

Fig. 2.57 Parallel multiplier (a) and serial parallel multiplier (b) designs.

Parallel multipliers are important devices for fast signal processing, but they are also used along with SRAMs to evaluate the real LSI capability of GaAs digital IC processes. Most of the parallel multipliers reported so far (4, 5, 6, 8, 12, 16 bits) very often implement simple algorithms in non-optimized architectures ([132], [133]).

In this section a review of current multiplication algorithms is presented, as well as a comparison between possible architectures for the adders array. Finally, the performance of GaAs multipliers is discussed.

2.5.4.2 Multiplication algorithms

(a) **Shift and Add alogrithm.** The shift and add alogirthm is the simplest way to perform a multiplication. When handling binary words, partial products may be equal either to the multiplicand, or zero. n partial products P_n are thus generated, shifted n bits to the left with respect to P1 and finally added together (Fig. 2.57).

It should be noted that this algorithm can only deal with unsigned numbers. Nevertheless, multiplication of signed numbers may be actually achieved by treating the sign bits and the mantisses separately, which results in extra combinational operation.

(b) **Modified Booth's algorithm [134].** The overall multiplication time achieved with parallel multipliers may be reduced in two ways: accelerating the addition of the partial products, which will be discussed in Section 2.5.4.3., or reducing the number of partial products to be added. This is the purpose of the modified Booth's algorithm. This algorithm is especially suited for signed two's complemented binary words.

The partial products to be generated from two input words A (a_i) and B (b_i) in a Booth's multiplier are: $PP_i = \alpha_i B$ for each odd value of i between 1 and $n-1$ where $\alpha_i = a_i - 2a_{i+1} + a_{i-1}$. α_i belongs to $-2, -1, 0, 1, 2$. The number of these PP_i is reduced by a factor of two, compared with the conventional shift and add algorithm. Nevertheless, the synthesis of partial products is no longer simply obtained by anding the corresponding bits of each word. In fact, the bits of each partial product must be selected within the set $b_{j+1}, b_j, 0, \overline{b_j}, \overline{b_{j+1}}$.

This selection results of course in a more complex synthesis of partial products and consequently the trade-off between the generation of the partial products and the number of the PP_is to be added together will be more and more in favour of Booth's algorithm when the number of bits increases, say when the gain in addition time will prevail over the synthesis of all the partial products, which is concurrently carried out.

2.5.4.3 *Architecture of the adders array*
The simplest method to add together n-bit words (in this case the multiplication of partial products) consists of implementing a ripple carry approach (Fig. 2.58).

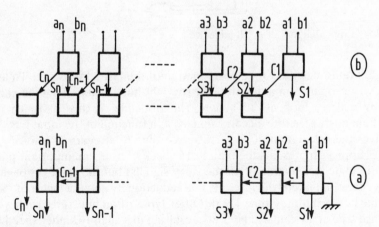

Fig. 2.58 Ripple carry (a) and carry save (b) architectures.

In this approach, the speed is governed by carry propagation delay.

To avoid this intrinsic phenomenon, the connections between each full adder may be slightly modified, resulting in improved parallelism. This approach is called a carry save architecture where the carry signals from each full adder are added at a later stage. Each row of the sub-adder array is also called a 'pseudo adder' for the simple reason that each row does not generate an $n+1$-bit word equal to the sum of two n-bit words but actually generates two n-bit words, the sum of which is equal to the final result. The multiply times for ripple carry or carry-save architectures associated with the shift and add algorithm, are given below:

$$t_m \text{ (carry save)} = 2\ (n\text{-}1)\ t_c$$

and

$$t_m \text{ (ripple carry)} = 2\ (n\text{-}1)\ t_c + nt_s$$

where t_c is the carry delay time and t_s the sum delay time of a full adder. As shown in Fig. 2.59, a full adder can be implemented using inverters, NOR gates and even pass transistors. For $t_c = 2t_{pd}$ (NOR gate), we then have t_m (carry save) $\approx 4\ (n\text{-}1)\ t_{pd}$.

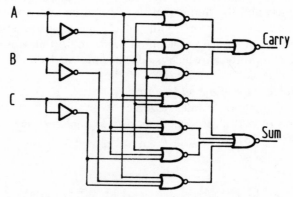

Fig. 2.59 Logic diagram of a full adder cell.

Addition of carries can be postponed until the last adder stage. To avoid the carry propagation along this last row, the final addition can be carried out by a carry look ahead adder. The principle of this approach is to simultaneously generate all carry signals, as a function of the input bits. This involves the calculation of all carries according to a recurrent law.

This approach presents the major drawback of using an important part of random logic (calculation of all the carry signals) but the carry propagation delays are eliminated which results in a reduction by about a factor of two in the maximum multiply time [135]. Other types of architectures such as the Wallace tree array can also be used, resulting in a slightly higher speed but, above all, in a reduced device count (–20%) in the adder array.

2.5.4.4 Booth's multiplier: partial products synthesis

As it has been previously pointed out, the generation of partial products for multipliers implementing a shift and add algorithm is very simple. Indeed, each bit of these PP_i can be generated by anding one bit of the multiplier and one of the multiplicand. When using the modified Booth's algorithm, the computation of the PP_i is no longer that simple: three bits of the multiplier are considered together as a substring and to each value of this substring, corresponds a special operation on the multiplicand to generate a partial product. As the caculated coefficient α_i *belongs to –2, 1–, 0, 1, 2, the partial product* $\alpha_i B$ can be obtained by complementing and/or shifting the multiplicand **B**.

An original way to implement this multiplexing operation was proposed in [135], using GaAs MESFETs as switches.

Provided Booth's encoders generate convenient control signals (corresponding to a given operation on the multiplicand) each bit of a partial product can be generated by a multiplexer as depicted in Fig. 2.60. Although some precautions must be taken on the control signals levels, this approach exhibits a very low power consumption.

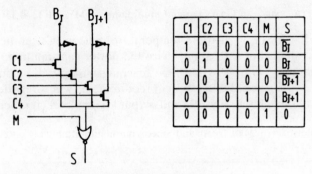

C1	C2	C3	C4	M	S
1	0	0	0	0	$\overline{B_J}$
0	1	0	0	0	B_J
0	0	1	0	0	$\overline{B_{J+1}}$
0	0	0	1	0	B_{J+1}
0	0	0	0	1	0

Fig. 2.60 Generator of Booth's partial products.

2.5.4.5 GaAs arithmetic circuits: state of the art

GaAs Booth multipliers ([135], [136]) which have been dealt with above are actually the first GaAs multipliers handling signed (2's complemented) binary words. Consequently, they are hardly comparable with other unsigned multipliers, which have been extensively reported recently ([74], [80], [132], [138]–[144]). However, a 4-bit implementation of the modified Booth's algorithm does not improve the overall multiplication time that much because the reduction of the partial products count is compensated by a longer synthesis of PP_is. The gain in multiplication speed will be much more important for 12- or 16-bit Booth's multipliers.

The main results for GaAs parallel multipliers are shown in Fig. 2.61 and in Table 2.8. To compare efficiently different size circuits, the performances are expressed as the multiplication time/bit (n) versus the power dissipation/ (bit)^2. It is clear that there is so far very little advantage to use E-HEMTs

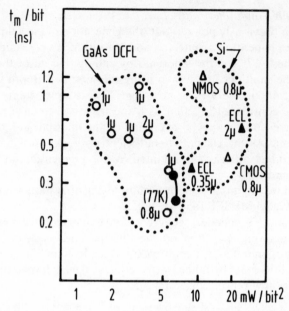

Fig. 2.61 Performance of GaAs parallel multipliers. ○ MESFET, ● HEMT.

instead of E-MESFETs at room temperature. Silicon ICs can be as fast as GaAs equivalents but with about twice the power consumption.

GaAs parallel multipliers are now reaching multiplication times in the range of 0.3–0.5 ns/bit. Future system requirements will surely demand on-chip integration of input registers and output accumulator. It is believed now

Table 2.8. Performances of GaAs and Silicon parallel multipliers (C = custom, GA = gate array)

		Bit number	Logic	Gate count	t_m (ns)	P (mW)	t_{pd} (ps)	P/gate (mW)
LEP/RTC	C	4 × 4	DCFL 1 µm	213	2.5	40	120	.2
Bell Labs	C	4 × 4	HEMT 1 µm	162	1.6	55	114	.34
Honeywell	C	5 × 5	MESFET 1 µm	343	2.6	79	87	.23
Honeywell	C	5 × 5	HEMT 1 µm	350	1.8	150	72	.43
Fujitsu	GA	8 × 8	HEMT 1 µm	888	4.9	5800		
					3.1	3200		
					(77 K)			
Sony	C	8 × 8	JFET 1 µm		5.6	900	170	0.5
Rockwell	C	8 × 8	SDFL 1 µm	1008	5.25	1000	160	1
Hughes	C	8 × 8	NMOS 0.8 µm	400	9.5	600	244	1.5
Fujitsu	C	16 × 16	DCFL 2 µm	3168	10.5	950	162	0.3
Matsushita	C	8 × 8	ECL-Si 2 µm		5	1400	160	
NTT	GA	16 × 16	ECL-Si 0.35 µm	1750	6	2050		
LEP	C	8 × 8	DCFL 0.8 µm	800	2	400	70	0.5
NEC	C	12 × 12	BFL 0.7 µm	1100	4		170	1.7

that ultimate digital circuit performances can be reached by implementing efficient algorithms in highly parallel, optimized architectures, using high performance basic logic blocks based on submicron devices. Using fully systolic mutipliers, multigiga bit/s data rates will be readily achieved with GaAs technologies.

2.6 THE NEXT STEP: GaAs MICROPROCESSOR

2.6.1 Introduction

In 1982, The Mayo Foundation was assigned by the United States Strategic Technology Office of the Defense Advanced Research Projects Agency (DARPA) the task of identifying a machine architecture that could be implemented with fewer custom ICs fabricated in GaAs rather than in silicon.

Silicon microprocessor designers have used the availability of large numbers of gates on VLSI chips to create relatively powerful architecture based upon parallelism and rich set of assembly language instruction types. Architectural complexity and the need for low power consumption have constrained the microcycle clock rate of these silicon-based processors to the range of 2 to 25 MHz.

Due to the present low fabrication yield, the gate count and packing density of GaAs digital ICs are currently considerably lower than in silicon ICs.

As a result, a microprocessor implemented in GaAs would have to contain less than 10 K to 20 K equivalent gates on a chip in order to obtain reasonable fabrication yield. A microprocessor architectural approach that seems adequate for a GaAs microprocessor is represented by the family of reduced instruction set computers (RISC). A RISC architecture can be defined by [144]:

- a small number reduced instructions,
- a fixed instruction format,
- hardwired rather than microcode architecture,
- simple cycle execution for most instruction,
- a load and store architecture.

2.6.2 GaAs microprocessor designs

The goal for a GaAs microprocessor is a chip of no more than 10 000 FETs, a 200 MHz clock rate, and a 100 200 MIPS. One of the major differences between CMOS or NMOS silicon technology and E-MESFET GaAs technology is the ratio of off-chip memory access time to on-chip access time. With the GaAs approach, it is necessary to minimize the number of off-chip accesses. As a result, when designing a GaAs microprocessor it is crucial to increase the size of on-chip storage to reduce the need for off-chip communication. One way to minimize the impact of off-chip memory access time is to utilize pipelining techniques for both memory and microprocessor [145]. The fixed instruction format in the RISC approach, leads to a highly simplified instruction decoder, reducing the number of transistors for its implementation and thus increasing the speed of the decoder. Therefore, the only significant control circuitry in the microprocessor will be the pipeline stage control which can represent 95% of the FET count for implementing the data path, as in the 32-bit RISC machine designed at McDonnell Douglas.

A difficulty is the choice of the number and type of execution pipeline stages. Two approaches are available: the four-stage pipeline or the six-stage pipeline [146]. The six-stage pipeline, which permits a pipelined memory to be accessed over two cycles (M1 and M2), consists of instruction fetch cycle 1 (IF1), instruction fetch cycle 2 (IF2), ALU execute memory access cycle (MI), memory access cycle M2, and writer-register file (WF). The 4-stage pipeline does not use IF2 and M2 instructions. Pipelined memory access increases the complexity of the design but can decrease the cycle time by about 40%. Furthermore a memory which is too slow for say a 5 ns cycle time required by a 4-stage pipeline could be operated in a synchronous mode over two cycles in a 6-stage pipeline.

The minimum microprocessor cycle time is given by the delay times across the critical path in the ALU pipe stage. Typically 40% of the cycle time is spent to precharge the internal buses and 60% in the adder and the register file access. Careful attention has to be paid to the design of the buses which must be as short as possible with minimum crossovers. As regards the adder, its design must achieve low gate fan-outs and then the number of gates in the longest path has to be kept to a minimum in order to increase the speed [146]. The ripple carry adder due to its very regular structure has a very low fan-out but the number of gates in the critical path (82 gates for a 34-bit adder) makes it inappropriate for the high speed. The carry-select adder has 37 gates in the critical path but still has too high a fan-out. The carry look ahead has only 11 gates in the critical path and a maximum fan-out of 8. But due to the number of transistors required and the irregular layout, the delay times associated with crossovers are similar to those due to fan-out. The adder which has been chosen by McDonnell Douglas for its 32-bit microprocessor is the Brent and Kung adder [147]. A maximum fan-out of 3 and a very regular layout make interconnection and crossover loading minimal.

2.6.3 State of the art in GaAs microprocessors

Today different design approaches are available with GaAs technologies. However a large amount of research work is still necessary to define special optimum architecture strategies for the GaAs environment. Some companies have developed GaAs RISC oriented machines. McDonnell Douglas has used the JFET technology approach to fabricate a 4-bit GaAs microprocessor circuit that contains 2K transistors and intends to develop a 32-bit RISC based microprocessor. Texas Instruments is implementing a 32-bit microprocessor which can support a 64 Mbyte virtual and real address space. The goal of the system is to operate using a 1 ns clock cycle, resulting in a 200 MIPs peak execution rate. This 32-bit microprocessor will be fabricated using GaAs heterojunction integrated injection logic (HI^2L) technology. The main advantage of HI^2L is the good control of logic gate threshold voltage over temperature and the fact that a NAND gate requires only one transistor, which permits a high gate density. The overall system would be built with a maximum gate count of 10 000 gates. RCA has demonstrated an 8-bit GaAs microprocessor which is a pipelined RISC machine [148]. The processor instruction set consists of 23 instructions: 19 of which are 8-bit and four of which are 16-bit. The pipeline permits simultaneous processing of several instructions. The memory consists of four 256 × 4-bit SRAMs with 3 ns access time in synchronized mode. The cycle time of the system is 10 ns with a 100 MIPS performance. The technology used is the GaAs E/D MESFET approach and the power consumption is 840 mW. RCA is also designing a single chip 32-bit GaAs processor.

Vitesse Electronics has used an E/D process to design and fabricate a GaAs 4-bit microprocessor slice compatible with 100K ECL I/O signal levels. The device consists of a 16 × 4 dual-port SRAM, a high-speed 4-bit ALU, shift register file, decoding and multiplexing circuits. The micro-instruction word is 9-bit wide. The device can execute an instruction every 14 ns, even though 31% of the total delay from input to output pins is used up in the input and output buffers. In order to demonstrate the minimum cycle time that can be achieved using the Vitesse family of bit-slice processors, a comparison has been done between a typical silicon bipolar 16-bit design and a design using GaAs circuits. Both control loop and data loop cycle time have been measured in the worst case configuration. The use of GaAs devices has reduced the control loop cycle time from 98 ns down to 22 ns, and the data loop cycle time from 87 ns down to 29 ns, which means a speed improvement by a factor of 3.

In future, E-mode logic will clearly lead the GaAs development to LSI and VLSI. However, because of the present low fabrication yield, the RISC approach will still be for the years to come the only suitable way for GaAs. RISC design strategy has to be greatly modified for GaAs compared with silicon. A great deal of research is still necessary to learn how to design and build a GaAs microprocessor, and above all to implement GaAs RISC-based chips in a mixed GaAs and silicon environment to build a GaAs VLSI computer.

2.7   CONCLUSION: GaAs DIGITAL ICs IN A WORLD OF SILICON ICs?

To conclude this review of GaAs digital design and performances, we have to discuss how GaAs digital ICs can complement silicon equivalents. This issue is vital to the future of GaAs digital ICs. Technical considerations show that as long as optimum interconnect technology and optimum architectures are used the speed/power performance of logic gates in VLSI environments is directly related to the high frequency response of active devices in the gates. This remark may seem contradictory to some silicon designers who have been used to carrying more for complexity than for speed. However, there is now a trend towards extensive parallelism in systems and, for instance, systolic arrays can be used to approach a maximum data rate of $1/t_p$ where t_p is the propagation delay time across one or a few gates placed between master slave registers.

In this respect, GaAs ICs have definitely been shown to be faster than silicon ICs or to exhibit lower power consumption, at the same system speed. Then comes the issue of the need for even higher performance ICs. These political and economical issues are certainly the main snags in the way of the rapid growth of a GaAs IC market and consequently of the availability of investment large enough to boost GaAs ICs development and production.

Recent scaling down of fast silicon processes (NMOS, CMOS, chilled CMOS, ECL) ([149]–[155]) and the impressive performances achieved have raised the question of the real advantage of GaAs over silicon for digital applications. Review papers concerning fast SRAMs, parallel multipliers, adders, multiplexers, prescalers showed that many laboratories are working on similar technologies with seemingly contradictory results.

The performance of a circuit ultimately depends on the system design and requirements. Various architectures, chip partitioning and algorithms can be used to improve the operating speed of the chips in a system. For example, circuits dedicated to an iterative algorithm are upgraded by parallel processing in a pipelined architecture, since in non-recursive structures the latency time is not that important. In semi-systolic arrays delay problems arise from the time constant associated with the charging and discharging of long data paths. In pure systolic arrays, these problems are transferred to the global distribution of the clock signals.

These various aspects of chip design make solid comparisons very difficult. Logic technologies can be described as the combination of microwave three-port devices with an interconnect technology.

We will try to come up with a realistic and significant comparison between technologies. This is a very difficult task, since the performances reported are often prejudiced or incomplete.

We will first consider the switching capabilities of silicon and GaAs transistors. Today's fast silicon bipolar transistors and MOSFETs are fabricated with self-aligned, fully implanted, double polysilicon processes with trench oxide isolation. Silicon bipolar transistors with 0.3 μm and even 0.1 μm emitter-to-base spacing have exhibited f_T as high as 18 and 26 GHz respectively at 1 mA collector current and $V_{CE} = 3$ V. Silicon MOSFETs with 0.3 μm effective gate lengths have also been reported with f_T of about 20 GHz. Such small feature sizes have been achieved in GaAs MESFETs and HEMTs resulting in f_T as high as 60 GHz and 80 GHz respectively. Scaling down of GaAs HBTs is not as advanced yet; however f_T of 68 GHz have already been observed.

We will now compare silicon and GaAs gates. The performance of logic gates is dependent on the transistor mean switching time and also on the gate loading (*FI*, *FO*, *C*), power consumption, design flexibility, gate noise margins and the logic swing. As long as sufficient stages are included in the ring oscillator, and as long as the inverters are designed with well-defined low and high logic levels and sufficient linear gain, the propagation delay times derived from RO measurements represent the behaviour of a logic gate. Significant performances are summarized in Table 2.9 for *FI* = *FO* = 1.

Table 2.9. Significant ring oscillator results ($FI = F\emptyset = 1$)

	Minimum feature size (μm)	t_{pd} (ps)	$n = t_{pd}/\tau_o$	P (mW)	$P \times t_{pd}$ (fJ)
GaAs MESFET	1.5	42	≈ 4	0.5	21
(DCFL)	0.4	16		1	16
GaAs HEMT	0.35	10	≈ 4	1	10
(DCFL)		5.8 (77 K)			5.8
Si BT (ECL)	0.1	50	≈ 6	3	150
	0.35	78	≈ 6	2.6	200
Si NMOS	0.3	30	≈ 4	1.5	30
GaAs HBT (CML)	< 1	27.6	≈ 6	5.8	160

Examples of the variation of RO propagation delay with various *FI*, *FO* for negligible metal delays are summarized in Table 2.10 for various technologies. It should be noted that these are only the best performances reported so far. However they are the only useful data for a fair comparison. Loaded gate performances are related to the transistor speed and it is clearly demonstrated that GaAs N-OFF logic gates are about two times faster than silicon BJT gates with equivalent feature sizes and large fan-in and fan-out (6 and 4 respectively). Below 0.5 μm feature size, silicon NMOS and CMOS gates (especialy at 77K) tend to be as fast as bipolar CML or ECL gates.

The metal delay contribution to the gate delay, becomes important when long lines have to be driven. It is then always better to use a buffered gate. Metal delays for three technologies are shown in Table 2.11. They are in complete agreement with the measured data reported for gate arrays. The metal delay in silicon bipolar ICs is about three times longer than in GaAs ICs at 1 mW/gate. For equivalent metal delay, the power consumption of silicon gates is about two to three times higher than their GaAs equivalents.

Table 2.10. Variation of logic gate t_{pd}s (FI, $F\emptyset$ for $C_W = 0$)

		Min. feature size (μm)	t_{pd} ($FI = F\emptyset = 1$) (ps)	t_{pd}/FI (ps)	$t_{pd}/F\emptyset$ (ps)	P/gate (mW)
	DCFL	1.5	42	11	16.5	0.5
GaAs MESFET	BFL	0.5	56	6	10	4.5
	BDCFL	0.7	22	–	9	3.4
	CML	0.7	40	5	6	7.5
GaAs HEMT	DCFL	0.8	40		22	1.6
Si BT	NTL	≤ 0.5	50	9.5	9.5	1.84
	ECL	≤ 0.5	78	14	19	2.6

Table 2.11. Metal delays of the main silicon and GaAs technologies for 2 μm lines

	P (mW)	V supply (V)	k	metal delay (ps/mm)	
				Without cross-talk	With max. cross-talk (3 coupled lines 2 μm apart)
DCFL 0.7 μm	1	1.5	1	45	60
BDCFL 0.7 μm	1	1.5	0.5	36	50
BDCFL 0.7 μm	3	2	0.5	16	22
Si ECL 0.5 μm	1	3	0.4	120	180
Si ECL 0.7 μm	3	3	0.4	40	60
Si ECL 0.7 μm	6	3	0.4	20	30

where: $k = P_{buffer}/P_{logic}$

However, when it comes to comparing all reported results on basic SSI, MSI and LSI ICs such as frequency dividers, SRAMs and logic circuits, the conclusions are more difficult to draw since the interconnect technology and the circuit configuration play an important role, and in many cases they are far from being optimum in GaAs demonstratros. Nevertheless, take frequency dividers for instance. Here the best circuit designs are used for GaAs and silicon technologies (1/2 t_{pd} flip-flops). The speed advantage of GaAs flip-flops is of about 2 for the best results reported to date: 20.1 GHz for GaAs against 10.4 GHz for silicon. At the maximum speed achieved with silicon, the power consumption of GaAs flip-flops is at least two times lower (Fig. 2.40).

Static random access memories are circuits with specifications and features of their own. A direct comparison is only meaningful if similar design rules are considered and the same memory architecture. For instance, bipolar 4 × 256-bit SRAMs cannot be directly compared with 1 × 1024-bit GaAs SRAMs.

0.6 ns 1 × 1K HEMT SRAMs and 1 ns 1 × 4-K bit MESFET SRAMs have been reported with 450 mW and 1.6 W power consumption respectively. This is definitely better than what is achieved with the best silicon bipolar SRAMs (Fig. 2.56). At present, a speed power advantage of about a factor of 3 has been demonstrated up to a capacity of 4 K bit.

The performance of LSI logic circuits is not only dependent on the technology, but also on the algorithm and architecture used. This is why it is fairly difficult to compare the performances of parallel multipliers. Equivalent speeds of 0.3 to 0.5 ns/bit have been achieved with 0.7 μm GaAs DCFL (MESFET, HEMT), 0.5 μm Si ECL and 0.8 μm CMOS (Fig. 2.61), the power advantage of the GaAs chips being of about 3, in terms of power/bit[2].

In Conclusion, while silicon ICs are reaching their limits, GaAs ICs are just getting out of their infancy and further performance improvements can be expected. GaAs MESFETs have been shown to be excellent for a first generation of fast digital ICs. But as described in the introduction, GaAs hetero MISFETs and HBTs are already exhibiting promising features for the next generation.

For performance improvement in any technology, silicon or GaAs, it will require painstaking developments, but GaAs ICs should eventually take full advantage of the best refinements in silicon process technology and CAD and really turn out to be top of range products in a silicon world of microelectronics.

REFERENCES

1. Jutzi, W. (1971) 'Direct coupled circuits with normally-off GaAs MESFET's at 4.3 K' *Arch. Elektron. und Ubertragungtech, (AEU)* **25** 595–598.
2. Van Tuyl, R., Rory, L. and Liechti, C. A. (1974) 'High speed integrated logic with GaAs MESFET's' *IEEE J. Solid-State Circuits* **SC–9** 269–276.
3. Yokoyama, N., Mimura, T., Kusakawa, H., Suyama K. and Fukuta M. 'Low-power high speed integrated logic with GaAs MOSFETs' *Digest of Tech. Papers* 11th Conf. (1979 International) on Solid State Devices, Tokyo.
4. Zuleeg, R., Notthoff J. K. and Troeger, G. L. (1984) 'Double implanted GaAs complementary JFET's' *IEEE Electron Device Letters* **EDL–5** No. 1 (Jan.).
5. Van Tuyl, R. L., Liechti, C. A., Lee, R. E. and Gowen, E. (1977) 'GaAs MESFET with 4 GHz clock rates' *IEEE J. Solid-State Circuits* **SC–12** 485.
6. Ishikawa, H., Kusakawa, H., Suyama, K. and Fukuta, M. (1977) 'Normally-off type GaAs MESFET for low power, high speed logic circuits' *1977 IEEE Intl. Solid-State Circuits Conference Tech. Digest* 200.
7. Jensen, J. F. *et al.* 'Ultra highspeed GaAs static frequency dividers' *Digest of the IDEM 86* 476–479.
8. Mimura, T., Hiyamizu, S., Fujii, T. and Nanbu, K. (1980) 'A new field-effect transistor with selectively doped GaAs/n-Al$_x$Ga$_{1-x}$As heterojunctions' *Japanese J. of Applied Physics* **19** (5) (May) L225–L227.

9. Takakuwa, H., Kato, Y., Watanabe, S. and Mori, Y. (1984) 'Low-noise HEMT fabricated by MOCVD' *Electronics Letters* (20 Dec.) 125.

10. Mishra, U.K., Palmateer, S. C., Chao, P. C., Smith, P. M. and Hwang, J. C. M. (1985) 'Microwave performance of 0.25 μM gate length high electron mobility transistors' *IEEE Electron Device Letters* **EDL-6** (3) (March) 142.

11. Abe, M., Mimura, T., Nishiuchi, K., Shibatomi, A. and Kobayashi, M. (1983) 'HEMT LSI technology for high speed computers' *Proc. GaAs IC symposium.* 158.

12. Sheng, N. H., Wang, H. T., Lee, S. J., Lee, C. P., Sullivan, G. J. and Miller, D. L. (1986) ' A high speed 1 K-bit high electron mobility transistor static RAM' *Proc. GaAs IC symposium*, 1986.

13. Katayama, Y., Morioka, M., Sawada, Y., Ueyanagi, K., Mishima, T., Ono, Y., Usagawa, T. and Shiraki, Y. (1984) 'A new two-dimensional electron gas field-effect transistor fabricated on undoped AlGaAs-GaAs heterostructure' *Japanese J. of Applied Physics* **23** (3) (March) L150–L152.

14. Mizutani, T., Fujita, S. and Yanagawa, F. (1985) 'Complementary circuit with AlGaAs/GaAs heterostructure MISFETs using high mobility two dimensional electron and hole gases' *GaAs and Rel. Compounds 1985* Inst. Phys. Conf. Ser. 79, Chap 13.

15. Kroemer, H. (1982) *Proc. IEEE*, **70** 13.

16. Ishibashi, T. *et al.* 'Self-aligned AlGaAs/GaAs heterojunction bipolar transistors for high speed digital circuits' *IDEM 86* 809–810.

17. Wang, K. C., Asbeck, P. M., Chang, M. F., Miller, D. L. and Sullivan, G. J. (1986) 'High speed MSI current mode logic circuits implemented with heterojunction bipolar transistors' *Proc. GaAs IC Symp.*

18. Yuan, H. T. (1982) 'GaAs by gate array technology' *Tech. Dig. GaAs IC Symp.* 100.

19. Alley, G. D., Bozler, C. O., Flanders, D. C., Murphy, R. A. and Lindley, W. T. (1980) 'Recent experimental results on permeable base transistors' *IEEE Proc.* 608.

20. Yokoyama, N., Imamura, K., Muto, S., Hiyamizu, S. and Nishi, H. (1985) 'A resonant tunnelling hot electron transistor (RHET)' *Proc. GaAs and Rel. Comp. 1985* Inst. Phys. Conf. Ser. 79.

21. Pande, K. P., Fathimulla, M. A., Gutierrez, D. and Messick, L. (1986) 'Gigahertz logic gates based on InP MISFET's with minimal drain current drift' *IEEE Electron Dev. Lett.* **EDL-7** (7) (July).

22. Schmitt, R., Steiner, K., Kaufmann, L. M. F., Brockerhoff, W., Heime, K. and Kuphal, E. (1985) 'InGaAs-JFETs with p+ gates diffused from spin-on films for Ka-band operation' *GaAs and Rel. Compounds 1985* Inst. Phys. Conf. Ser. 79.

23. Hirose, K., Ohata, K., Mizutani, T., Itoh, T. and Ogawa, M. (1985) '700 ms/mm 2 DEGFETs fabricated from high mobility MBE grown n-AlInAs/GaInAs heterostructures' *GaAs and Rel. Compounds 1985* Inst. Phys. Conf. Ser. 79.

24. Grote, N., Su, L. M. and Bach, H. G. 'Characteristics of double heterojunction InGaAsP/InP bipolar transistors' *GaAs and Rel. Compounds 1985* Inst. Phys. Conf. Ser. 79.

25. Eden, R. C. (1982) 'Comparison of GaAs device approaches for ultrahigh speed VLSI' *Proc. IEEE* **70** (1) (Jan).

26. Shockley, W. (1952) 'A unipolar field effect transistor' *Proc. IRE* **40** (Nov) 1365–1376.

27. Pucel, R. A. *et al.* (1974) 'Signal and noise properties of gallium arsenide microwave field effect transistors' *Advances in Electronics and Electron Physics* **38** 195–265.
28. Shur, M. S. (1976) 'Influence of non uniform field distribution on frequency limits of GaAs field effect transistors' *Elec. Letters* **12** (23) 615–616.
29. Onodera, T., Ohnishi, T. and Nishi, H. (1985) 'The role of piezoelectric effects in GaAs MESFET ICs' *GaAs and Rel. Compounds 1985* Inst. Phys. Conf. Ser. 79.
30. Grubene, A. B. and Ghandhi, S. K. (1969) 'General theory for pinched operations of the junction gate FET' *Solid State Electronics* **12** 573–589.
31. Drangeid, K. E. and Sommerhalder, R. 'Dynamic performance of Schottky-barrier field-effect transistors' *IBM J. Res. Develop.*
32. Cappy, A. (1986) 'Propriétés physiques et performances potentielles des composants submicroniques à efft de champ: structures conventionnelles et à gaz d'électrons bidimensionnelles Thèse, Université de Lille.
33. Zuleeg, R. and Lehovec, K. (1980) 'Radiation effects in GaAs junction field-effect transistors' *IEEE Transactions on Nuclear Science* **NS–27** (5) (Oct.).
34. Lehovec, K., Zuleeg, R. and Notthoff, J. K. (1984) 'Charge effects in GaAs semi-insulating substrates due to pulsed ionizing radiation' *GaAs IC Symposium 1984* 143.
35. Rocchi, M. (1985) 'Status of the surface and bulk parasitic effects limiting the performances of GaAs IC's' *Physica* **129 B** 119–138.
36. Jacob, G. (1982) 'How to decrease defect densities in LEC SI GaAs and InP crystals' *Semi-insulating III-V materials Book* Shiva Publ. Ltd.
37. Ozeki, M., Kodama, K. and Shibatomi, A. (1982) 'Surface analysis in GaAs MESFET by gm frequency dispersion measurement' *GaAs and Rel. Compounds, Osio, (1981)* Inst. Phys. Conf. Ser. 63, 323–328.
38. Tsironis, C. *et al.* (1984) 'Low frequency noise in GaAs MESFET's' *GaAs and Related Compounds Symposium* Biarritz.
39. Canfield, P. and Forbes, L. (1986) 'Suppression of drain conductance transcients, drain current oscillations, and low frequency generation-recombination noise in GaAs FET's using buried channels' *IEEE Trans. on ED* **ED–33** (7) (July).
40. Makram-Ebeid, S. *et al.* (1983) 'Effets parasites dans les transistors à effet de champ en GaAs: rôles de la surface et du substrat semi-isolant' *Acta Electronica* **25** (3) 241–260.
41. Lee, C. P. (1982) 'Influence of substrates on the electrical properties of GaAs FET devices' *Proc. Semi-insualting III–V materials* Conf. Evian 324–335.
42. Pichaud, B., Burle-Durbec, N., Minari, F. and Duseaux, M. (1985) 'Study of dislocations in highly In doped GaAs crystals grown by liquid encapsulation Czochralski technique' *J. Crystal Growth* **71** 648–654.
43. Pasqualini, F. *et al.* 'Piezo electric effects in GaAs IC's' *ESSDERC 85.*
44. Onodera, T. *et al.* (1985) 'The role of piezoelectric effects in GaAs MESFET IC's' *Symp. Gallium and Related Compounds* Karuizawa.
45. Schink, H., Packeiser, G., Maluenda, J. and Martin, G. M. (1986) 'GaAs substrate material assessment using a high lateral resolution MESFET test pattern' *Japanese J. Applied Physics*, **25** (5) (May) L369–L372.
46. Yamaguchi, K. and Kodera, H. (1976) 'Two dimensional numerical analysis of stability criteria of GaAs FET's' *IEEE Elect. Devices* **ED–23** (12) 1283–1290.
47. Grebene, A. B. and Ghandhi, S. K. (1969) 'General theory for pinched

operation of the junction gate FET' *Solid State Electronic* **12** 573–589.

48. Lehovec, K. and Zuleeg, R. (1970) 'Voltage current characteristics of GaAs J-FET's in the hot electron range' *Solid State Electron* **13** (Oct.) 1415–1426.

49. Curtice, W. R. (1980) 'A MESFET model for use in the design of GaAs integrated circuits' *IEEE Trans. Microwave Theory Tech.* **MTT–28** (5) 448–456.

50. Kacprzak, T. (1983) 'Compact D-model of GaAs FET's for large signal computer calculation' *IEEE J. SSC* **SC–18**, (April) 211–213.

51. Rocchi, M. (1980) 'Outil CAO pour circuits intégrés numériques sur AsGa' *Acta Electronica* **23** (3) 223–242.

52. Farrar, A. and Taylor Adams, A. (1972) 'Matrix methods for microstrip three dimensional problems' *IEEE* **MTT** (8) (Aug.).

53. Higashisaka, A. and Hasegawa, F. (1980) 'Estimation of fringing capacitance of electrodes on SI GaAs substrates' *Electr. Letter* **16** (11) (May).

54. Bakoglu, H. B. *et. al.* (1985) 'Optimal interconnection for VLSI' *IEEE* **ED–32** (5) (May).

55. Chilo, J., Angenieux, G. and Razban, T. (1985) 'CAD formulas for modelling the interconnections of fast GaAs integrated circuits' *EuMC.*

56. Yoshihara, K. *et al.* 'Cross-talk predictions and reducing techniques for high speed GaAs digital IC's' *GaAs IC Symposium 1984.*

57. Dang, R. L. M. and Shigyo, N. (1981) 'Coupling capacitances for two dimensional lines' *IEEE* **EDL–2** (8) (Aug.).

58. Rode, A. *et al.* 'A high yield GaAs gate array technology and applications' *1983 GaAs IC Symposium Digest.*

59. Kato, N., Hirayama, M., Asai, K., Matsuoka, Y., Yamasaki, K. and Ogino, T. 'A high density GaAs static RAM process using MASFET' *IEDEM 85.*

60. Rocchi, M. and Gabillard, B. (1983) 'GaAs digital dynamic IC's for applications up to 10 GHz' *IEEE J. Solid State Circuits* **SC–18** (03) (June 1983).

61. Meignant, D., Delhaye, E. and Rocchi, M. (1986) 'A 0.1– 4.5 GHz, 20 mW GaAs prescaler operating at 125°C' *GaAs IC Symp.* 129.

62. Csansky, G. and Warner, R. M. (1963) 'Put more snap in logic circuits with field-effect transistors' *Electronics* **36** 43–45.

63. Ohmori, M. (1981) 'Very low power logic circuits with enhancement mode GaAs MESFET's' *1981 Int. Microwave Symp., Los Angeles, Tech. Digest* 188–190.

64. Namordi, M. R. and White, W. A. (1982) *IEEE Electron Device Letters* **V–EDL–3** (9) (Sept.) 264–267.

65. Livingstone, A. W. and Mellor, P. J. T. 'Capacitor coupling of GaAs depletion mode FET's' *1980 GaAs IC Symposium Abstracts* paper No 10.

66. Eden, R. C., Lee, F. S., Long, S. I., Welch, B. M. and Zucca, R. (1980) 'Multi level logic gate implementation in GaAs IC's using Schokttky diode FET logic' *IEEE 1980 Int. Solid State Circuits Conf. California Tech. Digest* 122–123.

67. Eden, R. C. (1984) 'Capacitor diode FET logic (CDFL), circuit approach for GaAs D-MESFET IC's' *GaAs Symposium.*

68. Nuzillat, G., Bert, G., Danny-Kavala, F. and Arnodo, C. (1981) 'High speed, low-power logic ICs using quasi-Normally-off GaAs MESFETs' *IEEE J. Solid State Circuits* **SC–16**, (June) 226–232.

69. Barnard, A. and Liechti, C. (1979) 'Optimization of GaAs logic gates with subnanosecond propagation delays'. *IEEE J. SSC* **SC–14** (4) (Aug.).

70. Singh, H. P., Sadler, R. A., Geissberger, A. E., Fisher, D. G., Irvine, J. A. and Gorder, G. E. 'A comparative study of GaAs logic families using universal shift registers and self-aligned gate technology' *GaAs IC symposium, 1986.*

71. Ishikawa, H. (1977) 'Normally-off type GaAs MESFET for low power, high speed logic circuits' *ISSCC* 200.

72. Wallmark, J. T. and Marcus, S. M. (1959) 'Integrated devices using direct coupled unipolar transistor logic'. *IRE Trans. Electron. Comp.* **EC–8** 98–107.

73. Chung, H. K. 'High speed ultra low power GaAs MESFET 5 × 5 multipliers' *GaAs IC Symposium, 1986* 15.

74. Watanabe, Y. (1986) 'A high electron mobility transistor 1.5 K gate array' *ISSCC* 80.

75. Nakamura, N. (1985) 'A 390 ps 1000 gate array using GaAs super buffer FET logic' *ISSCC*.

76. Katsu, S., Nambu, S., Shimano, S. and Kano, G. (1982) 'A GaAs monolithic frequency divider using source coupled FET logic' *IEEE Electron Device Letters* **EDL-3** (8) (Aug.).

77. Yang, L., Yen, A. T. and Long, S. I. (1985) 'A simple method to improve the noise margin of III–V DCFL digital circuit coupling diode FET logic' *IEEE Electron device letters* **EDL–7** (3) (March) 145–148.

78. Vogelsang, C. H., Hogin, J. L. and Notthoff, J. K. (1983) 'Yield analysis methods for GaAs IC' *Digest of GaAs IC Symposium.*

79. Gabillard, B., Rocher, C. and Rocchi, M. (1985) 'Theoretical and experimental temperature dependence of GaAs N-OFF IC's over 120 K to 400 K' *Physica* **129 B** 403–407.

80. Toyoda, N. *et al.* (1985) 'A 42 ps 2 K Gate GaAs Gate array' *ISSCC* 206.

81. Davenport, W. H. (1986) 'Macro evaluation of a GaAs 3000 gate array' *GaAs IC symposium* 19.

82. Cathelin, M., Gavant, M. and Rocchi, M. (1980) 'A 3.5 GHz self-aligned single clocked binary frequency divider on GaAs' *IEE Proc.* **127** (5) (Oct.) 270.

83. Frey, P., Gabillard, B. and Rocchi, M. '0.25 mW, 1.9 GHz, GaAs T flip-flops' *GaAs IC Symposium 1983* 62.

84. Rocchi, M. and Gabillard, B. (1983) 'GaAs digital dynamic ICs for applications up to 10 GHz' *IEEE J. Solid State Circuits*, **SC–18** (3) (June) 369.

85. Van Tuyl, R. L., Liechti, C. A., Lee, R. E. and Gowen, E. (1977) 'GaAs MESFET logic with 4 GHz clock rate' *IEEE J. Solid State Circuits* **SC–12**, (Oct.).

86. Ohmori, M. *et al.* (1981) 'Very low power gigabit logic circuits with enhancement mode GaAs MESFETs' *MTT Symposium* 188.

87. Flahive, P. G., Clemetson, W. J., Connor, P. O., Dori, A. and Shunk, S. C. (1984) 'A GaAs DCFL chip set for multiplex and demultiplex applications at gigabit/sec data rates' *GaAs IC Symposium* 7.

88. Jensen, J. F., Salmon, L. G., Deakin, D. S. and Delaney, M. J. 'Ultrahigh speed GaAs static frequency dividers' *IEDM 86* 476.

89. Takada, T., Kato, N. and Ida, M. (1986) 'An 11 GHz GaAs frequency divider using source coupled FET logic' *IEEE Electr. Dev. Letters* **ED–7** (1) (Jan.).

90. Osafune, K. and Ohwada, K. (1987) 'An ultra high speed GaAs prescaler using a dynamic frequency divider' *IEEE Trans. on MTT* **MTT–35** (1) (Jan.).

91. Asbeck, P. M. *et al.* (1986) *GaAs IC Symp. Tech. Dig.* (Oct.)

92. Sakai, T. *et al.* (1985) *IEDM Tech. Dig.* 18–21.

93. Nakazato, K., Nakamura, T., Nakagawa, J. I., Okabe, T. and Nagata, M.

(1985) 'A 6 GHz ECL frequency divider using sidewall base contact structure' *IEEE ISSCC.*

94. Fraser, D. L., Boll, H. J., Byruns, R. J., Wittwer, N. C. and Fuls, E. N. 'Gigabit logic circuits with scaled nMOS' *ESSCIRC 81* Digest of technical papers 202.

95. Osafune, K. *et al.* (1987) '20 GHz dynamic frequency divider with GaAs advanced saint and air-bridge technology' *Electronics Letters* **23** (6) (March).

96. Chantepie, B. *et al.* (1986) 'Packaged 7 mW, 1–2 GHz dynamic 60/61 GaAs prescaler' *Electronics Letters* **22**(7) (March) 355–356.

97. Anderson, C. J. 'GaAs MESFET 4 × 4 crosspoint switch' *GaAs IC Symposium 1986* 155.

98. Takada, T., Shimazu, Y. Yamasaki, K., Togashi, M., Hoshikawa, K. and Idda, M. (1985) 'A 2 Gb/s throughput GaAs digital time switch LSI using LSCFL' *IEEE 1985 microwave and millimeter-wave monolithic circuits symposium* St-Louis.

99. Carney, J. K., Helix, M. J. and Kolbas, R. M. (1985) 'Gigabit optoelectronic transmitters' *IEEE 1985 microwave and millimeter-wave monolithic circuits symposium* St-Louis.

100. Hasegawa, K., Tezuka, A., Uenoyama, T., Nishii, K., Bando, K., Utsumi K. and Onuma, T. (1985) 'High yield and low power multiplexer/demultiplexer by SCFL' *IEEE 1985 microwave and millimeter-wave monolithic circuits symposium* St-Louis.

101. Reiman, R. *et al.* (1986) *ISSCC Dig. Tech. Papers* (Feb.) 186–187.

102. Hughes, J. B., Coughlin, J. B., Harbott, R. C., van den Hurk, T. H. J. and van den Bertgh, B. J. (1979) 'A versatile ECL multiplexer IC for the Gbit/s range' *IEEE J. Solid State Circuits* **SC–14** (3) (Oct.).

103a. Flahive, P. G., Clemetson, W. J., Connor, P. O., Dori, A. and Shunk, S. C. (1984) 'A GaAs DCFL chip set for multiplex and demultiplex applications at gigabit/sec data rates' *GaAs IC Symposium 1984* 7.

103b. Takada, T., Nozawa, K., Ida, M. and Asai, K. (1985) 'A Gigabit rate GaAs multi-functional LSI with half micron gate buried p-layer saint FETs' *IEEE 1985 microwave and millimeter-wave monolithic circuits symposium* St-Louis.

104. Hickling, R. M., Argyroudi, P., Lai, H., Chow, J., Lee, F. S. and Eden, R. C. (1985) 'Monolithic 1.6 Gbit/s 8:1 multiplexer and 1:8 demultiplexer subsystems using CDFL' *GaAs IC Symposium* 79.

105. McCormack, G. D., Rode, A. G. and Strid, E. W. (1982) 'A GaAs MSI 8–bit multiplexer and demultiplexer' *GaAs IC Symposium* 25.

106. Nakamura, H., Tanak, T., Inokuchi, K., Saito, T., Kawakami, Y., Sano, Y., Akiyama, M. and Kaminishi, K. (1986) 'Multiplexer and demultiplexer using DCFL/SBFL circuit and precise Vth control process' *GaAs IC Symposium* 151.

107. Liechti, C. A., Baldwin, G. L., Gowen, E. G., Joly, R., Namjoo, M. and Podell, A. F. (1982) 'A GaAs MSI word generator operating at 5 Gbits/s data rate' *IEEE Trans. on MTT* **MTT–30** (7) (July).

108. McDonald, M. A. and McCormarck, G. (1986) 'A 12:1 multiplexer and demultiplexer chip set for use in a fiber optic communication system' *GaAs IC Symposium* 229.

109. Clawin, D. and Langmann, U. 'Multigigabit/second silicon decision circuit' *Proc. ISSCC 85* 222.

110. Peltier, M., Nuzillat, G. and Gloanec, M. 'A monolithic GaAs decision circuit for Gbit/s PCM transmission systems'.

111. Connor, P. O., Flahive, P. G., Clemetson, W., Panock, R. L., Wemple, S. H., Shunk, S. C. and Takahashi, D. P. (1984) 'A monolithic multigigabit/second DCFL GaAs decision circuit' *IEEE Elect. Dev. Lett* **EDL–5** (7) (July).

112. Ducourant, T. (1986) '3 GHz, 150 mW 4bit GaAs analogue to digital converter' *GaAs IC symposium*.

113. Lee, C. P. (1985) 'GaAs/GaAlAs high electron mobility transistors for analogue
to digital converter application; *IEDM*.

114. Wang, K. C. (1985) 'High speed high accuracy voltage comparators implemented with GaAs/(GaAl)As heterojunction bipolar transistors' *GaAs IC Symposium*.

115. Fawcett, K. (1986) 'High speed high accuracy self-calibrating GaAs MESFET voltage comparator for A/D converters' *GaAs IC Symposium*.

116. Hsieh, K. C. (1985) 'A GaAs 12 bit D/A converter' *GaAs IC Symposium*.

117. Wliss, F. (1986) 'A 1Gb/s 8 bit GaAs DAC with on chip current sources' *GaAs IC Symposium*.

118. de Graff, K. (1986) 'GaAs technology for analogue to digital conversion' *GaAs IC Symposium* 205–208.

119. Fiedler, A., Chun, J., Eden, R. and Kang, D. (1986) 'A GaAs 256 × 4 static self-timed random access memory' *GaAs IC Symposium* 89.

120. Hirayama M. *et al.* (1986) 'A GaAs 16 Kbit static RAM using dislocations free crystal' *IEEE* **Ed–33** (1) (Jan.) 104.

121. Tanino, N. *et al.* (1986) 'A 2.5 ns 200 mW GaAs 4Kb SRAM' *GaAs IC Symposium* 101.

122. Hayashi, T. *et al.* (1984) 'ECL compatible GaAs SRAM circuit technology for high performance computer application' *GaAs IC symposium* 111.

123. Gabillard, B. *et al.* (1987) 'A 1 K GaAs SRAM with 2 ns cycle time' *ISSCC Technical Digest*.

124. Rocher, C. *et al.* (1986) 'Evaluation of the theoretical maximum fabrication yield of GaAs 1K bit SRAM's' *International Symposium on Gallium Arsenide and Rel. Compounds*.

125. Hayaski T. *et al.* (1985) 'Small access time scattering GaAs SRAM technology using bootstrap circuits' *Digest of 1985 GaAs IC Symposium* 199–202.

126. Mizo-guchi, T. *et al.* (1984) 'A GaAs 4K bit static RAM with normally-on and off combination circuit' *GaAs IC Symposium* 117.

127. Yokoyama, N. *et al.* (1985) 'A 3 ns GaAs 4 K × 1 bit static RAM' *IEEE* **ED–32** (9) (Sept).

128. Graham, A. and Sando, S. (1984) 'Pipe-lined static cache memories with 1 ns speed electronic design' (Dec.).

129. Sheng, N. H. (1986) 'A high speed 16 KBit HEMT SRAM' *GaAs IC Symposium* 97–99.

130. Takano, S. *et al.* (1987) 'A GaAs 16 Kbit static RAM' *ISSCC Technical Digest*.

131. Mitonneau, A., Rocchi, M., Talmud, I., Mauduit, J. C. and Henry, M. (1984) 'Direct experimental comparison of submicron GaAs and Si NMOS MSI digital IC's' *GaAs IC Symposium*.

132. Lee, F. S. *et al.* (1982) 'A high speed LSI GaAs 8 × 8 bit parallel multiplier' *IEEE J. of Solid State Circuits* **SC–17** (4) (Aug.).

133. Nakayama, Y. *et al.* (1983) 'A GaAs 16 × 16 bit parallel multiplier' *IEEE J. Solid State Circuits* **SC–18** (5) (Oct.) 599.

134. Mac Sorley, O. L. (1961) 'High speed arithmetic in binary computers' *Proc. IRE* **49** (Jan.) 67–91.

135. Delhaye, E. *et al.* (1986) 'A 2.5 ns, 40 mW GaAs 4 × 4 multiplier in 2's complement mode' *ESSCIRC 1986* Delft 56. *IEEE J. Solid State Circuits* (to be published).

136. Furutsuka, T., Takahashi, K. *et al.* (1984) 'A GaAs 12 × 12 bit expandable parallel multiplier LSI using sidewall-assisted closely-spaced electrode technology *Proc. IEEE International Electron Devices Meeting* (Dec. 9–12).

137. Arch, D. K. *et al.* (1986) 'A self-aligned gate superlattice (Al, Ga)As/n+ GaAs MODFET 5 × 5 bit parallel multiplier' *IEEE Electron Device Letters* **EDL–7** (12) (Dec.).

138. Chung, H. K. *et al.* (1986) 'High speed and ultra-low power GaAs MESFET (× 5 multipliers)' *GaAs IC Symposium.*

139. Wada, M. *et al.* (1985) 'GaAs JFET Technology for DCFL LSI' *GaAs IC Symposium.*

140. Horiguchi, S. *et al.* (1985) 'An 80 ps 2500 gate bipolar macrocell array' *ISSCC 85.*

141. Schlier, A. R. *et al.* (1985) 'A high speed 4 × 4 bit parallel multiplier using selectively doped heterostructure' *GaAs IC Symposium.*

142. Miller, B. E. and Owen, R. E. (1986) 'A sub 10 ns low power bipolar 16 × 16 bit multiplier' *IEEE 1986 Custom Integrated Circuits Conference.*

143. Lee, J. Y. *et al.* (1985) 'A 8 × 8 b parallel multiplier in submicron technology' *ISSCC 85.*

144. Patterson, D. A. and Ditzal, D. R. (1980) 'The case for the reduced instruction set computer' *Computer Architectures News* **8** (6) (Oct.) 25–32.

145. Gilbert, B. K. (1984) 'Design and performance trade-offs in the use of VLSI and Gallium Arsenide in high clockrate signal processors' *Proc. IEEE ICCD 84* 260–266.

146. Mikitinovic, V. *et al.* (1986) 'An introduction to GaAs microprocessor architecture for VLSI' *Computer* (March) 30–42.

147. Breut, R. P. and Kung, H. T. (1982) 'A regular lay-out for parrallel adders' *IEEE Trans. Computers* **C 31** (3) (March) 260–264.

148. Helbig, M. *et al.* (1985) 'The design and construction of a GaAs technology demonstration microprocessor' *Proc. MIDCON 85* 1–6.

149. Miyanaga, H. *et al.* 'A 0.85 ns 1Kb bipolar ECL RAM'.

150. Washio, K. *et al.* (1987) 'A 48 ps ECL in a self-aligned bipolar technology' *ISSCC 87.*

151. Bayruns, R. J. *et al.* (1986) 'A 3 GHz 12 channel time division multiplexer-demultiplexer chip set' *ISSCC 86.*

152. Oowaki, Y. *et al.* (1987) 'A 7.4 ns CMOS 16 × 16 multiplier' *ISSCC 87.*

153. Hanamura, S. *et al.* (1985) 'Low temperature CMOS 8 × 8 b multipliers with sub 10 ns speeds' *ISSCC 85.*

154. Tang, D. D. and Solomon, P. M. (1979) 'Bipolar transistor design for optimized power delay logic circuits' *IEEE J. Solid State Circuits* **SC–14** (4) (Aug.).

155. Knorr, S. G. (1981) 'The potential of bipolar devices in LSI gigabit logic' *IEEE Circuits and Systems* **3** (1).

Chapter 3

Digital Integrated Circuit Technologies

Kimiyoshi YAMASAKI, Naoki KATO and
Yutaka MATSUOKA

3.1 INTRODUCTION

Gallium arsenide is an excellent semiconductor for use in very high-speed integrated circuits. This is because GaAs has the following superior electrical properties in comparison with silicon: (1) a very high low-field mobility (six times that of silicon) resulting in very high-speed operation of intrinsic devices, (2) an easily obtainable high resistivity substrate that decreases parasitic capacitances. Therefore, it is expected that GaAs ICs will be applied to high-speed computers, telecommunications and measurement systems.

However, GaAs also has some disadvantages which restrict IC processing. Developing high-quality GaAs crystals is difficult because GaAs is a binary semiconductor. Processing temperatures should be kept relatively low to suppress arsenic decomposition. Since GaAs does not have a stable native oxide, sophisticated methods are necessary for surface passivation.

The first GaAs digital IC constructed with MESFETs was reported in 1974. Since then, progress has been made in decreasing switching time and dissipation power and in increasing circuit complexity and process yield. As a result, LSI circuits such as 16 Kb static RAM consisting of about 100 000 MESFETS [1] and a 6 K-gates GaAs gate array [2] have been developed. These IC technologies have been applied to commercial SSIs, MSIs and LSIs.

The key technologies leading to the rapid progress are as follows:

(1) selective implantation to form a uniform and reproducible active layer
(2) self-aligned FETs with high performances
(3) high quality liquid encapsulated Czochralski (LEC) substrates, and
(4) advanced circuit configurations with low power and large operation margins.

In this chapter, the present state of the art for GaAs digital IC technologies is reviewed and discussed. In particular, device technologies, ion implantation and IC fabrication processes and performance will be considered.

Prior to discussing each technology in detail, the historical development of

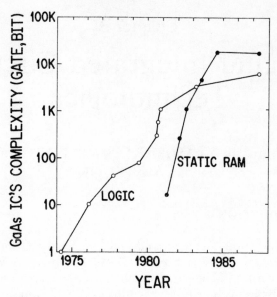

Fig. 3.1 Evolution of GaAs IC complexity; logic and memory (static RAM) ICs.

GaAs digital IC technologies is briefly reviewed. The trend of GaAs IC complexity is illustrated in Fig. 3.1.

The first GaAs IC was reported by Hewlett-Packard in 1974 [3]. This was to demonstrate its potential for very high-speed digital ICs. The circuit was a buffered FET logic (BFL) with 1 μm gate length normally-on (depletion mode) MESFETs. The active layer was formed by liquid phase exitaxy (LPE) and mesa-isolation. This technology was used in fabricating MSI circuits, such as an 8-bit multiplexer/data generator with 60–gates operating up to 3 Gb s^{-1} in 1977 [4]. However, this technology was not suitable for complex circuits. The LPE grown active layers were not reproducible, and the BFL circuits dissipated large amounts of power.

Rockwell researchers overcame these problems in 1977 with selective ion-implantation and Schottky diode FET logic (SDFL) circuits [5]. Selective ion-implantation was used to form n-channel and n$^+$-source/drain layers, resulting in planar GaAs surfaces. The uniformity and reproducibility of the threshold voltage were remarkably improved; the standard deviations of the threshold voltages were below 100 mV. SDFL circuit configurations reduced device dimensions and power dissipation. This technology led to development of the first LSI, an 8 × 8–bit multiplier, containing about 1 K gates [6].

In order to increase circuit complexity further, direct coupled FET logic (DCFL) was investigated, mainly in Japan. The advantages of this kind of circuit are very low power dissipation and simple gate construction. However, DCFLs required normally-off (enhancement mode) FETs with much more uniform threshold voltages because of low logic swings of 0.5 V. In order to satisfy this requirement, several self-aligned FET technologies and the growth technology of high quality crystals with low dislocation densities were developed. In 1983, Fujitsu fabricated a 3 K gate 16 × 16–bit parallel multiplier [7] and a 1 Kb RAM [8] using refractory gate self-aligned MESFETs [9]. NTT advanced the complexity of static RAM. They fabricated a 1 Kb static RAM in 1982 [10], a 4 Kb static RAM in 1984 [11], and a 16 Kb static RAM in 1984 [1], using self-aligned implantation for n+-layer technology (SAINT) FETs [12].

3.2 FET TECHNOLOGIES

3.2.1 Reviews of GaAs FETs for digital ICs

Since the first GaAs digital IC was reported in 1974, many GaAs FET structures have been proposed and developed for improving operation speeds and increasing circuit complexity and process yields. Schematic cross-sections of some typical devices are shown in Fig. 3.2.

The mesa-epitaxially fabricated MESFET, similar to ones used as discrete microwave devices, was applied to digital ICs for the first time in 1974 [3]. This device is shown in Fig. 3.2 (a). It successfully demonstrated the

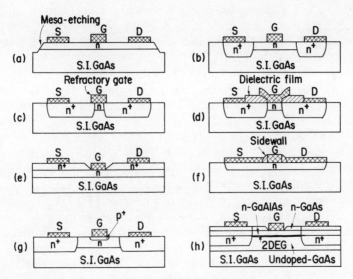

Fig. 3.2 Schematic cross sections of various GaAs FET structures developed for GaAs digital ICs: (a) mesa-epitaxial MESFET, (b) planar MESFET with selective ion implantation, (c) refractory gate self-aligned MESFET, (d) self-aligned implantation for N+-layer technology (SAINT) MESFET, (e) selectively grown n+ source/drain MESFET, (f) sidewall-assisted self-aligned technology (SWAT) MESFET, (g) junction FET (JFET), and (h) high electron mobility transistor (HEMT).

potential of GaAs in making high-speed logic circuits. However, this FET was not suitable for large integration, because its threshold voltage was scattered within wafers and from wafer to wafer due to poor epitaxial techniques (LPE) at that time. Moreover, the mesa-isolation technique restricted the development of fine line fabrication. These disadvantages were overcome by selective ion implantation FET technology [5]. The GaAs surface is planar, as shown in Fig. 3.2 (b). The n-channel layer and source/drain n^+-layers are formed by selective ion implantation directly into the GaAs semi-insulating substrate. The unimplanted semi-insulating substrate isolates each device.

Normally-off (enhancement-mode) FETs were required to reduce dissipation power in order to increase circuit complexity. For this purpose, various self-aligned structures were developed. Some schematic examples are shown in Figs 3.2 (c)–(f). Reduction of source resistance and very uniform threshold voltages were obtained by using these self-aligned technologies. For refractory gate self-aligned FETS (Fig. 3.2 (c)), the source and drain n^+-layers are embedded close to the gate by n^+-implantation with the gate refractory metal used as a mask [9]. In self-aligned implantation for N^+-layer technology (SAINT) FETs (Fig. 3.2 (d)), the n^+-layers are formed by n^+-implantation with a T-shaped multilayer resist, which also defines the gate contact. The fabrication procedures and performances of these self-aligned FETs are described in detail in the next subsection.

In Fig. 3.2 (e), n^+-layers are grown by selective epitaxial techniques with refractory gate metal as a mask [13]. Another type of self-aligned FET is shown in Fig. 3.2 (f). The gate and source/drain electrodes are narrowly separated by the sidewall [14].

To apply sub-micron gate length FETs to digital ICs for the purpose of achieving much faster operation, we must suppress fatal short channel effects. Some short channel FETs were developed and are discussed in Section 3.3.

One of the disadvantages of MESFETs is low circuit operation margins caused by low-barrier heights (0.7–0.8 V) of Schottky junctions. A p-n junction gate enlarges the built-in potential and improves the circuit margin, especially in low power circuits constructed of normally-off FETs [15]. A schematic cross-sectional view of a junction FET (JFET) is shown in Fig. 3.2 (g). However, the p-n junction depth should be controlled precisely to obtain an appropriate reproducible threshold voltage. This increases the difficulty of development.

Heterostructure FETs, such as a high electron mobility transistor HEMT (Fig. 3.2 (h)), have several advantages over MESFETs. The two-dimensional electron gas, induced at the surface of undoped-GaAs, has larger electron mobility than three-dimensional electron gas in highly doped MESFET channels. This advantage is significant at low temperatures and for scaled-down sub-micron gate length FETs.

3.2.2 Self-aligned MESFET technologies

A normally-off (enhancement-mode) GaAs MESFET's n-channel layer is thin enough to be pinched off by only the built-in potential of the Schottky gate junction. The channel is turned on by forward gate voltage. As a result, the n-layer between the source and gate electrodes is fairly depleted by large surface state density, as schematically shown in Fig. 3.3. This surface depletion increases the external source resistance and decreases the transconductance and switching speed of conventional planar MESFETs. Surface states with high densities are inherent characteristics of GaAs surfaces covered with dielectric films. Therefore, it is necessary to develop sophisticated processing techniques to decrease the source resistance rather than attempting to reduce the surface states.

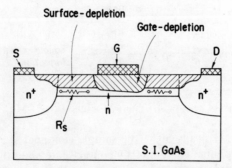

Fig. 3.3 Schematic cross section of a planar GaAs MESFET. The surface depletion layer induced by surface states decreases the n-channel between the source and gate electrodes resulting in increased source resistance.

Many n^+ self-aligned technologies, in which the n^+-source/drain layers are embedded close to gates, have successfully reduced the undesirable source resistances of normally-off MESFETs [16]. High-speed and low-power LSIs, constructed with direct coupled FET logics (DCFLs) or low-power source coupled FET logics (LSCFLs), were achieved after the development of self-aligned technologies. In this sub-section, fabrication techniques, performance and comparison of typical self-aligned FETs, such as refractory gate type and SAINT type FETs, are described.

(a) **Refractory gate technology**. The refractory gate self-sligned MESFET was pioneered by Fujitsu [9]. The fabrication steps are shown in Fig. 3.4.

The first step of the process is to selectively implant donor impurities, typically silicon ions, for the n-channel layer, followed by activation annealing. Next, refractory metal, typically WSi, is deposited over the wafer, and is patterned by reactive ion etching (RIE) to define the gate electrode. Then, n^+-implantation is performed. At that time, the gate electrode acts as an implantation mask. Consequently, n^+-layers are embedded outside the gate region in a self-aligned manner. After deposition

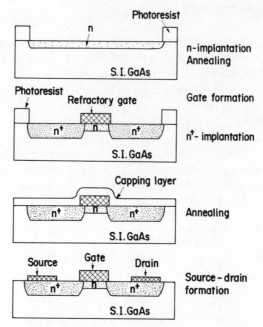

Fig. 3.4 Process steps of a refractory gate self-aligned GaAs MESFET.

of a capping layer, typically SiO_2, the sample is annealed at about 800°C to activate the n^+-dopants. The Schottky contact gate should be stable during this high-temperature process. Finally, source and drain ohmic electrodes are formed with AuGe-based alloyed contacts.

The key point of this technology is to develop a temperature stable refractory gate material. Following the first metal of TiW [9], various materials including TiWSi [17], WSi [18], WAl [19], WN [20] and TaSi [21], were investigated as suitable candidates from the standpoint of high-temperature stability and reproducibility.

One of the disadvantages of refractory gate technology is that the resistivities of refractory materials are higher by a factor of tens to hundreds than those of conventional Au-based gate metals, typically Ti/Pt/Au. In order to reduce gate resistance, which is necessary for monolithic microwave IC (MMIC) applications, the Au/TiW/WSi gate electrode was developed [22].

(b) **Dummy gate technology**. In the SAINT process developed by NTT [12], a T-shaped dummy gate was used as a mask in place of a refractory gate during n^+-implantation. Although the process steps were relatively complicated, there were many advantages.

The fabrication steps of a SAINT FET are illustrated in Fig. 3.5. At first, an n-channel layer is formed by selective implantation. Next, the whole surface is covered by a plasma enhanced CVD, PECVD, SiN film. Then, a T-shaped multilayer resist is formed on the SiN film. This consists of a bottom resist, an intermediate sputtered-SiO_2 layer and a top photoresist.

The top resist pattern, defined by photolithography or electron-beam-lithography, is replicated in the intermediate SiO_2 layer by RIE. Then, the bottom resist is shaped by RIE in an O_2 plasma discharge. By controlling the O_2 pressure and rf-power, the bottom resist is undercut with respect to the upper SiO_2 pattern. The vertical and lateral etching characteristics of the bottom resist are shown in Fig. 3.6. The lateral etching rate is about one-seventh of the vertical etching rate. The reproduction of the undercut is controlled to within a deviation of 0.02 μm. A cross-sectional SEM photograph of a T-shaped multilayer resist is also shown in Fig. 3.6. With the T-shaped resist, an n^+-implantation is performed to embed the source and drain n^+-layers. The intermediate SiO_2 layer acts as an ion stopping mask.

Then, the second SiO_2 layer is deposited by sputtering. The SiO_2 film is lifted off by using the T-shaped resist after slight etching of SiO_2 deposited on the side of the bottom resist. Consequently the SiO_2 film remains above the n^+-implanted layers, and extends somewhat beyond the n^+-projection edges by the amount of the T-shaped resist undercut. After that, activation of the implanted ions is done by high-temperature annealing. The source and drain ohmic electrodes are formed by AuGe-based alloyed contacts. Finally, gate metallization is accomplished by patterning photoresist and etching SiN (opening the gate window), and evaporating gate metals, typically Ti/Pt/Au, followed by lift-off. Consequently, the edge of the Schottky gate contact is automatically separated from the n^+-projection edge

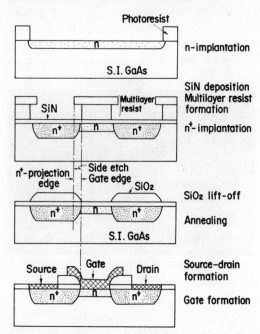

Fig. 3.5 Process steps of a GaAs SAINT MESFET.

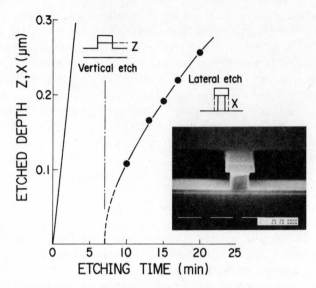

Fig. 3.6 Vertical and lateral etching characteristics of the bottom resist in the T-shaped multilayer resist used in the SAINT process as a dummy gate. The etching is performed by RIE in O_2 discharge. The insert is a SEM photograph of the finished multilayer.

by a distance equal to the distance of the undercut of the T-shaped resist. This is accomplished with the aid of the lifted-off SiO_2 film.

Typical current-voltage (I-V) characteristics for a SAINT FET and a conventional planar MESFET without n^+-layers are compared in Fig. 3.7. The conventional FET was fabricated by the same process as the SAINT FET, with the exception of the n^+-implantation. The drain conductance in

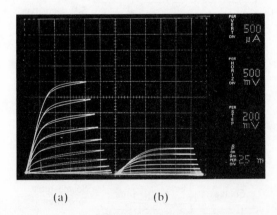

(a)   (b)

Fig. 3.7 Comparison of static current-voltage (I-V) characteristics between (a) a SAINT FET and (b) a conventional MESFET without n^+-layers. The gate lengths and widths are 1 and 10 μm, for both FETs. Horizontal axis: 0.5 V/div; vertical axis: 0.5 mA/div; gate voltage: 0.2 V/step from –1.0 to 0.8 V.

the linear region and the transconductance are dramatically improved by the self-aligned embedding of n+-layers. The maximum transconductance of the SAINT FET is 280 mS/mm, which is about three times the value (90 mS/mm) of the conventional FET.

The optimization of n+-gate spacing in a MESFET employing n+-layers self-aligned with its gate is of great importance in improving the switching speed. When the n+-projection edges are too close to the gate, the gate contact overlaps the n+-layers which spread laterally due to ion recoiling and thermal diffusion. As a result, gate parasitic capacitance increases. This situation also occurs in simpler refractory gate self-aligned FETs, because the n+-projection edges coincide with the gate edges. On the other hand, n+-placement too far from the gate causes increases in external source resistance. SAINT has made it possible for the first time to optimize n+-gate spacing.

An experimental result of the influence of n+-gate spacing on source resistance is shown in Fig. 3.8 [23]. This result verifies that significant improvements in source resistance cease at an n+-gate spacing of about 0.3 µm. This 0.3 µm length might correspond to a lateral spreading of the n+ implants. The influence of the spacing on gate capacitance was estimated by using a two-dimensional simulator [24]. In this calculation, lateral diffusion during high-temperature annealing was not taken into account. Nevertheless, gate capacitance for a 0.2 µm n+-gate spacing FET with 1 µm gate length decreases by a factor of two in comparison with the case of non-spacing. The optimum n+-gate spacing from the trade-off of source resistance and gate capacitance is 0.2–0.3 µm.

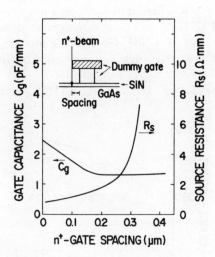

Fig. 3.8 Effects of n+-gate spacing on gate capacitance and external source resistance for normally-off n+ self-aligned MESFETs with 1 µm gate length. The gate capacitances were calculated with two-dimensional simulation and source resistance was determined by experiment.

A T-gate structure has also been developed using the refractory gate self-aligned technology [25]. The n⁺-implantation process step is illustrated in Fig. 3.9. The gate metal consists of two metals, an upper mask layer and a lower refractory gate layer. A T-shaped gate is formed by undercutting the lower metal with respect to the upper metal. After that, n⁺-implantation is performed, and the n⁺-projection edges are separated from the lower gate contact edges by the extent of the undercut.

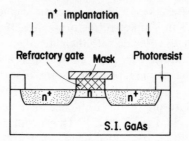

Fig 3.9 Schematic cross section of T-shaped refractory gate self-aligned GaAs MESFET.

An important advantage of the SAINT process is that conventional Au-based gate metals with low resistivities or unique gate materials with large barrier heights can be used. The barrier heights of ordinary n-GaAs Schottky junctions are about 0.7 to 0.8 V. They do not depend much on the work function of metals [26]. The barrier height limits forward gate voltages and logic swings, especially in DCFL circuits. Enlargement of the barrier height improves the operation margin for high complexity circuits. A large barrier height of 1.0 V is achieved by using an amorphous SiGeB (a-SiGeB) as a gate material [27]. An a-SiGeB was deposited on GaAs surface by thermal decomposition of a SiH_4-GeH_4-B_2H_6 mixture in a low pressure furnace at 450°C. The barrier height was controlled by the B_2H_6 flow rate, as shown in Fig. 3.10.

One of the disadvantages of the early SAINT process is that the gate

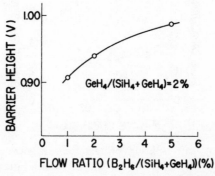

Fig 3.10 Barrier height of amorphous SiGeB versus deposition condition (flow ratio of reactive gases).

metal overlaps the dielectric film. This excess gate metal increases the parasitic capacitance and degrades the switching speed. Although the excess capacitance is 10–20% of the intrinsic gate capacitance for 1 μm gate length SAINT FETs, the ratio of excess capacitance over intrinsic gate capacitance inherently increases with decreases in gate length. This structural disadvantage was successfully eliminated by utilizing a planarisation technique using ion milling with a large incident angle [28]. With this technique, the gate electrode is automatically constrained only within the gate opening region of SAINT FETs.

A simplified version of SAINT was developed by Rockwell and is referred to as a dual-level double-lift-off substitutional gate (DDS) technique [29]. The T-shaped trilevel resist is replaced by a dual level photoresist, which consists of a top layer of AZ-type photoresist and a bottom layer of PMMA. The top resist is delineated by regular UV exposure. After that, undercutting of the bottom resist is obtained by deep UV flood exposure followed by chlorobenzene development.

3.2.3 Short channel FET technologies

One of the most effective methods for increasing the operation speed of FET circuits is to reduce the gate length. However, short channel effects restrict gate length reduction. With decreases in gate length, the threshold voltage shifts to a negative voltage direction, and current turn-off characteristics deteriorate. One example of the threshold shift is shown in Fig. 3.11. This was obtained using ordinary SAINT FETs. In the sub-micron gate region the deviation of the threshold voltage increases rapidly with decreasing gate length. Small gate length non uniformity induced by the photolithography process would also induce a large threshold voltage scatter. These effects are undesirable for LSI applications because LSIs require rigorous control of reproducibility and uniformity of threshold voltage.

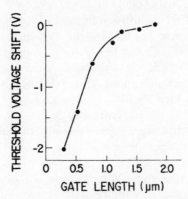

Fig. 3.11 Gate length dependence of threshold voltage determined for conventional SAINT FETs.

In this subsection, sub-micron gate length FET structures that suppress short channel effects, such as a buried P-layer SAINT (BP-SAINT) FET and a lightly doped drain-source (LDD) FET, are described. Scaling law and heterostructure FETs are also discussed.

(a) **Buried p-layer structure**. Short channel effects for n^+ self-aligned GaAs MESFETs are caused by substrate leakage current and n^+ lateral spreading. Lateral spreading due to thermal diffusion during high-temperature activation can be suppressed by capless annealing without surface stress [30] or by rapid thermal annealing [31]. The substrate current, which flows in the semi-insulating substrate between the adjacent source-drain n^+-layers, is a problem for GaAs MESFETs on semi-insulating substrates. This leakage current increase with reductions in gate length. This is because the spacing of the source-drain n^+-layers is reduced in proportion to reductions in the gate length. The substrate leakage current is illustrated in Fig. 3.12.

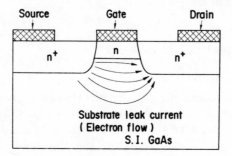

Fig. 3.12 Schematic cross section of an n^+ self-aligned GaAS MESFET to represent the substrate leak current.

An effective and direct method to suppress the substrate current is to intentionally bury a barrier layer between the active layer and the substrate. A p-layer, formed by Be ion implantation in a SAINT FET functions as an electrical barrier [32]. A schematic cross-sectional view of a buried p-layer SAINT (BP-SAINT) FET is shown in Fig. 3.13. The buried p-layer does not lead to instability or increases in parasitic capacitance. This is because it includes mainly shallow acceptor levels and is completely depleted of free holes by its built-in potential against the upper n-layer. Little difficulty is caused by the fabrication procedure, since only one simple process step, Be ion implantation, is added. Similar barrier layers are formed by Mg ion implantation [33] or C/O ion implantation [34].

To suppress the substrate current successfully, the Debye length of the p-layer should be shorter than about one-tenth that of the gate length. Also the depletion layer created by the built-in potential against the n-channel layer must extend across the whole p-layer. Otherwise, an increase in parasitic capacitance and instability will occur. These two conditions determine the thickness and impurity density of the p-layer.

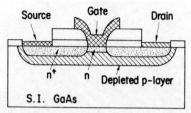

Fig. 3.13 Schematic cross section of a BP-SAINT FET.

Static I-V characteristics of a BP-SAINT FET are compared with a conventional SAINT FET without p-layer in Fig. 3.14. Both types have the same gate length of 0.4 μm. The triode-like and poor current turn-off characteristics of the conventional SAINT FET are improved with the buried p-layer. The threshold voltage shift is also reduced from 2.45 V (without p-layer) to 0.9 V (with p-layer) for 0.4 μm gate length FETs. For BP-SAINT FETs with a half micron gate length, the short channel effects are suppressed enough for LSI applications [35].

The switching speeds of BP-SAINT FETs were evaluated by a ring oscillator with an E/D DCFL configuration. The gate lengths are 0.4 μm. The delay time of BP-SAINT FETs was reduced by a factor of 1.6 in comparison with that of the conventional SAINT FETs. The minimum delay time observed was 9.9 ps/gate.

(b) **LDD MESFET technology**. Reduction of n^+-layer thickness is an effective way of suppressing short channel effects [73]. Separating the source and drain n^+-layers would also decrease substrate currents [36]. These methods by themselves are not useful, because they tend to increase the source resistance. The combination of these methods is however a sophisticated way of suppressing substrate leakage currents, and short channel effects, with only a small increase in source resistance. Thick and heavily doped n^+ source drain layers can be slightly separated from the gate by thinner and/or lower doped n'-layers embedded next to the gate. This structure is referred to as LDD MESFET [37] or a double n^+-layer SAINT FET [38].

The fabrication process steps for an LDD MESFET are illustrated in Fig. 3.15. First, the n-channel layer is formed by selective ion implantation into a semi-insulating GaAs substrate. Then, refractory gate metal (WSi/W) is deposited and patterned by RIE. After that, the whole surface is coated with SiO_2 film, followed by anisotropic dry etching for sidewall formation. Then, n^+-implantation is performed with the gate metal and sidewalls as a mask. After removal of the sidewalls, n'-implantation is performed. Finally, source and drain ohmic contacts are made by alloying a AuGe/Ni film.

The threshold voltage shift is successfully reduced from 0.35 to 0.04 V for an LDD MESFET with 0.8 μm gate length. The K-value is increased by a factor of 1.4 in spite of a slight increase in the source resistance.

(c) **Scaling law**. Scaling-down vertical dimensions along with horizontal dimensions is required for suppressing undesirable short channel effects and

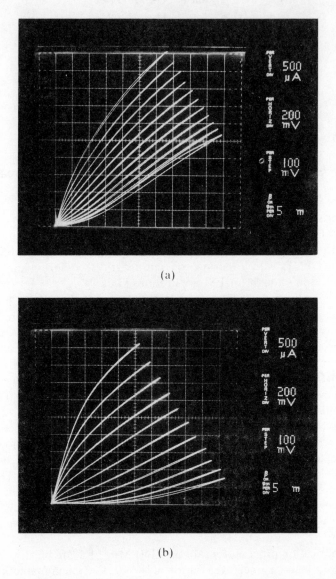

(a)

(b)

Fig. 3.14 Comparison of static I-V characteristics of (a) a conventional SAINT FET and (b) a BP-SAINT FET with 0.4 μm gate length. Gate voltage: 0.1 V/step from –0.3 to 0.7 V.

for improving transconductance and switching speed. However, few investigations have been made yet because short channel effects due to substrate currents strongly depend on the residual impurity density and thickness of semi-insulating substrate assumed in two-dimensional simulation. The buried p-layer structure removes the uncertainty of the substrate in simulation and makes it possible to discuss the scaling law of MESFETs.

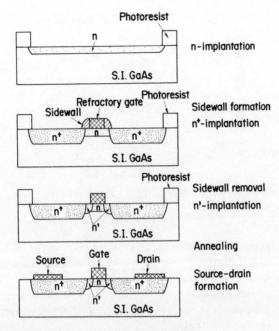

Fig. 3.15 Process steps of an LDD GaAs MESFET.

The scaling law of BP-SAINT FETs was evaluated using a two-dimensional device simulator [39]. The scaling factors and standard values are summarized in Table 3.1. A constant voltage scaling was chosen, because the threshold voltage should be kept constant for circuit operation margin, and because the constant Schottky barrier height is comparable to supply voltage in low pinch-off FET circuits. Therefore, the doping densities

Table 3.1. A scaling table for GaAs MESFETs

Device parameter	Scaling factor		Standard value
n thickness	t_n	$1/K$	0.1 μm
p thickness	t_p	$1/K$	0.4 μm
n^+ thickness	t_n^+	$1/K$	0.1 μm
n doping	N_D	K^2	1.3×10^{17} cm^{-3}
p doping	N_A	K^2	1.6×10^{17} cm^{-3}
n^+ doping	N^+	K	1.0×10^{18} cm^{-3} (peak value)
n^+-gate spacing	L_{n^+-g}	$1/K$	0.1 μm
Gate length	L_g	$(1/K)^{1.2}$	0.52 μm

n and p layers: uniform dopings (step distributions)
n^+ layer: Gaussian distribution

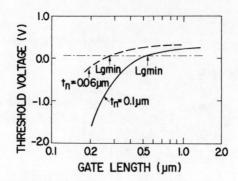

Fig. 3.16 Calculated gate length dependence of threshold voltages for two n-channel thickness, 0.1 and 0.06 μm.

of the n and p layers were expected to increase by a factor of K^2 with a $1/K$ reduction in vertical dimensions. However, according to simulation the doping density of n^+-layers was only increased by a factor of K. This was because the carrier density of n^+-layers was restricted to below 1×10^{19} cm^{-3}. Under these scaling factors, the sheet resistivity of the n^+-layer was kept almost constant.

When the gate length is reduced with constant vertical dimensions, the threshold voltage shifts to the negative direction (Fig. 3.16). From our experience in GaAs LSI design, we defined the boundary between long channel and short channel FETs as the gate length at which the negative threshold shift is 0.2 V. The minimum gate lengths applicable to LSIs are defined by this boundary. When the vertical dimension is reduced, the minimum gate length is decreased. The obtained relationship between n-channel thickness, t_n, and minimum gate length, L_{gmin}, is $L_{gmin} \propto t_n^{1.2}$ (Fig. 3.17). The solid line shows the boundary between the short and long

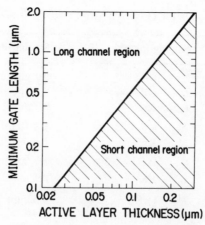

Fig. 3.17 Relationship between n-channel thickness and minimum gate length for BP-SAINT FETs, calculated by two-dimensional simulation.

channel regions. The slope is slightly larger than one. The reason for this is that the density of the n^+-layer is increased by a factor of only K. When the device is scaled down along the boundary, the threshold voltage remains constant and transconductance increases in proportion to $1/t_n$. A large value of 850 mS mm^{-1} is predicted for a 0.2 μm gate length FET.

For a 0.2 μm gate length, the n-channel thickness should be less than 0.045 μm. This value equals the effective thickness of Si ion distribution implanted at 25 keV. Therefore, it would be possible to realize 0.2 μm gate length FETs by using implantation in combination with low diffusion annealing techniques.

(d) **Heterostructure FETs**. According to the scaling law, a very thin n-channel 0.025 μm thick with a high doping density of 2×10^{18} cm^{-3} is required for a 0.1 μm gate length MESFET applicable to LSIs. It is very difficult at present to achieve this thin channel layer by the ion implantation technique ordinarily used in LSI fabrication. Moreover, the heavily doped channel layer degrades the electron mobility due to ionized impurity scattering.

These scaling-down related problems are resolved by heterostructure FETs (HFETs). A cross sectional view of a typical HFET, HEMT, is illustrated in Fig. 3.2 (h). Very thin layers are obtained by current epitaxial techniques, MBE and MOCVD, and carriers induced in the undoped heterointerface drift without impurity scattering. Therefore, the HFET is one of the most promising candidates for very short gate length, very high-speed devices. A record propagation delay time of 5.8 ps/gate was measured at 77 K with a ring oscillator integrated with 0.35 μm gate length HEMTs [40].

One serious problem with HEMTs is difficulty in threshold voltage reproducibility, caused by strong dependence of threshold voltage on the doping density of the AlGaAs layer. Another problem is related to deep levels (D–X centres) in n-AlGaAs, which are induced by donors.

The simplest and most effective method to overcome these problems is to use undoped-AlGaAs. From this point of view, MIS-like HFETs (MIS-HFETs) have been proposed [41]. One such MIS-HFET is schematically shown in Fig. 3.18 [42]. The threshold voltage is essentially determined by the electron affinity difference between Ge and GaAs and is expected to be 0.1 V. Transconductances up to 430 mS mm^{-1} have been obtained with AlGaAs thickness of 0.01 μm [43].

3.3 ACTIVE LAYER FORMATION

Direct, selective ion implantation into semi-insulating GaAs substrate is one important technology which has promoted GaAs digital IC developments. In the early stages of GaAs device fabrication, epitaxial growth was used in conjunction with the etching technique to control the FET channel layers. It was difficult to make reproducible, uniform active layers employing this

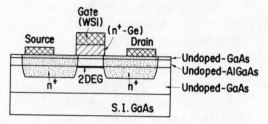

Fig. 3.18 Schematic cross section of n⁺-Ge gate MIS heterostructure FET (MIS-HFET).

technique because of difficulties in controlling the growth and the etch precisely. The etched mesa structure was also not suitable for LSI fabrication, due to photolithography limitations. Ion implantation made it possible to overcome these disadvantages. In this section, semi-insulating substrate characteristics and active layer formation by ion implantation suitable for GaAs ICs are described.

3.3.1 Ion implantation

Considerable work has been done with silicon, sulphur and selenium in the formation of an n-type layer [44]. Carrier concentration levels required for high speed GaAs device channel layer are of the order of 10^{17}–10^{18} cm^{-2}, with depth of 0.05–0.2 μm.

An example of a carrier concentration profile is shown in Fig 3.19 [45]. In

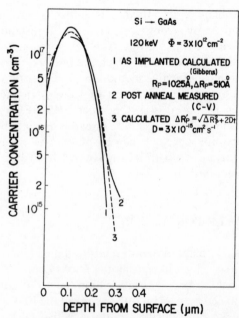

Fig. 3.19 A carrier concentration profile of Si implantation into GaAs.

this profile, ^{28}Si ions are implanted into GaAs at an energy of 120 keV and a dose of 3×10^{12} cm^{-2}. The three types of curves are shown: (1) as implanted Si ion profile calculated with LSS [46], (2) measured carrier profile with C–V after annealing, and (3) calculated profile using a diffusion constant. The experimental sample for obtaining the C–V profile was coated with a PECVD silicon nitride film (from a gaseous mixture of silane and ammonia) subsequently annealed in a furnace anneal system at 800°C. Meanwhile the post-annealing theoretical profile using a diffusion constant was given by the following equation [47]

$$\Delta R_{p}'^{2} = \Delta R_{p}^{2} + 2Dt$$

where $\Delta R_p'$ is standard deviation of post-anneal implanted atoms
$\quad \Delta R_p$ is standard deviation of as implanted atoms
$\quad t$ is annealing time
$\quad D$ is diffusion constant.

The diffusion constant D was calculated to be 3×10^{-15} cm^2 sec^{-1} by fitting the LSS curve to the C–V profile. Annealing time was 1200 s.

Typical ion implantation conditions for GaAs LSI are:

- use of Si icons
- 30–60 keV for channel layer formation
- selective implantation using patterned mask (usually photo-resist).

The most important parameter in realising high-speed ICs is FET transconductance (g_m). It is widely known that g_m can be increased by reducing the channel depth. Recently most of the successful integrated circuits employ an energy level of around 30 keV ion implantation to reduce the channel depth.

The following example illustrates the advantage of low energy ion implantation [48]. Device structure was buried p-layer SAINT FET (BP-SAINT). The original version was fabricated with 67 keV Si ion implantation and a more recent version with 30 keV. The g_m–V_T relationship for the two energy levels is shown in Fig. 3.20. The g_m was taken under the

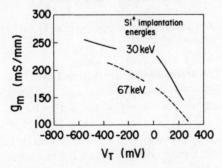

Fig. 3.20 FET g_m and V_T relationship for two ion implantation energies.

condition of 1 V drain voltage and 0.5–0.6 V gate voltages. This figure clearly demonstrates g_m increase from 170 mS mm^{-1} to 230 mS mm^{-1} at the V_T of 0 V, or an improvement factor of 1.3. The improvement factor was less than expected from the ideal scale-down law where g_m is inversely proportional to channel depth. Further improvements are discussed in more detail in a later section.

3.3.2 Activation annealing

Post-implantation annealing is carried out to activate implanted ions and provide carriers for conduction in semi-conductors. Presently the most widely utilized technique in IC fabrication consists of holding wafers in a heated furnace for a fixed duration. To protect GaAs wafer during annealing, a dielectric cap is usually provided. SiO_2 or SiN with thicknesses of 500–2000 Å are used.

The relationship between sheet resistivity and annealing temperature is shown in Fig. 3.21 [48]. Here the experimental sample was implanted with 200 keV Si ions, and a dose of 4×10^{13} cm^{-2} through a 0.15 µm thick SiN film. Annealing time was fixed at 10 minutes. This figure indicates an optimum annealing temperature of around 800°C. This temperature provides the highest activation efficiency for relatively high doses for GaAs. A decline in sheet resistivity at temperatures higher than 825°C can be ascribed to Si ions being positioned in the As site.

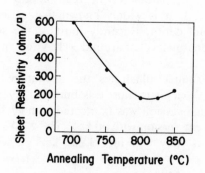

Fig. 3.21 Sheet resistivity vs annealing temperature of Si implanted GaAs layer.

3.3.3 Electrical Characteristics of the Active Layer

(a) **Electrical evaluation methods of the active layer**. Electrical characteristics of GaAs active layers can be evaluated by several methods. Carrier mobility and sheet carrier concentration are determined by Hall effect measurements. A large chip, usually several mm square, is conventionally needed for the measurement. Active layers used in the FET channel, especially in the enhancement mode FET channel, are difficult to measure by the Hall effect because of surface depletion. So this method is suitable for macroscopic distribution investigation of sheet carrier concentration and/or activation efficiency for relatively high dose or high implantation energy, although micro Hall chip measurements [49] make it possible to investigate areas as small as 40 microns square.

Carrier concentration profiles of the active layer can be measured by the capacitance–voltage (C–V) method. The method also requires large area contact to maintain an accuracy by avoiding periphery effects, and the depth of the active layer measured should be deeper than that in the FET channel region. It is impossible to measure active layers with carrier concentrations near 1×10^{18} cm^{-3} or higher because reverse leakage current through the Schottky contact rapidly increases with carrier concentration in such high concentration regions. Ohmic contact resistivity and sheet resistivity of the active layer can be determined at the same time using the transmission line method. An E-FET channel layer cannot be measured by this method due to surface depletion, for the same reason as for the Hall effect measurement. The transmission line method is often adopted for n$^+$-layer evaluation.

The most direct characterization of the channel layer is made by FET characteristic evaluation. Although more complicated procedures must be used for test device fabrication than with other methods, the smallest area can be investigated with highest sensitivity by this method. Both substrate-induced and process-induced inhomogeneities can be evaluated by using LSI-directed FETs with a self-aligned, short-length gate.

(b) **Origins of FET characteristic scattering**. Maintaining high uniformity in high-performance FETs is one of the greatest demands of GaAS ICs, especially digital ICs, because low power consumption with high-speed operation cannot be attained without highly uniform characteristics of component devices. There are several origins of the FET characteristic scattering. They are classified as follows.

(1) Threshold voltage scattering arising from gate length non uniformity related to short-channel effects.
(2) Drain current scattering with a constant threshold voltage arising from non-uniformity of external source resistance.
(3) Microscopic threshold voltage scattering introduced by dislocation effects.
(4) Macroscopic sloped threshold voltage distribution arising from channelling effects.

An example of the relationship between ungated drain current, which is measured with a passivation layer of Si$_3$N$_4$ over the channel region, and threshold voltage measured after gate formation and two-level interconnection processes is shown in Fig. 3.22. Threshold voltage is uniquely determined according to ungated current normalized by gate width. This indicates that threshold voltage scattering is introduced at the stage of implanted ion activation. All processes including Schottky gate formation after activation introduce negligibly small threshold voltage scattering, and Schottky gate barrier height uniformity is very high.

Relationships between threshold voltage and drain current at fixed bias voltages are shown in Fig. 3.23. The square root of drain current at a fixed bias is almost linear with threshold voltage. The unique relation is confirmed for wafers that include FETs with small deviation of external source resistance. However, drain current scatterings are observed in some cases

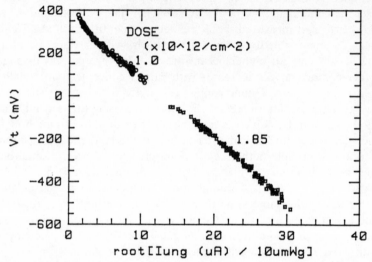

Fig. 3.22 Relationship between the square root of ungated drain current measured before gate Schottky contact formation and threshold voltage for SAINT FETs.

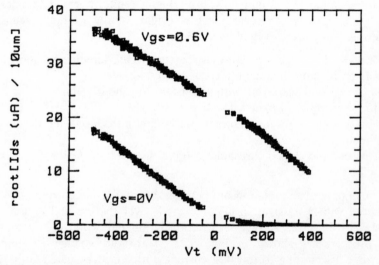

Fig. 3.23 Relationship between the square root of saturation drain current and threshold voltage. Saturation drain current was measured at 0.6 V and 0 V gate bias.

using a conventional gate structure. These are concerned only with scattering in source resistance depending on Schottky gate formation process or gate-to-source distance. Fig. 3.23 also indicates that scatterings in FET characteristics other than threshold voltage are independent of substrate effects.

Gate length non-uniformity brings about threshold voltage scattering of LSI-directed FETs with short gate length, because threshold voltage depends strongly on gate length. The contribution of substrate effects and gate length non-uniformity can be separated analytically by sub-threshold characteristic measurement. Short channel effects appear in threshold voltage shifts in negative direction, drain conductance increase and sub-threshold current increase. Sub-threshold drain current I_{DS} is characterized by sub-threshold factor N_G which is defined by the following expression, same as the Schottky contact ideality factor,

$$I_{DS} \propto \exp\left[\frac{qV_{GS}}{N_G kT}\right]$$

where, q is the electron charge, V_{GS} is the gate voltage, k is Boltzman constant, and T is the temperature in °K. Typical dependences of sub-threshold factor N_G on gate length are shown in Fig. 3.24 [50]. N_G increases rapidly according to gate length reductions in the short gate length region. This means that factor N_G is a good figure for indicating the degree of short-channel effects.

Individual contributions of substrate effects and gate length non-uniformity are separated by plotting threshold voltage versus sub-threshold factor N_G as

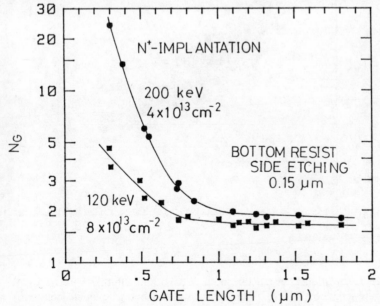

Fig 3.24 Dependence of sub-threshold factor N_G on gate length.

shown in Fig. 3.25. The line in Fig. 3.25 is determined by the least-squares method assuming a linear relation between threshold voltage and sub-threshold factor N_G. From the N_G sensitivity to gate length, the threshold voltage standard deviation σV_{th} can easily be divided into two components. The standard deviation of threshold voltage attributable to the substrate inhomogeneity, denoted by $\sigma V_{th}(sub)$, can be obtained by evaluating the vertical deviations of threshold voltage from the lines determined by the least squares method, and that attributable to gate length non-uniformity denoted by $\sigma V_{th}(L_G)$, is the product of the slope of the line and standard deviation of sub-threshold factor N_G. The relation

$$\sigma V_{th}^2 = \sigma V_{th}(sub)^2 + \sigma V_{th}(L_G)^2$$

is easily obtained from basic statistics. In the example in Fig. 3.25, σV_{th}, $\sigma V_{th}(sub)$ and $\sigma V_{th}(L_G)$ are calculated at 87, 43 and 74 mV, respectively. Suppressing the short-channel effects leads to a small deviation of subthreshold factor N_G, resulting in reduction of $\sigma V_{th}(L_G)$.

(c) **Dislocation effects**. Substrate effects have been observed in carrier concentration distribution in an implanted and annealed active layer along a wafer diameter. For conventional LEC-grown GaAs substrate, the dislocation density has a W-shaped distribution, and the macroscopic distribution of the implanted layer carrier concentration, as measured by Hall

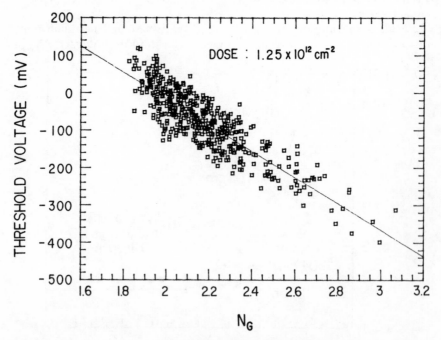

Fig 3.25 Correlation between threshold voltage and sub-threshold factor N_G for 1 μm gate-length SAINT FETs on a 2 in. wafer. Then gate length non-uniformity was reflected on the N_G scattering through short-channel effects.

measurement, is similar. This fact suggests that dislocations affect the activation efficiency in some manner.

Direct observation of dislocation effects was made by NTT using simple FETs without an n$^+$-layer [51], and also using SAINT FETs with a 1 μm gate length [50]. A dielectric passivation layer of Si$_3$N$_4$ was used to prevent As evaporation from the GaAs surface in the experiments. Figure 3.26 shows dependence of threshold voltage on the distance between the FET gate and the nearest neighbour dislocation revealed by KOH etching after the electrical measurement. There is a threshold voltage shift to the negative direction within 40–50 microns of the dislocation.

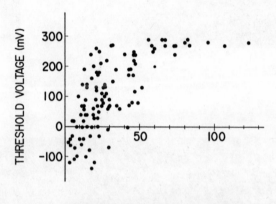

DISTANCE FROM AN ETCH PIT (μm)

Fig 3.26 Correlation between threshold voltage and distance from the nearest neighbour dislocation pit.

It is not always correct to take the distance between the FET gate and the nearest neighbour dislocation as a dislocation effect parameter because dislocations often form networks or cells in conventional LEC-grown substrates. The effects should be understood as a superposition of plural dislocations. Threshold voltage distribution of 100 SAINT FETs with a 1 μm gate arrayed in a 400 μm square and dislocation distribution as revealed by molten KOH etching are shown in Fig. 3.27. It is evident that the threshold voltage shifts towards the negative direction in the region having high dislocation density. Isolated dislocations have little effect on threshold voltage. The dislocation network in the substrate must be reduced to improve GaAs IC yield.

Typical threshold voltage distribution in a 2-inch diameter, conventional, slightly Cr-doped substrate is compared to other kinds of substrates. The same type of SAINT FETs with a 1 μm gate length are used in the three cases in Fig. 3.28 (a)–(c).

Effects of dislocation network bring about a short-range correlation in threshold voltages. The threshold voltage correlation between a D-FET and

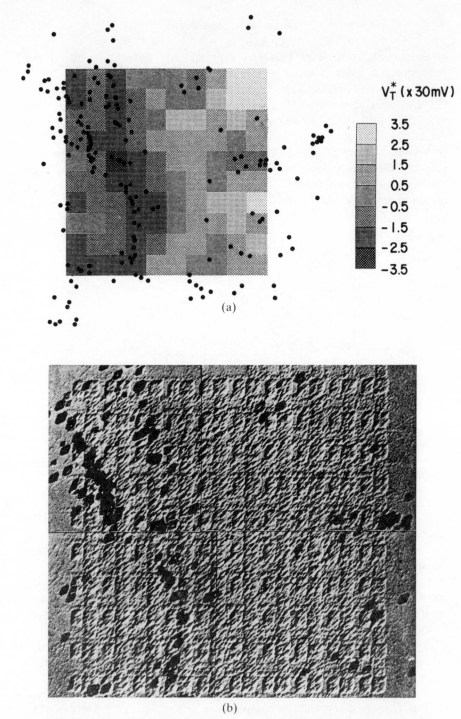

Fig. 3.27 (a) Threshold voltage distribution of 100 FETs in an array. The FET distance is 40 μm in both horizontal and vertical directions. Dots indicate dislocation etch pits. (b) Microphotograph of the GaAs surface after molten KOH etching in the same FET array region.

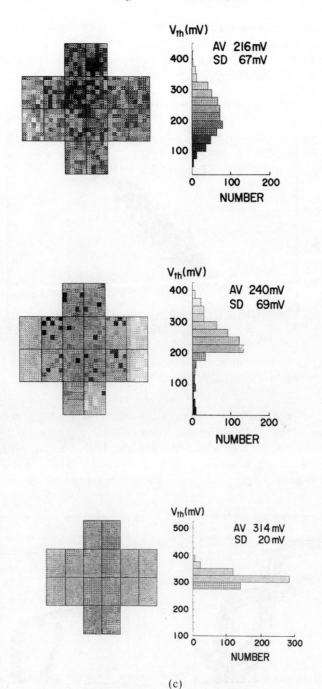

Fig. 3.28 Examples of threshold voltage distributions in 2 in. wafers. (a) Conventional, slightly Cr-doped substrate, (b) Ingot annealed substrate (at 800°C for 12 hours) and (c) In-doped low-EPD substrate.

an E-FET 12 μm apart is shown in Fig. 3.29. Owing to this distinct correlation, transfer characteristics are very uniform, as shown in Fig. 3.30 (a) for a direct coupled FET logic (DCFL) configuration using D-FETs placed close to E–FETs for load resistance.

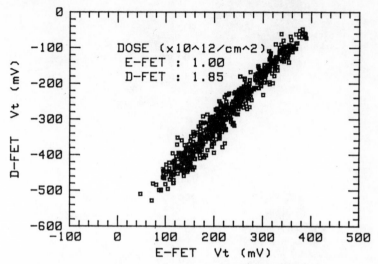

Fig. 3.29 Short-range correlation of threshold voltage of FETs arranged at 8 μm distance on a conventional substrate.

Threshold voltage shift in the negative direction or high activation efficiency caused by dislocations are related to a decrease in the Si_{As} acceptor in Si-implanted and activated substrates [52]. There is an inverse correlation between sheet carrier concentration in small Hall chips and the photoluminescence (PL) peak intensity ratio of $I(Si)/I(ex.)$, as shown in Fig.

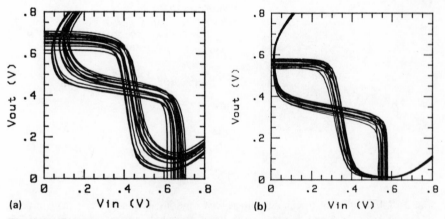

Fig. 3.30 Transfer characteristic uniformity for the DCFL configuration obtained on (a) a conventional substrate and (b) an In-doped low-EPD substrate.

3.51. Here, $I(\text{ex.})$ is the intensity of a donor bounded exciton peak related to donor concentration, and $I(\text{Si})$ is the intensity of a Si-related donor-acceptor transition peak proportional to the product of donor (Si_{Ga}) and acceptor (Si_{As}) concentrations. Fig. 3.31 indicates that sheet carrier concentration decreases with Si_{As} acceptor concentration.

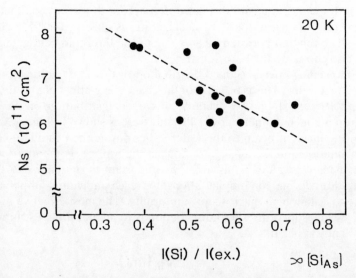

Fig. 3.31 Relationship between carrier concentration measured on small Hall chips and photoluminescence intensity ratio of Si-related donor-acceptor transition peak to donor bounded exciton peak.

It should be mentioned here that dislocation effects on threshold voltage depend on annealing conditions such as the passivation layer and annealing atmosphere. The Oki group reported that dislocation effects are suppressed under arsenic vapour pressure when an SiO_2 layer is used as an annealing cap [53].

Dislocation effects on threshold voltage are reduced if the substrates are annealed at temperatures around 800°C for about 10 hours [54]. The improvement in homogeneity by annealing is also confirmed by taking a scanning cathodoluminescence (CL) intensity profile. An example of threshold voltage on a 2 inch wafer prepared from a conventional LEC-grown GaAs ingot annealed at 800°C for 12 hours is shown in Fig. 3.28 (b). Uniformity is improved, but there are some exceptional FETs with large negative threshold voltages. Those FETs exhibit anomalous short-channel effects. The existence of such abnormal FETs is probably observed only in the case of high-performance FETs with self-aligned, short length gate.

In recent years, dislocation density in semi-insulating GaAs crystal has been steadily reducing, and completely dislocation-free wafers have been produced using a vertical magnetic field and fully encapsulated Czochralski

(VMFEC) method in combination with indium doping [55]. An example of threshold voltage distribution on In-doped low-dislocation density, or low-etch pit density (low-EPD), substrate is shown in Fig. 3.28 (c). The standard deviation in the entire 2 inch wafer is 20 mV, which is about ⅓ that of conventional substrate. Transfer characteristics on the low-EPD substrate are shown in Fig. 3.30 (b).

Except for uniformity, no distinct difference, such as transconductance, between conventional and low-EPD substrates is observed in FET characteristics. The relation between mean value of the threshold voltage and the dose is also the same for both substrates.

(d) **Channelling effects**. Sloped distribution of threshold voltage is often observed in a wafer. This is because of the channelling effect of ions near the GaAs <100> axis [56]. An example of sloped distribution of threshold voltage on a 2-inch diameter low-EPD substrate is shown in Fig. 3.32. The angle of the incident beam to the wafer varies during the beam scanning in the implantation and the angle variations in the wafer become more severe in large diameter wafers. Thus it is very important to optimize coordinate configuration during ion implantations for each individual implantation machine to obtain high macroscopic uniformity. The most effective method of suppressing sloped threshold distribution is to adopt parallel scanning of the incident ion beam.

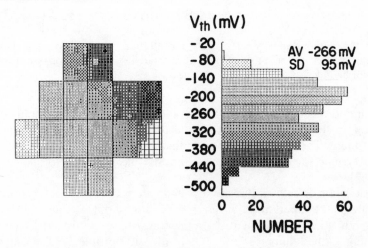

Fig. 3.32 An example of sloped threshold voltage distribution on a 2 inch wafer caused by the ion implantation channelling effect.

3.4   IMPROVEMENTS IN ACTIVE LAYER CHARACTERISTICS

There are three key requirements for the development of high performance GaAs FETs: source resistance reduction, gate length reduction without short-channel effects and channel layer thinning combined with increasing carrier concentration. All of these are concerned with improvements of active layer characteristics.

Thinning the channel layer is very important. The *K*-value, that is, the slope of the square root of drain current versus gate voltage, is inversely proportional to channel layer thickness. Short-channel effects are also suppressed by channel layer thinning. One approach for thinning the channel-layer and increasing carrier-concentration is reducing ion-implantation energy. For example, fitting parameters of I-V characteristics of a 0.5 μm gate BP-SAINT FET are listed in Table 3.1 [57] for channel layer implantation energies of 67 keV and 30 keV. The parameters in Table 3.2 are used to express saturation drain current I_{DS} as an equation.

$$I_{DS} = K (1 + \lambda V_{DS}) (V_{GS} - V_{th} + \gamma V_{DS} - I_{DS} \cdot R_s)^2$$

The *K*-value is increased from 210 mA/V^2 mm to 292 mA/V^2 mm by the reduction of implantation energy from 67 keV to 30 keV. It has already been recognized that the γ-value is approximately equal to the ratio of drain-conductance, g_D, to transconductance, g_m. Therefore the decrease in γ-value from 0.066 to 0.045 indicates a decrease in g_D due to the implantation energy reduction. It should be noted that the γ-value for the 30 keV-implanted BP-SAINT FET is, in spite of its short gate length, smaller than that for the conventional 1 μm gate length SAINT FET used in the 16 Kbit SRAM [11]. Due to the ion implantation energy reduction, the N_G-value is also decreased from 1.5 to 1.4, which is the smallest ever reported for submicron gate-length FETs.

Table 3.2. I-V curve fitting parameters for three kinds of FETs.

FET	L_G (μm)	E_n (keV)	dose (x10^{12} cm^{-2})	K (mA/V^{2}mm)	λ (1/V)	V_{th0} (mV)	γ	R_s (Ω)
BP-SAINT	0.5	30	4.0	2.93	0	100	0.045	40
BP-SAINT	0.5	67	2.2	2.10	0	268	0.066	40
SAINT	1	67	1.1	1.85	0.065	155	0.051	38

Threshold voltage uniformity for FETs with a self-aligned half-micron gate is also improved by short-channel suppression due to implantation energy reduction. The standard deviation of the threshold voltage of the 0.5 μm-gate BP-SAINT FET implanted with 30 keV Si ions in the channel layer was only 30 mV for the entire 2-inch diameter, low-EPD wafer, while that implanted with 67 keV Si ions was 49 mV. When a conventional wafer with dislocations of the order of 10^4/cm^2 was used as a substrate, threshold voltage scattering was not decreased by the implantation energy reduction, so dislocation effects are not suppressed by thinning the channel layer.

Besides implantation energy reduction, there are other implantation techniques for channel-layer thinning. Fujitsu [58] reported that FETs fabricated by implantation through an AlN layer have excellent characteristics. Heavy-ion implantation was reported to be effective for making thin channel layers and improving FET characteristics [14].

The above-mentioned methods are effective for obtaining a steep impurity profile. However, it is difficult to maintain the as-implanted steep impurity profile and obtain a thin channel layer reflecting as-implanted profile. The disadvantage of conventional furnace annealing becomes significant as the as-implanted layer becomes thinner, activation efficiency becomes poorer with increases in carrier concentration and thermal diffusion of implanted impurities becomes relatively large. Rapid thermal annealing (RTA) has been investigated to overcome such disadvantages. The annealing is characterized by high temperature (> 900°C) and short annealing time (a few seconds).

Examples of carrier concentration profiles obtained by C–V measurement for furnace annealing and rapid thermal annealing are shown in Fig. 3.33. Although the implantation energy reduction effects can be observed for the furnace annealing, the effects are not as conspicuous as expected. Owing to the rapid thermal annealing, the active layer became shallower and the carrier concentration became higher. In this example, activation efficiency and diffusion length are calculated as 53% and 0.025 µm for the 30 keV-implanted sample activated by rapid thermal annealing, and 35% and 0.037 µm for the 30 keV-implanted sample activated by conventional furnace annealing.

The effect of rapid thermal annealing of FET characteristics is shown in Fig. 3.34. Here, K-values of long-gate FETs are plotted as a function of threshold voltage. The change from furnace annealing to rapid thermal annealing shifts threshold voltage to the negative direction as a result of high activation efficiency, and also increases the K-value as a result of diffusion suppression. The K-value increase produced by the annealing method

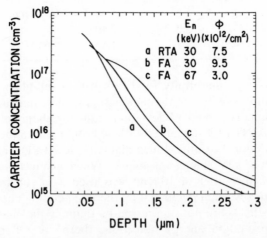

Fig. 3.33 Carrier concentration profiles obtained by C–V measurements for rapid thermal annealing (RTA) and furnace annealing (FA).

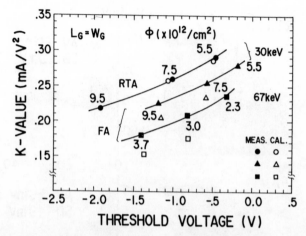

Fig. 3.34 *K*-value versus threshold voltage obtained for wafers using rapid thermal annealing (RTA) and furnace annealing (FA). *K*-value is normalized for a square gate FET. Unfilled points were calculated using C–V data assuming constant electron mobility.

change is, in this case, nearly equal to that caused by implantation energy reduction from 67 keV to 30 keV for the furnace annealing. In this figure, threshold voltage and *K*-value calculated analytically using C–V data are also plotted to confirm the consistency of C–V data with FET characteristics.

Rapid thermal annealing was reported to have the additional beneficial effect of suppressing undesirable dislocation effects. An example of threshold voltage uniformity observed in a conventional substrate with high dislocation density is shown in Fig. 3.35. Here, the standard deviation of long gate-length FET threshold voltage is as small as 18 mV for rapid thermal annealing while it is 49 mV for conventional annealing. High-temperature short-time annealing may arrange the silicon ions preferentially at a Ga site independent of small fluctuations in As-interstitial or Ga-vacancy concentration.

Rapid thermal annealing is also effective for low-resistivity n⁺-layer fabrication. Kuzuhara et al. [59] reduced n⁺-layer sheet resistivity to 20 ohms/square using SiO_xN_y capped rapid thermal annealing and formed non-alloyed ohmic contacts to the Si-implanted layer.

3.5 KEY PROCESSING TECHNOLOGIES FOR GaAs LSI

3.5.1 Photo-lithography

The lithographic process plays a central role in LSI fabrication. Previously the most popular technique was contact printing with an alignment made by a machine operator. This was common when the GaAs wafer processed was oval-shaped or cut-out. However recent advances in wafer technology made round-shaped wafers possible. If these round wafers can be supplied on a

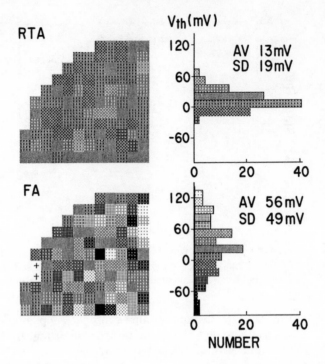

Fig. 3.35 Comparison of threshold voltage uniformity of long-gate FETs on conventional substrate between rapid thermal annealing (RTA) and furnace annealing (FA).

regular basis, the GaAs process engineers can purchase the most advanced fabrication machines which would increase wafer throughput and yield.

The most advanced lithographic system warranted for production is a reduction projection aligner or step-and-repeat (stepper). The typical commercial stepper has an NA of 0.35 to 0.42, with 0.8 μm resolution, alignment accuracy of 0.2–0.25 μm.

The main features of this lithographic process are (i) there are no resist stripping defects found in contact printing, (ii) highly uniform pattern generation, (iii) extremely precise automatic alignment using laser-assisted stage positioning.

An example of the steppers's superior uniformity is shown in Fig. 3.36. A histogram of gate length distribution fabricated with conventional contact lithography and 10:1 reduction projection, g-line stepper is presented. Note that the FET process is SAINT. The SAINT permits sub-micron gate lengths since gate is defined by etching. Standard deviation in gate lengths of 0.03 μm in 2 inch wafers for an average gate length of 0.5 μm can be achieved.

This high uniformity in gate length also permits high uniformity in FET electrical characteristics. The relationship between closely arranged enhancement-mode FETs and depletion-mode FETs is shown in Fig. 3.37. A stronger correlation when using a stepper indicates high uniformity in

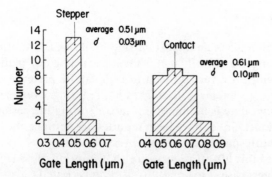

Fig. 3.36 Histogram of gate length distribution comparing contact printing and 10:1 reduction stepper.

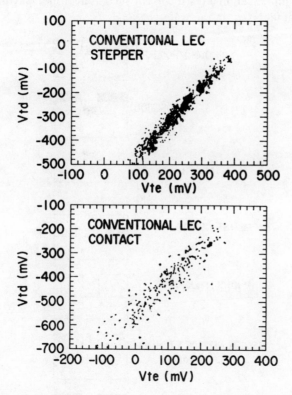

Fig. 3.37 Relationship between closely arranged enhancement-mode FET and depletion-mode FET.

pattern delineation. Direct delineation of 0.5 μm resist pattern cannot be achieved with a g-line stepper. On the other hand, carefully arranged processing can produce 0.5 μm with contact lithography, however the reproducibility is relatively poor.

3.5.2   Lift-off process

The lift-off technique is an important step in GaAs LSI fabrication. This is because in using the lift-off, metal films such as gold, platinum or gold-germanium, which are very difficult to etch can be patterned easily. There are no specific problems with regard to realising even sub-micrometer patterns. Lithography is an important factor setting the resolution limit.

A variety of techniques have been reported for refining the lift-off process to meet linewidth/spacing demands of recent LSIs.

Variations of this lift-off process are shown in Fig. 3.38 (a)–(d). Version (a) is the simplest case. First photo-resist film is coated on a wafer and then patterned with standard lithography. Then metal is deposited over the whole wafer covering both the resist area and the patterned resistless area. Next, the wafer is immersed in a resist solvent such as acetone, leaving behind the patterned metal layer on the wafer.

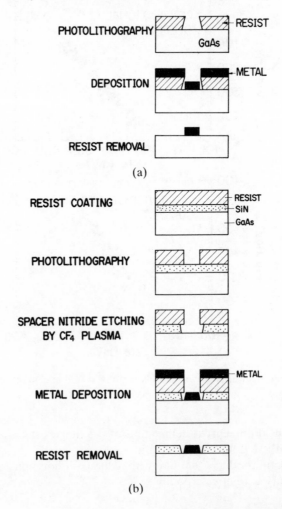

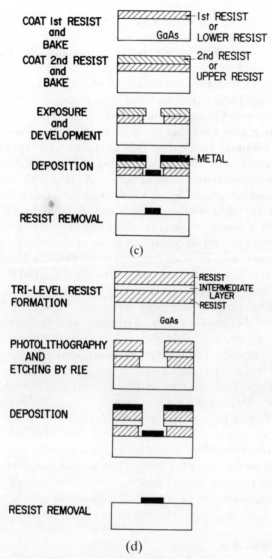

Fig. 3.38 Lift-off process sequences (a) simple version, (b) dielectric-assisted, (c) double-layer resist, (d) tri-level resist.

However the drawback to this simple version is the poor success rate in small dimension line-and-space patterns. This is because the metal over the photo-resist and the metal attached to the wafer are not fully separated. The next versions can be applied to dimensions as small as 1.5 μm by realising intentional separation of these metals.

Version (b) is a dielectric-assisted lift-off [60]. First, a wafer is deposited with dielectric film such as silicon nitride. Then a photo-resist film is coated and patterned. Next, the dielectric film is etched with a slight undercutting.

Metal is deposited over the whole wafer and the resist is removed along with the unnecessary metal. The advantages of this version are the high success rate of even 1.5 μm line-and-space and built-in planarizing treatment. This version can be applied to embedding metal in via holes.

Version (c) is a simultaneously developed double-layer resist system. The first resist is coated and baked to avoid intermixing with the second resist. Then the second resist is coated and baked. If the sensitivity of the second resist is slightly higher than that of the first resist, the opening of the first (lower) resist becomes larger than that of the second (upper) resist. This causes undercutting in the first resist ensuring high reproducibility of the lift-off process. This version looks simple but selection of resist materials can be difficult. A good combination of resists requires the following factors: (i) realisation of resist systems which can be developed in the same developer with a slightly different sensitivity, (ii) no intermixing when coating the second resist, (iii) controllability of undercutting depth. As far as the authors know, there is no such resist system that can be easily usable in production level, but several combinations such as PMMA/MMA [61], double-layer poly (phenyl methacrylate-co-methacrylic acid) [62] for electron-beam lithography have been reported.

Version (d) of the lift-off employs a tri-level resist technique. For the tri-level, bottom organic materials such as resist or polyimide, intermediate layer usually composed of dielectric material, and a top pattern defining resist, are common. The top resist is patterned precisely as standard lithography. Then the intermediate dielectric layer is etched using the top resist as a mask. The bottom resist is etched through the opening produced by the preceding etching. The top resist is also etched with the etching process of the bottom resist resulting in diminished thickness. Undercutting in the bottom resist assists successful lift-off by separating unnecessary metal from patterned metal. The drawback of this version is the larger number of time-consuming process steps. However, reproducibility of this version is excellent and application to LSI-level fabrication is definitely recommended.

3.5.3 Interconnection

The present minimum interconnection reported to date is 1.5 μm/1.5 μm line-and-space and 1.5 μm × 1.5 μm via hole. The first example is the interconnection technology employed in 16 Kb static RAMSs [63]. Main features of this technology are dielectric-assisted lift-off, homogeneous dry etching using reactive ion etching (RIE), and 1.5 μm line-and-space lithography. RIE etching characteristics for PECVD SiN film is shown in Fig. 3.39. The etching rate in the vertical direction was approximately 0.1 μm/min, while the etching rate for the lateral side was ¼ of the vertical direction. This etching rate ratio is appropriate for the dielectric-assisted lift-off process.

A cross sectional view of the interconnection with the first-level metal, via

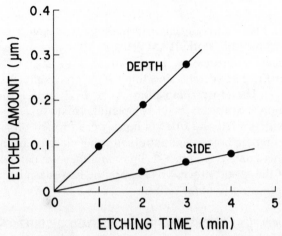

Fig. 3.39 RIE etching characteristics of plasma-assisted CVD silicon nitride. Vertical and side etching are shown.

metal and second-level metal is shown in Fig. 3.40. First level metal as well as the via metal are embedded in the PECVD SiN film through the dielectric-assisted lift-off mentioned in the preceding section. Next, the second level interconnection located on the cylindrical via metal can be seen. This interconnection was patterned using ion milling. Interconnection resistance measurements revealed very low via resistance of 0.3 ohms for 1.5 μm square mask pattern.

Recently NTT has proposed a tri-level interconnection technology [62]

Fig. 3.40 Cross-sectional view of double-layer interconnection.

which is expected to play an important role in realizing LSIs of higher complexities. This interconnection technology was developed to meet the demands to reduce the static RAM cell size.

The fabrication process is presented in Fig. 3.41. The first interconnection is the gate metal connecting FET ohmic contacts over relatively short distance. Therefore completion of the FET process means the first interconnection has already been completed. Next, first dielectric, silicon dioxide, was deposited and first via holes were formed on gate and ohmic metals. The first via holes were patterned using electron beam lithography. Via size was as small as 0.8 μm × 1.3 μm, and they are embedded with gold as a result of the lift-off process. These are the first sub-micrometer vias for GaAs LSIs reported to date.

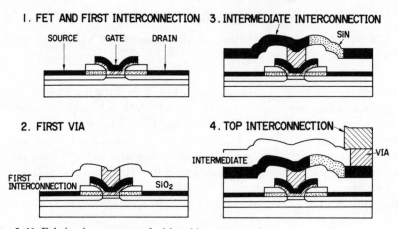

Fig. 3.41 Fabrication process of tri-level interconnection.

The next fabrication step is deposition of the second dielectric, silicon nitride. Then intermediate interconnection metal titanium/platinum/gold is embedded in the dielectric once again by the lift-off technique. Next, the third dielectric, silicon dioxide, was deposited, and the second via holes were formed by etching and embedding gold in the same manner as used with the first vias. The second via holes were 1.3 μm × 1.3 μm. Finally top interconnection was formed by ion milling. Interconnection and via holes were fabricated using planarizing technology by a dielectric-assisted lift-off except for the ion-milled top interconnection.

A SEM view of the partially processed tri-level interconnection is shown in Fig. 3.42. The process stage is at the completion of the intermediate interconnection. The most significant phase is the step over the gate, whose height is approximately 0.7 μm. However no disconnection at the step was observed if metal thickness exceeded 0.5 μm.

This new interconnection techology can be applied in the near future for the fabrication of 64 Kb static RAMs, for example, or other highly complex GaAs LSIs.

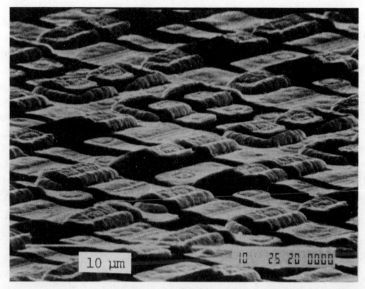

Fig. 3.42 Partially processed tri-level interconnection. Viewed after intermediate interconnection.

3.5.4 Control of FET threshold voltage

A GaAs 16 Kb static RAM [63] consists of more than 100 000 FETs each of which must operate successfully. The most important parameter determining successful operation of LSIs is scatter in FET electrical parameters, especially threshold voltage. Dislocation densities in low Cr LEC grown crystal generally exceed 10^4 cm^{-2}. As described previously, the dislocations are known to affect threshold voltages of FETs in close proximity [64]. Therefore the use of dislocation-free In-doped wafers are desirable for producing high uniformity LSI circuits.

Threshold voltage control from wafer to wafer is another important factor. The variation of threshold voltage result with ion implantation dose is shown in Fig. 3.43. ^{29}Si ions were used to implant active layer with 67 keV

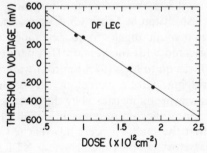

Fig. 3.43 Threshold voltage against ion implantation doses. 67 keV Si into In-doped GaAs.

acceleration energy. The four points in the figure lie on a straight line, indicating the wafer-to-wafer reproducibility in the threshold voltages. The dose-threshold voltage relationship is almost identical with Cr-doped wafers.

There are cases where threshold voltages do not result in expected values. For GaAs this is not unusual because of crystal impurities, fluctuations in Cr concentration or dislocations. Several efforts have been reported to overcome this uncontrollable factor. Experiments aimed at controlling the threshold voltage will be reviewed.

One uses a process named SLICE (self limiting ion assisted chemical etching) [65]. This is a dry-recess type of etching using reactive ion etching. After opening the gate window, GaAs is etched using CF_4/H_2 plasma. The etched recess reaches a constant depth after a certain plasma on-time. This depth can be controlled by changing the plasma power levels.

Another interesting report concerns the possibility of trimming after FET completion [66]. The control of threshold voltage was performed with a small dose of boron through the gate metal. The gate metal was 0.1 μm thick W-Al, while the boron acceleration energy was 200 keV. Linearity between the threshold voltage shift and the boron dose was fairly good. It was reported that the threshold voltage can be controlled to within 30 mV. Furthermore, it was claimed that the process did not increase the standard deviation of the threshold voltage.

3.5.5 Yield problems

Divide by 4 frequency dividers using low power source coupled FET logic [35] were fabricated to evaluate the yield status for GaAs IC processing. This circuit was composed of an ECL-compatible input buffer, 2 flip-flops, and an output buffer. The total number of FETs, diodes and resistors was 76. A photomicrograph of the circuit is shown in Fig. 3.44. Chip size was 0.6 mm × 0.85 mm. FET gate length and width were 0.5 μm and 20 μm respectively. Frequency dividers were measured at their designed supply voltages on four wafers. Toggle frequency histogram comparing the process using 67 keV and 30 keV ion implantation energies is shown in Fig. 3.45. Process versions are listed in Table 3.3. Circuits using process version II operated at the maximum toggle frequency range of 5.2–6 GHz. In contrast, circuits fabricated with the version III operated in the 6–7 GHz range. The chip yield was quite high (97%) in four wafers. Chip failure was mainly due to a bridging pattern defect in the second level of metal interconnection. This was probably caused by the presence of a dust particle caught by the wafer and replicated into the interconnection layer.

The authors feel that the GaAs process has reached a level of maturity that can provide satisfactory circuit yield. Thus far NTT has published yield data for several MSI–LSI level complexities ([35], [67]). In a general sense, the yield can only be discussed for a mass production line which has a large

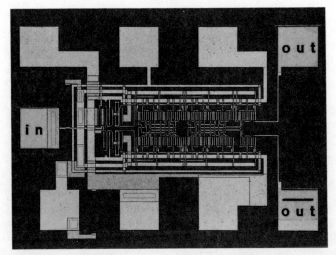

Fig. 3.44 Divide by 4 frequency divider in LSCFL. Chip size is 0.6 mm × 0.85 mm.

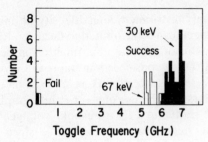

Fig. 3.45 Toggle frequency histogram of ¼ frequency dividers. The circuits were fabricated using 67 keV and 30 keV ion implantation energies for channel.

Table 3.3. List of process features

Version	Process	Lg	Lithography	n-Layer	Wafer
I	SAINT	0.8 μm	Contact + EB	67 keV	Conventional LEC
II	BP–SAINT	0.55 μm	Stepper	67 keV	Low EPD LEC
III	BP–SAINT	0.5 μm	Stepper	30 keV	Low EPD LEC

process tolerance. However the yield can be used to evaluate the process improvements at the laboratory level.

Published yield data are shown in Fig. 3.46. Points indicating a chip active area at 3.8 mm² represent a 4-channel time switch LSI [35]. The point at

GaAs Integrated Circuits

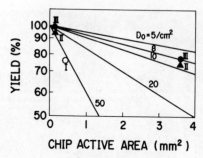

Fig. 3.46 Published yield data for divide by 4 dividers (process versions II and III), divide by 2 divider (I), 4-channel time switch LSIs (II and III).

0.4 mm^2 represents a divider circuit. All points plotted here used the LSCFL configuration. Circuits with larger complexities naturally have lower yields. To clarify the improvement steps, the solid lines shown in Fig. 3.46 were calculated by assuming a Poisson distribution.

$$Y = \exp(-D_O A)$$

where D_O is randomly distributed defect density of any type, and A is the chip active area. Defects can be attributed to dust, scratches, non-uniformity in threshold voltages or other local abnormalities in device parameters. A process with a small D_O is considered an improved process.

Advances from version I to version II were based on the use of In-doped, dislocation-free wafers and the use of stepper lithography. In-doped wafers effectively reduced scatter in FET threshold voltages and ion implanted resistor values. The stepper was also effective since it eliminates resist stripping defects common in contact printing. No significant difference was observed for versions II and III. However, version III is superior since it is able to produce faster ICs.

3.6 DIGITAL LSIs

GaAs IC technology offers attractive prospects for the implementation of giga-bit rate digital data/signal processing components. Specifically, this technology may be used in static random access memories (static RAMs), multipliers, arithmetic logic units (ALUs), and adders. This section considers GaAs digital LSI performance, focusing their suitability of use in fabrication technologies.

3.6.1 Memory LSIs

The first LSI level GaAs memory circuit, a 1–Kb static RAM, was fabricated in 1982 by NTT [10]. Technology related to fabrication processes and substrate materials developed rapidly, leading to an integration complexity evident in the 16 Kb static RAM [1]. Fujitsu fabricated a 1 Kb static RAM in 1983 [8] and a 4 Kb static RAM in 1984 [68]. Toshiba [69], Hitachi [70] and NEC [71] also fabricated 4 Kb static RAMs while Mitsubishi [72] fabricated a 16 Kb static RAM.

Fig. 3.47 is a photomicrograph of a 16 Kb static RAM chip having 102 028 MESFETs and 256 diodes in a 7.2 mm × 6.2 mm area [1]. An E/D-DCFL circuit was used as it features the essential advantages of low power dissipation and circuit simplicity. Fig. 3.48 is a photomicrograph of a memory cell composed of E/D type cross-coupled flip-flop and two transfer gates. The cell size is 41 μm × 32.5 μm. The MESFETs, having 1 μm gate length, were fabricated using the SAINT process. All FET gates were aligned in the same direction in order to avoid the orientation dependence of threshold voltage [73]. These orientation effects are considered to be caused by a piezoelectric charge induced by elastic stress [74]. A 1/10 reduction step and repeat mask aligner was used to achieve 1.5 μm/1.5 μm for the first-level metal line/space, 3 μm/3 μm for the second and 1.5 μm square for the via hole size. The first-level and via metals were formed by the dielectric-assisted lift-off technique. The dielectric films were etched with a slight undercut by RIE.

Threshold voltage uniformity is of great importance for achieving high speed and low power dissipation with high yield. For this, In-doped, low

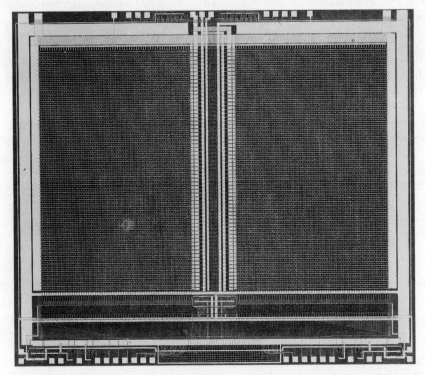

Fig. 3.47 Photomicrograph of a 16 Kb GaAs static RAM chip having 102 028 MESFETs and 256 diodes in an 7.2 mm × 6.2 mm area.

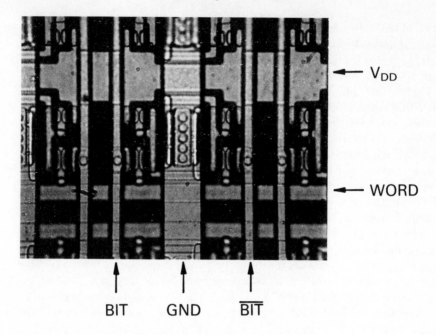

Fig. 3.48 Photomicrograph of a 16 Kb GaAs static RAM memory cell composed of six MESFETs. The cell size is 41 μm × 32.5 μm.

dislocation GaAs wafers were used. The stepper also contributed to the threshold voltage uniformity, because it suppresses gate length scattering. The standard deviation of the threshold voltage for E-FETs was 20 mV in a 2 in. full wafer. This value is about one-third that for those fabricated on a conventional LEC substrate. An interesting result for RAM performance, showing dependence on the substrate dislocation density, can be observed in Fig. 3.49 [63]. Though the minimum address access time for both substrates was 4.1 ns, power dissipation for the conventional LEC was 1.7 times larger than that for the low-dislocation LEC.

A number of problems remain to be solved for the production of large GaAs static RAMs at a commercial level. Such problems include the process yield, access time scattering, chip size, and power dissipation.

NTT first developed a fully functional 4 Kb static RAM in 1984. The measured access times, however, ranged from 4 to 16 ns [75]. Since the practical performance of a RAM is limited by the slowest access time, it is very important to reduce this scattering. Hitachi researchers proved that bootstrap circuits, introduced to shorten the read signal transient time, are effective in reducing the access time scattering by more than 50% [70]. It was shown that the delay time scattering, which is strongly related to the data-line signal amplitude variation, was reduced by the bootstrap effect. This technique led to development of fully functional two 1 Kb blocks of a 4 Kb static RAM having small access time scattering of 2.2–3.0 ns. Further work by Hitachi led to a 4 Kb static RAM having 1 ns access time [76]. Gate length reduction to 0.7 μm also improved the performance.

Very low power dissipation is another feature of the GaAs static RAM. McDonell Douglas researchers fabricated a 256–b static RAM having a low dissipation power of 1–2 μW/cell by using normally-off JFET drivers and resistive loads [77]. Power dissipation was further decreased by developing complementary memory cells constructed of n-channel and p-channel, normally-off GaAs JFETs [78]. Both FETs were fabricated by selective implantation. From measurements of an 8-b memory array chip, ultra-low-power operation in the range of 50–200 nW/cell was confirmed. As hole mobility is smaller than electron mobility by a factor of about ten, the W_g/L_g (W_g: gate width; L_g: gate length) ratio for the p-channel JFET must be increased by this factor in comparison with the n-channel device to achieve the same drain current. Unfortunately, this results in an increase in gate capacitance and a decrease in switching speed. Therefore, a complementary design for GaAs LSIs is useful only when a power dissipation reduction is mandatory.

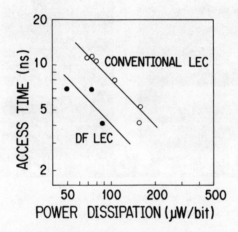

Fig. 3.49 Measured results between 16 Kb static RAMs fabricated on low-dislocation LEC and conventional LEC substrates.

3.6.2 Logic LSIs

An 8 × 8 bit parallel multiplier, the first GaAs LSI circuit, containing about 1 K gates was achieved in 1980 using an SDFL circuit approach [6]. Fujitsu fabricated a more complex logic: a 16 × 16 bit parallel multiplier containing about 3 K gates of DCFL circuits [7]. Toshiba achieved a DCFL gate array of 6 K gate complexity [2].

NTT fabricated a digital time switch LSI [35] and an 8-b ALU LSI [79] using an LSCFL circuit approach. The MESFETs involved were BP-SAINT FETs having a half-micron gate length. A microphotograph of a fabricated

8-b ALU is shown in Fig. 3.50. Chip size is 3.34 mm × 2.52 mm. The LSI containing an equivalent of 250 gates consists of 1426 devices (MESFETs, diodes and resistors). An airgap technique was also incorporated to reduce interconnection capacitance. The critical path delay time was determined to be 0.8 ns, which corresponds to an equivalent loaded delay time of 57 ps/gate.

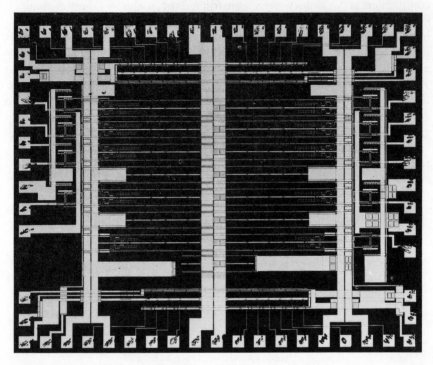

Fig 3.50 Photomicrograph of a GaAs 8 b ALU, containing 1426 devices (250 equivalent gates). The chip size is 3.34 mm × 2.52 mm.

3.7 FUTURE ISSUES AND CONCLUSIONS

Normally-on GaAs IC technology with 1 μm gate length has reached production levels. Some SSI, MSI, and a few LSI circuits are commercially available. However, the GaAs IC's market share is much smaller than that of Si-based ICs. How large an application field GaAs digital ICs will acquire is not clear at present, because they must compete directly with giant Si-based ICs. Is the GaAs IC a competitor of or complementary to the Si IC? The comparison of GaAs with Si ICs has been discussed in technical papers and international conferences ([80]–[82]).

One comparison is shown in Fig. 3.51 [83]. The performances of static RAMs were compared based on a circuit simulation. The smallest access

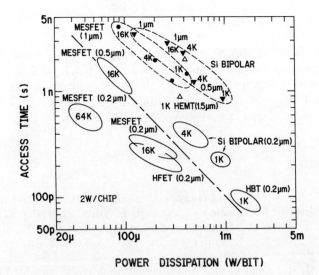

Fig. 3.51 Comparison of GaAs and Si static RAMs. Triangles and circles represent experimental results. Solid ellipses are the calculated results based on circuit simulation.

time of 0.85 ns was realised for a 1 Kb static RAM using 0.5 μm emitter Si-bipolar transisters [84]. The minimum obtained access time of a GaAs 1 Kb static RAM with 1 μm gate length MESFETs was 1 ns [68]. It should be noted that the device size (gate length) of the GaAs IC was twice as large as the size (emitter width) of the Si IC, and that the power dissipation of the GaAs IC was about one-third that of the Si IC. The simulated access times are plotted in this figure as solid ellipses. With reduction of device size (gate length and emitter width), other parameters, such as interconnection line width and spacing, are also scaled down. By using 0.2 μm gate length GaAs MESFETs, a 64 Kb static RAM operating with access time below 1 ns and 16 Kb static RAM below 0.5 ns is predicted. These values are half or one-third those of Si bipolar static RAM of the same device size.

The technical requirements for realization of such high speed LSIs are:

(1) gate length reduction down 0.2 μm with suppression of short channel effects,
(2) uniformity improvement in threshold voltage (standard deviation below 10 mV),
(3) establishment of a submicron interconnection technology.

It can be expected that the application of GaAs ICs will certainly expand with further advances in technologies.

ACKNOWLEDGMENTS

The authors wish to thank M. Fujimoto, T. Ikegami and T. Sugeta for their encouragement and support. They also are much indebted to several other members of NTT Electrical Communications Laboratories, especially to M. Hirayama, S. Miyazawa, M. Ino, T. Mizutani, M. Suzuki, T. Takada, H. Yamasaki and T. Ogino for many valuable discussions.

REFERENCES

1. Ishii, Y., Ino, M., Idda, M., Hirayama, M. and Ohmori, M. (1984) 'Processing technologies for GaAs memory LSIs' *GaAs IC Symp. Tech. Dig.* 121–124.
2. Terada, T., Ikawa, Y., Kameyama, A., Kawakyu, K., Sasaki, T., Kitaura, Y., Ishida, K., Nishihori, K. and Toyoda, N. (1987) 'A GaAs 6K gate array' *ISSCC Dig. Tech. Papers* 144–15.
3. Van Tuyl, R. L. and Liechti, C. A. (1974) 'High speed integrated logic with GaAs MESFETs' *ISSCC Dig. Tech. Papers* 114–15.
4. Van Tuyl, R. L., Liechit, C. A., Lee, R. E., and Gowen, E. (1977) 'GaAs MESFET logic with 6–GHz clock rate' *IEEE J. Solid State Circuits* **SC–12** 485–496.
5. Welch, B. M. and Eden, R. C. (1977) 'Planar GaAs integrated circuits fabricated by ion implantation' *IEDM Tech. Dig.* 205–207.
6. Lee, F. S., Eden, R. C., Long, S. I., Welch, B. M. and Zucca, R. (1980) 'High speed LSI GaAs integrated circuits' *Proc. IEEE Int. Conf. Circuit and Computor* 697–700.
7. Nakayama, Y., Suyama, K., Shimizu, H., Yokoyama, N., Shibatomi, A. and Ishikawa, H. (1983) 'A GaAs 16 × 16 b parallel multiplier using self-alignment technology' *ISSCC Dig. Tech. Papers* 48–9.
8. Yokoyama, N., Onodera, H., Shinoki, T., Ohnishi, H., Nishi, H. and Shibatomi, A (1983) 'A GaAs 1K static RAM using tungsten-silicide gate self-alignment technology' *ISSCC Dig. Tech. Papers* 44–5.
9. Yokoyama, N., Mimura, T., Fukuta, M. and Ishikawa, M. (1981a) 'A self-aligned source/drain planar device for ultrahigh-speed GaAs MESFET VLSIs' *ISSCC Dig. of Tech. Papers* 218–19.
10. Ino, M., Hirayama, M., Ohwada, K. and Kurumada, K. (1982) 'GaAs 1Kb static RAM with E/D MESFET DCFL' *GaAs IC Symp. Tech. Dig.* 2–5.
11. Hirayama, M., Ino, M., Matsuoka, Y. and Suzuki, M. (1984) 'A GaAs 4 Kb SRAM with direct coupled FET logic' *ISSCC Dig. Tech. Papers* 46–7.
12. Yamasaki, K., Asai, K., Mizutani, T. and Kurumada, K. (1982a) 'Self-align implantation for n^+-layer technology (SAINT) for high-speed GaAs ICs' *Electron. Lett.* **18** 119–121.
13. Imamura, K., Yokoyama, N., Ohnishi, T., Suzuki, S., Nakai, K., Nishi, H. and Shibatomi, A. (1984) 'A WSi/TiN/Au gate self-aligned GaAs MESFET with selectively grown n^+-layer using MOCVD' *Japan. J. Appl. Phys.* **23** L342–L345.
14. Higashisaka, A., Ishikawa, M., Katano, F., Asai, S., Furutsuka, T. and Takayama, Y. (1983) 'Side-wall-assisted closely spaced electrode technology for high speed GaAs LSIs' *Extended Abstracts 15th Conf. Solid State Devices and Materials* 69–72.

15. Zuleeg, R., Notthoff, J. K., Friebertshauser, P. E., and Troeger, G. L. (1977) 'Femto joule, high-speed planar GaAs E-JFET logic' *IEDM Tech. Dig.* 198–200.

16. Bland, S. W., Wood, D. and Mun, J. (1987) 'Self-alignment techniques for GaAs MESFET IC's' *Journal of IERE*, **57** (1) 84–91.

17. Yokoyama, N., Ohnishi, T., Odani, K., Onodera, H. and Abe, M. (1981b) 'Ti/W silicide gate technology for self-aligned GaAs MESFET VLSIs' *IEDM Tech. Dig.* 80–83.

18. Ohnishi, T., Yokoyama, N., Onodera, T., Suzuki, S. and Shibatomi, A. (1983) 'Characterization of WSi$_x$/GaAs Schottky contacts' *Appl. Phys. Lett.* **43** 600–602.

19. Nakamura, H., Sato, Y., Nonaka, T., Ishida, T. and Kaminishi, K. (1983) 'A self-aligned GaAs MESFET with W–A1 gate' *GaAs IC Symp. Tech. Dig.* 134–17.

20. Uchitomi, N., Kitaura, Y., Mizoguchi, T., Ikawa, Y., Yoyoda, N. and Hojo, A. (1984) 'Refractory WN gate self-aligned GaAs MESFET technology and its application to gate array ICs' *Extended Abstracts 16th Int. Conf. Solid State Devices and Materials* 383–6.

21. Tseng, W. F. and Christou, A. (1982) 'Stable high temperature tantalum silicide Schottky barrier on gallium arsenide' *IEDM Tech. Dig.* 174–6.

22. Imamura, K., Ohnishi, T., Shigaki, M., Yokoyama, N. and Nishi, H. (1985) 'Au/TiW/WSi-gate self-aligned GaAs MESFETs using rapid thermal annealing method' *Electron. Lett.* **21** 805–804.

23. Yamasaki, K., Yamane, Y. and Kurumada, K. (1982b) 'Below 20ps/gate operation with GaAs SAINT FETs at room temperature' *Electron. Lett.* **18** 592–3.

24. Yamasaki, K., Asai, K. and Kurumada, K., (1982c) 'N$^+$ self-aligned MESFET for GaAs LSI' *Proc. 14th Int. Conf. Solid State Devices, Japan J. Appl. Phys.* **22** Suppl. 22–1 381–4.

25. Lee, R. E., Levy, H. M. and Matthews, D. S. (1982) 'Material and device analysis of self-aligned gate GaAs ICs' *GaAs IC Symp. Tech. Dig.* 177–179.

26. Sze, S. M. (1981) *Physics of Semiconductor Devices* (2nd edition) John Wiley & Sons 275.

27. Suzuki, M., Murase, K., Asai, K. and Kurumada, K. (1983) 'A large barrier height Schottky contact between amorphous Si-Ge-B and GaAs' *Japan J. Appl. Phys.* **22** L709–L711.

28. Enoki, T., Yamasaki, K., Osafune, K. and Ohwada, K. (1985) 'Advanced GaAs SAINT FET fabrication technology and its application to above 9 GHz frequency divider' *Extended Abstracts 17th Conf. on Solid State Devices and Materials* 413–16.

29. Chang, M. F., Ryan, F. J., Vahrenkamp, R. P. and Kirkpatrick, C. G. (1985) 'Self-aligned MESFETs by a dual level double-lift-off substitutional gate (DDS) technique for high-speed low-power GaAs ICs' *Electron. Lett.* **21** 354–6.

30. Sadler, R. A. and Eastman, L. F. (1983) 'Orientation effect reduction through capless annealing of self-aligned planar GaAs Schottky barrier field-effect transistors' *Appl. Phys. Lett.* **43** 865–7.

31. Ohnishi, T., Yamaguchi, Y., Onodera, T., Yokoyama, N. and Nishi, H. (1984) 'Experimental and theoretical studies on short channel effects in lamp-annealed WSi$_x$-gate self-aligned GaAs MESFETs' *Extended Abstracts 16th Int. Conf. Solid State Devices and Materials* 391–394.

32. Yamasaki, K., Kato, N. and Hirayama, M. (1984) 'Below 10 ps/gate operation with buried p-layer SAINT FETs' *Electron. Lett.* **20** 1029–1031.

33. Matsumoto, K., Hashizume, N. and Atode, N. (1984) 'Sub-micron-gate self-aligned GaAs FET with p-layer barrier layer fabricated by ion implantation' *42nd Annual Device Research Conf., paper VI B–5.*

34. Nakamura, H., Tsunotani, M., Sano, Y., Nonaka, T., Ishida, T. and Kaminishi, K. (1984) 'The effect of substrate purity on short-channel effect of GaAs MESFETs' *Extended Abstracts 16th Int. Conf. Solid State Devices and Materials* 395–398.

35. Takada, T., Shimazu, Y., Yamasaki, K., Togashi, M., Hoshikawa, K. and Idda, M. (1985) 'A 2 Gb/s throughput GaAs digital time switch LSI using LSCFL' *IEEE Microwave and Millimeterwave Monolithic Circuits Symp. Dig. Tech. Papers* 22–26.

36. Kato, N. Matsuoka, Y., Ohwada, K. and Moriya, S. (1984) 'Influence of n^+-layer-gate gap on short-channel effects of GaAs self-aligned MESFETs (SAINT)' *IEEE Electron Device Lett.* **EDL–4** 417–419.

37. Asai, S., Goto, N., Kanamori, M., Tanaka, Y. and Furutuka, T. (1986) 'A high performance LDD GaAs MESFET with a refractory metal gate' *Extended Abstracts 18th Int. Conf. Solid State Devices and Materials* 383–386.

38. Yamasaki, K. and Hirayama, M. (1986a) 'Determination of effective saturation velocity in n^+ self-aligned GaAs MESFETs with submicrometer gate lengths' *IEEE Trans. Electron Devices* **ED–33** 1652–1658.

39. Yamasaki, K. and Hirayama, M. (1986b) 'Scaling law of GaAs MESFETs' *National Convention Record of IECE (Japan)* **S6–9** 361–362.

40. Shah, N. J., Pei, S. S., Tu, C. W. and Tiberio, R. C. (1986) 'Gate-length dependence of the speed of SSI circuits using submicrometer selectivity doped heterostructure transistor technology' *IEEE Trans. Electron Devices* **ED–33** 543–547.

41. Drummond, T. J., Kopp, W., Fischer, R., Arnold, D. and Morkoc, H. (1983) 'Enhancement-mode metal/(Al, Ga)As/GaAs buried-interface field-effect transistor (BIFET)' *Electron. Lett.* **19** 986–988.

42. Arai, K., Mizutani, T., Yanagawa, F. (1985) 'An n^+-Ge gate MIS-like heterostructure FET' *Int. Symp. GaAs and Related Compounds, Inst. Phys. Conf. Ser.*, 79 631–636.

43. Maezawa, K., Mizutani, T., Arai, K. and Yanagawa, F. (1986) 'Large transconductance n^+-Ge gate AlGaAs/GaAs MISFET with thin gate insulator' *IEEE Electron Device Lett.* **EDL–7** 454–456.

44. Eisen, F. H. (1980) 'Ion implantation in III–V compounds' *Rad. Effects* **47** 99–166.

45. Yamazaki, H., Honda, T. and Ishii, Y. (1985) 'Si ion implantation for GaAs IC fabrication' *Review ECL* **33** (1) 130–135.

46. Gibbons, J. F., Johnson, W. S., and Myloie, S. W. (1975) *Projected range statistics* (2nd ed.), Halsted Press.

47. Dearnaley, G., Freeman, J. H., Nelson, R. S. and Stephen, J. (1973) 'Ion implantation' *Defects in Crystalline Solids* **8** (499) North Holland Publishing Company.

48. Kato, N., Takada, T., Yamasaki, K. and Hirayama, M. (1985a) '6–7 GHz GaAs ICs with high yield 0.5 µm gate-length SAINT' *Extended Abstracts 17th Conf. Solid State Devices and Materials* 417–20.

49. Hyuga, H. (1985) 'Effect of dislocations on sheet carrier concentration of Si-implanted, semi-insulating, liquid-encapsulated Czochralski grown GaAs' *Japan. J. Appl. Phys.* **24** L160–L162.

50. Matsuoka, Y., Ohwada, K., and Hirayama, M. (1984) 'Uniformity evaluation of MESFETs for GaAs LSI fabrication' *IEEE Trans. Electron Devices* **ED–31** 1062–1067.

51. Miyazawa, S., Ishii, Y., Ishida, S. and Nanishi, Y. (1983) 'Direct observation of dislocation effects on threshold voltage of GaAs field-effect transistor' *Appl. Phys. Lett.* **43** 853–856.

52. Watanabe, K., Hyuga, F., Nakanishi, H., and Hoshikawa, K. (1986) 'Inhomogeneity of electrical properties around dislocations in In-doped/ alloyed, semi-insulating GaAs' *Inst. Phys. Ser.* **79** 277–282.

53. Egawa, T., Sano, Y. and Nakamura, H. (1985) 'The dependence of threshold voltage scattering of GaAs MESFET on annealing method' *Japan. J. Appl. Phys.* **24** L35–L38.

54. Miyazawa, S., Honda, T., Ishii, Y. and Ishida, S. (1984) 'Improvement of crystal homogeneity in liquid-encapsulated Czochralski grown, semi-insulating GaAs by heat treatment' *Appl. Phys. Lett.* **44** 410–412.

55. Kohda, H., Yamada, K., Nakanishi, H., Kobayashi, T., Osaka, J. and Hoshikawa, K. (1985) 'Crystal growth of completely dislocation-free and striation-free GaAs' *J. Crystal Growth* **71** 813–816.

56. Kasahara, J., Sakurai, H., Suzuki, T., Arai, M. and Watanabe, N. (1985) 'The effect of channeling on the LSI-grade uniformity of GaAs-FETs by ion implantation' *GaAs IC Symp. Tech. Dig.* 37–40.

57. Matsuoka, Y., Sugitani, S., Kato, N. and Yamazaki, H. (1987) 'Effects of thin and high-carrier-concentration active layer for GaAs MESFET performance' *13th Int. Symp. GaAs and Related Compounds*.

58. Onodera, H., Yokoyama, N., Kawata, H., Nishi, H. and Shibatomi, A. (1984) 'A high-transconductance self-aligned GaAs MESFET using implantation through an AlN layer' *Electron. Lett.* **20** 45–47.

59. Kuzuhara, M. and Nozaki, T. (1985) 'Nonalloyed ohmic contacts to Si-implanted GaAs activated using SiO_xN_y capped infrared thermal annealing' *J. Appl. Phys.* **58** 1204–1209.

60. Eden, R. C. and Welch, B. M. (1982) 'GaAs FET principles and technology' (Ed. Dilorenzo) 696–698.

61. Grobman, W. D., Lohn, H. E., Donohue, T. P., Speth, A. J., Wilson, A., Hatzakis, M. and Chang, T. H. P. (1979) '1 µm MOSFET VLSI technology: Part VI' *IEEE J. Solid State Circuits* **SC–14** 282–290.

62. Kato, N., Hirayama, M., Asai, K., Matsuoka, Y., Yamasaki, K. and Ogino, T. (1985b) 'A high density GaAs static RAM process using MASFET' *IEDM Tech. Dig.* 90–93.

63. Hirayama, M., Togashi, M., Kato, N., Suzuki, M., Matsuoka, Y. and Kawasaki, Y. (1986) 'A GaAs 16 Kbit static RAM using dislocation-free crystal' *IEEE Trans. Electron Devices* **ED–33** 104–110.

64. Yamazaki, H., Honda, T., Ishida, S. and Kawasaki, Y. (1984) 'Improvement of field-effect transistor threshold voltage uniformity by using very low dislocation density liquid encapsulated Czochralski-grown GaAs' *Appl. Phys. Lett.* **45** 1109–1111.

65. Ryan, F. J., Chang, M. F., Vahrenkamp, R. P., Williams, D. A., Fleming, W. P. and Kirkpatrick, C. G. (1985) 'New dry recess etching technology for GaAs digital ICs' *GaAs IC Symp. Tech. Dig.* 45–48.

66. Nakamura, H., Tanaka, K., Inokuchi, K., Saito, T., Kawakami, Y., Sano, Y., Akiyama, M., and Kaminishi, K. (1986) '2 GHz multiplexer and demultiplexer using DCFL/SBFL circuit and precise V_{th} control process' *GaAs IC Symp. Tech. Dig.* 151–154.

67. Takada, T., Togashi, M., Kato, N. and Idda, M. (1984) 'A GaAs HSCFL 4 GHz divider with 60/70 ps transition time' *Extended Abstracts 16th Conf. Solid State Devices and Materials* 403–406.

68. Yokoyama, N., Onodera, H., Shinoki, T., Ohnishi, H., Nishi, H. and Shibatomi, A. (1984) 'A 4K × 1b static RAM' *ISSCC Dig. Tech. Papers* 44–45.

69. Mizoguchi, T., Toyoda, N., Kanazawa, K., Ikawa, Y., Terada, T., Mochizuki, M. and Hojo, A. (1984) 'A GaAs 4K bit static RAM with normally-on and -off combination circuit' *GaAs IC Symp. Tech. Dig.* 117–120.

70. Hayashi, T., Tanaka, H., Yamashita, H., Masuda, N., Doi, T., Shigeta, J., Kotera, N., Masaki, A. and Hashimoto, N. (1985) 'Small access time scattering GaAs SRAM technology using bootstrap circuits' *GaAs IC Symp. Tech. Dig.* 199–202.

71. Takahashi, K., Maeda, T., Katano, F., Furutsuka, T., and Higashisaka, A. (1985) 'A CML GaAs 4 Kb SRAM' *ISSCC Dig. Tech. Papers* 68–9.

72. Takano, S., Makino, H., Tanino, N., Noda, M., Nishitani, K. and Kayano, S. (1987) 'A 16K GaAs SRam' *ISSCC Dig, Tech. Papers* 140–141.

73. Yamasaki, K., Asai, K. and Kurumada, K., (1982d) 'GaAs LSI-directed MESFETs with self-aligned implantation for n^+-layer technology (SAINT)' *IEEE Trans. Electron Devices* **ED–29** 1772–1777.

74. Asbeck, P. M., Lee, C. P. and Chang, M. F. (1984) 'Piezoelectric effects in GaAs FETs and their role in orientation-dependent device characteristics' *IEEE Trans. Electron Devices* **ED–31** 1377–1380.

75. Idda, M., Yamazaki, H., Kato, N. and Ohmori, M. (1984) 'A 2 ns GaAs 4 Kb SRAM using a dislocation free LEC crystal' *Late News Abstracts 16th Int. Conf. Solid State Devices and Materials* 30–31.

76. Tanaka, H., Yamashita, H., Masuda, N., Matsunaga, N., Miyazaki, M., Yanazawa, H., Masaki, A. and Hashimoto, N. (1987) 'A 4K GaAs SRAM with 1 ns access time' *ISSCC Dig. Tech. Papers* 138–139.

77. Troeger, G. L. and Nottoff, J. K. (1983) 'A radiation-hard low-power GaAs static RAM using E-JFET DCFL' *GaAs IC Symp. Tech. Dig.* 78–81.

78. Zuleeg, R., Notthoff, J. K. and Troeger, G. L. (1984) 'Double-implanted GaAs complementary JFETs' *IEEE Electron Device Lett.* **EDL–5** 21–23.

79. Ino, M., Takada, T., Suto, H., Kato, N. and Ida, M. (1986) 'A sub-nanosecond GaAs 8b ALU by LSCFL' *Extended Abstracts 18th Int. Conf. Solid State Devices and Materials* 371–374.

80. Bosch, B. G. (1979) 'Gigabit electronics – a review' *Proc. IEEE* **67** 340–379.

81. Solomon, P. M. (1982) 'A comparison of semiconductor devices for high-speed logic' *Proc. IEEE* **70** 489–509.

82. Welbourn, B. A. (1982) 'Gigabit logic – a review' *IEE Proc.* **129** Pt. I 157–172.

83. Sugeta, T., Mizutani, T., Ino, M. and Horiguchi, S. (1986) 'High speed technology comparison – GaAs vs Si –' *Proc. GaAs IC Symp.* 3–6.
84. Miyanaga, H., Konaka, S., Yamamoto, Y. and Sakai, T. (1984) 'A 0.85 ns 1Kb bipolar ECL RAM' *Extended Abstracts 16th Int. Conf. on Solid State Devices and Materials* 225–228.

Chapter 4
Monolithic Microwave Integrated Circuit Design

James M. SCHELLENBERG and Thomas R. APEL

4.1 INTRODUCTION

The last ten years have witnessed the emergence of a new and exciting microwave technology which promises to revolutionize military and commercial microwave systems by offering new levels of performance at an affordable price. As the silicon chip revolutionized the electronic industry, this technology holds the promise of producing microwave ICs performing complicated functions and costing only a dollar or so per chip. It is this new microwave approch, monolithic microwave integrated circuits (MMICs), which is the subject of this chapter.

MMICs consist of both active and passive circuit elements integrated on a single semi-insulating substrate or chip forming a component or subsystem as shown in Fig. 4.1. The active devices consist typically of MESFETs and Schottky barrier diodes while the passive elements include both lumped and distributed matching elements, bias networks and interconnections between active and passive elements and ground. While currently at the multistage amplifier level of complexity, this technology holds the promise of integrating whole subsystems, such as a transmit/receiver (T/R) module, on

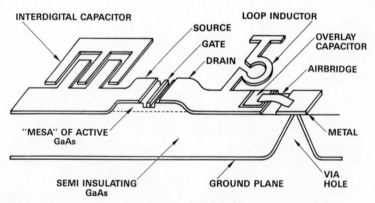

Fig. 4.1 Schematic of a monolithic microwave integrated circuit (MMIC).

a single chip of GaAs. The full implication of this approach underscores the potential advantages of monolithic circuits.

(1) Low cost circuits due to batch processing.
(2) Improved reliability due to the elimination of wire bonds and discrete components.
(3) Small size and weight.
(4) Circuit design flexibility and multi-function performance on a single chip.

The circuits are fabricated in a batch mode with anywhere from several hundred to several thousand circuits on a single wafter. With 3-inch wafers, now in common use, there is a useful area of 3850 mm^2. Current costs to fabricate a 3-inch wafer in six wafer lots are approximately $4000 per wafer. Hence, with a chip area of 2 mm^2, 1925 potential chips are possible with an unyielded cost of $2 per chip. This promise of inexpensive microwave circuits in large quantities has dramatically changed the way in which designers approach microwave system configurations and has attracted relatively large amounts of government and commercial investment.

The salient feature of the monolithic approach is the elimination of all wire-bonding within the IC itself. This results in cost effectiveness and improved reliability. Wire bonds have always been a serious factor in reliability and reproducibility. For example, at 20 GHz, a change of one mil in the bond wire length produces a reactance change of 3 ohms – not an insignificant change for a large power device. Bond wires represent a source of uncontrolled parasitic elements. Furthermore, wire bonding, being labour-intensive, is often a significant factor in the cost of a hybrid circuit.

Small size and volume are intrinsic properties of the monolithic approach. Small size allows batch processing of hundreds of circuits per wafer or substrate. Since the essence of batch processing is that the cost of fabrication is determined by the cost of processing the entire wafer, it follows that the processing cost per chip is proportional to the area of the chip. Thus the smaller the chip, the lower its cost. Hence the importance of minimizing the chip size. The small circuit size, intrinsic to the monolithic approach, will enable circuit integration on a chip level, ranging from the lowest degree of complexity such as an amplifier, oscillator, or mixer, to a next higher 'functional block' level, for example a receiver front end or a phase shifter. A still higher level of circuit complexity, for example, a transmit-receiver module, will probably by integrated in multichip form.

Prominent among the potential applications are active phased array rader and direct broadcast television receivers. Both systems require literally millions of ICs at low cost (several dollars per chip) if they are to be successful.

The concept of a microwave monolithic IC originated with a government funded program at Texas Instruments in 1964. This initial effort, based on silicon technology, failed due to the difficulty of maintaining a high resistivity silicon semi-insulating substrate as required for microwave circuitry [1]. In 1964, Mehal and Wacker [2] revived the idea by using semi-insulating gallium arsenide (GaAs) as the base material and Schottky barrier diodes and Gunn devices to fabricate a 94 GHz receiver front end. However, it was not until Plessy applied the monolithic idea to an X-band amplifier [3] using the MESFET as the key active element that the current wave of MMIC activity began around the world.

4.2 MONOLITHIC CIRCUIT ELEMENTS

In this section the passive and active monolithic circuit 'building blocks' are presented. The passive elements include the traditional lumped elements (resistors, inductors and capacitors), distributed elements (transmission lines) and circuit interconnections (via holes and air bridges). The active elements consist predominately of MESFET devices and, to a lesser degree, Schottky barrier diodes. Due to its versatility, and as a consequence, dominance in MMIC applications, we will focus on the three-terminal FET device. The properties of the Schottky barrier diode and its integration into MMICs is adequately covered elsewhere [4] and will not be repeated here.

4.2.1 Substrate thickness

The substrate thickness has a major impact on the MMIC design and its subsequent performance. The thickness of the GaAs wafer or substrate is a compromise between many factors, some of which demand a thick substrate while others dictate a substrate that is as thin as possible. Thermal considerations, via holes, material costs and the propagation of higher order modes all dictate the thinnest possible substrate, while circuit losses, wafer fragility and ground plane effects demand a thick substrate. The impact of these factors on the selection of a substrate thickness is considered in the following discussion.

Thermal considerations, in the design of a power MMIC, generally require a substrate thickness of less than 150 μm (6 mils). This is due, in part, to the rather poor thermal conductivity (0.46 W/cm °K) of GaAs, which is a factor of three worse than silicon. For typical MMIC device geometries and substrate thicknesses (100 μm), there is little lateral heat flow, and hence, to first order, the active device thermal resistance and temperature rise, ΔT, increase linearly with the substrate thickness. For example, with a 100 μm thick substrate, the FET thermal resistance normalized to one millimeter of gate width is approximately 40°C/watt producing a ΔT of 60°C for typical power FET bias conditions. Increasing the substrate thickness of 150 μm results in a thermal resistance of ~60°C/watt and a typical ΔT of 90°C. This temperature rise represents a potential reliability problem if the ambient temperature is not controlled. For many high reliability applications, the active device channel temperature must be maintained below 120°C, and therefore with this design the ambient temperature cannot exceed 30°C. For most applications, this is an impractical environment restriction.

Wafer fragility limits the substrate thickness to approximately 50 μm. At the thickness the wafer is difficult to handle and must be waxed down to a silicon carrier wafer. After sawing or scribing, the resultant chips are also difficult to handle and die attach. It is likely that automated equipment developed to handle MMICs will require a chip thickness of greater than 50 μm.

Via holes limit the substrate thickness to less than approximately 125 μm. With current via hole technology, the via hole is etched through the substrate from the backside as described in Section 4.2.3. As the substrate thickness increases, the via hole opening on the backside increases requiring more substrate area. For example, a typical via hole etched through a 100 μm thick substrate produces a backside opening of typically 150 × 200 μm. Also, the inductance associated with the via hole increases with the substrate thickness at the rate of ~0.5 pH/μm further limiting the thickness. While new methods of forming the via hole, namely ion milling and plasma etching, hold promise of reducing the required via hole area, the inductance limitation still constrains the substrate thickness to probably 125 μm or less.

For MMICs employing the microstrip configuration, the substrate thickness determines circuit losses (element Q), the microstrip line width and the upper frequency limit due to higher order modes. The cut-off frequency for the lowest order (TE) surface mode as a function of the substrate thickness is given as

$$f_c = \frac{c}{4 \sqrt{(\varepsilon_r - 1)} \, h} = \frac{75}{\sqrt{(\varepsilon_r - 1)} \, h \, (\text{mm})} (\text{GHz})$$

where c is the velocity of light (30×10^9 cm/s), ε_r is the relative dielectric constant of the substrate and h is the substrate thickness. For example, for a substrate thickness of 100 μm, the 'safe' operating frequency range is up to 200 GHz. For currently contemplated circuit applications, it appears that surface mode propagation is not a limiting factor in the choice of the substrate thickness.

Microstrip conductor losses are inversely proportional to the substrate thickness as described in Section 4.2.2. For example, the conductor loss for a 50 ohm line at 10 GHz increases from 0.35 dB/cm with a substrate thickness of 100 μm to 0.7 dB/cm for a 50 μm thick substrate. This latter case implies a microstrip element Q of only 38. Microstrip radiation losses, on the other hand, are directly proportional to the square of the product of frequency and substrate thickness. For example, the radiation Q of 30 GHz of a 50 ohm quarter-wave microstrip transmission line decreases from 1000 with a 100 μm thick substrate to 160 with a substrate thickness of 250 μm. Hence, the substrate thickness cannot be increased arbitrarily to decrease conductor losses without first considering the frequency and the consequences of radiation losses.

In summary, for monolithic ICs requiring via holes, the substrate thickness is limited on the high side by thermal resistance and via hole size and inductance to perhaps no more than 125 μm, and on the low side by substrate fragility and circuit losses to more than 50 μm. A good compromise, for power applications up to 30 GHz, is a substrate thickness of 100 μm. For low noise applications below 18 GHz not requiring via holes, substrate thicknesses up to 250 μm may be useful due to the lower circuit losses (higher Q).

4.2.2 Transmission line elements

For monolithic ICs operating at microwave frequencies, transmission line (TL) elements play an important role. TLs act as interconnecting elements between the active and passive devices on the chip and are used as impedance matching elements. Also, the TL configuration determines the substrate thickness circuit geometry, thermal impedance and packaging considerations. These and other factors relative to TLs are considered in the following discussion.

Four basic TL configurations, as illustrated in Fig. 4.2, are possible on a planar substrate. The first structure, microstrip (MS), shown in Fig. 4.2 (a), requires a ground plane on the backside of the substrate. Its 'inverse', slot line (SL), is shown in Fig. 4.2 (b). The third configuration is coplanar waveguide (CPW) shown in Fig. 4.2 (c). It consists of a central 'hot' conductor separated by a slot on each side and two coplanar ground planes. Its 'inverse', the coplanar stripline (CS), is illustrated in Fig. 4.2 (d). With the exception of MS, all the configurations generally assume an infinitely thick substrate. In practice, of course, this condition cannot be met, and the presence of a backside ground plane results in a hybrid mode. The features of these four TL configurations are summarized in Table 4.1.

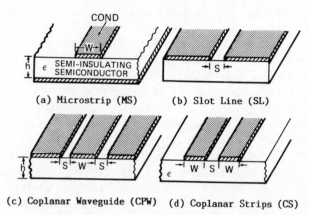

(a) Microstrip (MS) (b) Slot Line (SL)

(c) Coplanar Waveguide (CPW) (d) Coplanar Strips (CS)

Fig. 4.2 Four candidate transmission line configurations for monolithic circuits.

Table 4.1. Properties of candidate transmission line configurations

Property	Microstrip (MS)	Coplanar waveguide (CPW)	Coplanar strips (CS)	Slot line (SL)
Attenuation loss in 50 Ω line (dB/cm)	0.33	0.5	1.7	> 2
Dispersion	Low	Medium	Medium	High
Z_0 range (ohms)	10–100	25–125*	40–250*	50–300*
Connect shunt elements	Difficult	Easy	Easy	Easy
Connect series elements	Easy	Easy	Easy	Difficult

* Infinitely thick substrate

Of the four TLs, only slot line supports a non-TEM-like mode of propagation and therefore is quite dispersive. It also does not interface well with FET devices, and does not allow a great deal of versatility in the circuit layout. For these reasons, it is not considered a viable candidate for monolithic circuits, except perhaps in special cases.

The line losses of MS, CPW and CS on 0.1 mm thick GaAs substrates are compared in Fig. 4.3. For GaAs, the principle source of losses are ohmic conductor losses as opposed to dielectric losses. The coplanar structures (CPW and CS) are 'edge-coupled' tranmission lines with high current concentrations at the strip edge. As a result, the losses tend to be somewhat higher than for MS.

While the MS structure has the lowest loss, the lack of a coplanar ground

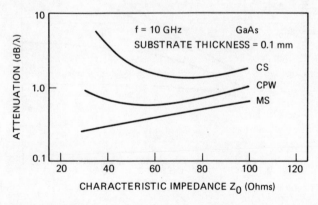

Fig. 4.3 Comparison of line losses for microstrip, coplanar waveguide and coplanar strips (after Pucel).

plane with microstrip is a considerable disadvantage when shunt elements are required. Since ground is on the backside of the substrate, shunt elements can only be connected to ground either by 'wrap-round' or 'via' hole connections. The inductance of these ground connections represents a serious limitation particularly at higher frequencies. This is discussed in greater detail below.

With CPW and CS, the characteristic impedance range is somewhat greater than for MS, particularly at the high end of the impedance scale. This is true only for an infinitely thick substrate. This range is reduced considerably when practical substrate thicknesses are used and the back side of the chip is metallized. Fig. 4.4 illustrates how the high impedance end of the scale is lowered when substrates of the order of 0.1 to 0.25 mm thick are mounted on a metal base for heat sinking purposes. The sensitivity of Z_0 to the presence of a ground plane, particularly at the high end of the impedance scale, makes the design of monolithic circuits with CPW and nearly as dependent on substrate thickness as with MS.

Weighing all of these factors, of the four candidate modes, MS and CPW are the most suitable for GaAs monolithic circuits, with preference toward MS for low frequency operation and CPW for frequencies about 40 GHz where the microstrip via hole impedance becomes significant. Indeed, there will be

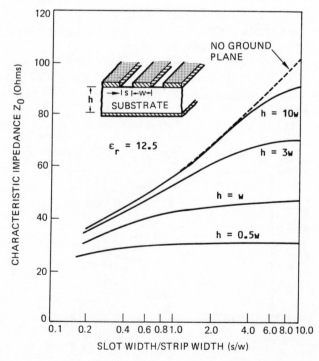

Fig. 4.4 Effect of the ground plane on the characteristic impedance of coplanar waveguide.

instances where both modes may be used on the same chip to achieve some special advantage.

4.2.2.1 Microstrip lines

The microstrip configuration, shown in Fig 4.2 (a), consists of a strip conductor of width w and thickness t on a dielectric (GaAs) substrate with the back side metallized to form a ground plane. Microstrip is the most popular transmission line configuration for monolithic IC applications due to the following:

(1) Passive and active elements are easily inserted in series with the MS strip conductor on the surface of the chip.
(2) The metallized ground plane on the back of the substrate can be used both as the mounting surface and the heat sink for heat generated by active devices on the surface.
(3) A large body of theoretical and experimental data exists for the MS configuration.
(4) The losses and dispersion are low while the Z_0 range is moderate.

The only serious disadvantage of MS is due to its noncoplanar geometry as described above, which makes it difficult to connect elements in shunt to ground.

While there are many analyses of microstrip and its discontinuities, including rigorous 'full-wave' solutions, these numerical solutions are not very useful for circuit design purposes. Generally, design equations based on calculator functions are more useful. In the following discussion, MS design equations are presented in terms of calculator functions. These equations are derived by numerically fitting the simplified expressions to the rigorous solutions. The accuracy is generally better than 1% for MS parameters in popular ranges ($10 \, \Omega < Z_0 < 100 \, \Omega$).

The characteristic impedance of a MS line with width w and substrate thickness h and negligible metallization thickness t (a good assumption with MMICs) is given as follows [5]:

For $w/h \leq 1$,

$$Z_0 = \frac{60}{\sqrt{\varepsilon'_r}} \, \ln\left(\frac{8h}{w} + \frac{w}{4h}\right) \tag{1}$$

where the effective dielectric constant ε'_r is given as

$$\varepsilon'_r = \frac{\varepsilon_{r+1}}{2} + \frac{\varepsilon_{r-1}}{2}\left[\left(1 + \frac{12h}{2}\right)^{-1/2} + 0.04\left(1 - \frac{w}{h}\right)^2\right] \tag{2}$$

For $w/h \geq 1$

$$Z_0 = \frac{120\pi/\sqrt{\varepsilon'_r}}{w/h + 1.393 + 0.667 \, \ln(w/h + 1.444)} \tag{3}$$

where

$$\varepsilon'_r = \frac{\varepsilon_{r+1}}{2} + \frac{\varepsilon_{r-1}}{2}\left(1 + 12\frac{h}{w}\right)^{-1/2} \tag{4}$$

Since the fields between the strip and the ground plane are not contained entirely in the substrate, the wave propagating along the strip is not strictly transverse electromagnetic (TEM) but is quasi-TEM. The phase velocity is given by

$$v_p = \frac{c}{\sqrt{(\varepsilon'_r)}} \tag{5}$$

where c is the velocity of light. The wavelength λ in the microstrip medium is given as

$$\lambda = \frac{v_p}{f} = \frac{c}{\sqrt{(\varepsilon'_r)f}} = \frac{30}{\sqrt{(\varepsilon'_r)f\,(\mathrm{GHz})}} \;(\mathrm{cm}) \tag{6}$$

where f is the frequency.

The impedance range of MS is limited by the minimum line width on the high side and higher order modes on the low side. The highest achievable impedance is determined by the smallest line width w that can be realised with acceptable integrity over a long length, say a quarter of a wavelength. A minimum line width of 5 μm is reasonable. The maximum line width must be well below a quarter-wavelength, say, one-eighth-wavelength, in order to prevent transverse modes and unpredictable performance. With these restrictions and a substrate thickness of 0.1 mm, the characteristics impedance as a function of frequency is constrained within the range indicated by Fig. 4.5.

Microstrip losses on GaAs are generally dominated by conductor losses. With good quality high resistivity GaAs substrates, the dielectric loss is usually negligible. Also, for popular substrate dimensions ($h \sim 0.1$ mm) and operating frequencies ($f < 30$ GHz), the radiation loss is also negligible in comparison to the dominant conductor loss.

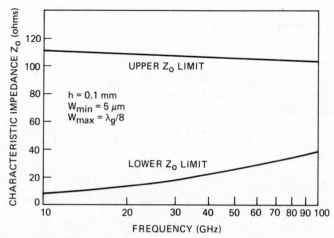

Fig. 4.5 Microstrip characteristic impedance range.

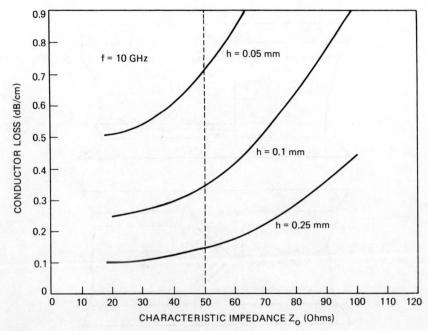

Fig. 4.6 Microstrip conductor loss (after Pucel).

The microstrip conductor loss is illustrated in Fig. 4.6 as a function of the characteristic impedance and substrate thickness at 10 GHz. The conductor loss for a 50 ohm line of a 0.1 mm thick substrate is approximately 0.35 dB cm^{-1} at 10 GHz. Due to skin effect, the loss in decibels per centimeter increases as the square root of frequency to 0.86 dB cm^{-1} at 60 GHz.

4.2.2.2 Coplanar waveguide

The CPW configuration, shown in Fig. 4.2 (c), is a coplanar geometry with a central conductive strip and two adjacent grounds planes on the same surface of the substrate. While an infinitely thick substrate is usually assumed for analysis, in practice the backside of the substrate is usually metallized and mounted to a heat sink for heat extraction. The presence of this backside ground plane significantly reduces the available impedance range as described previously. However, due to its coplanar nature, this configuration has other unique properties which make it particularly useful at millimeter wave frequencies. The advantages of CPW are summarized below:

(1) The coplanar geometry makes it easy to insert shunt elements to ground.
(2) The planar geometry of CPW interfaces directly with the planar FET geometry resulting in minimum parasitic elements. Three possible device geometries in a CPW configuration are shown in Fig. 4.7. The

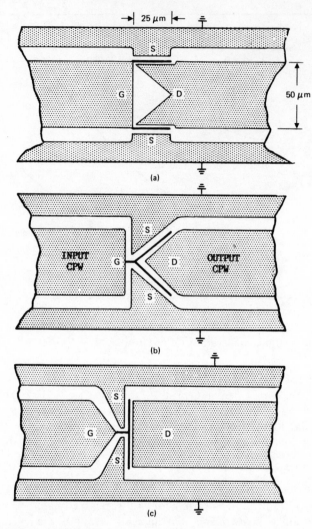

Fig. 4.7 Three GaAs MESFET designs in CPW configuration for minimizing L_S and gate and drain parasitics.

parasitic source inductance, particularly troublesome at millimeter wave frequencies, is virtually eliminated.

(3) The losses, dispersion and Z_o range are moderate.

On the negative side, CPW is a difficult topology for complex layouts, due to its three conductor patterns (central strip and two ground planes). Also, in order to avoid slot line modes, symmetry must be maintained between the two ground planes. For example, if we require an inductance, L, in shunt to ground, we must instead use an inductance of $2L$ to each ground plane, utilizing four times the surface area. To avoid this problem, we must interconnect the ground planes at regular intervals (every $\lambda/8$).

The properties of CPW on an infinitely thick substrate are given by Wen [6], and the effect of a backside ground plane is treated in reference [7]. From Wen the characteristic impedance of CPW on an infinitely thick substrate is given by

$$Z_o = \frac{30\pi}{\sqrt{\varepsilon'_r}} \frac{K'(k)}{K(k)}$$

where

k $= w/(w + 2s)$
$K(k)$ $=$ complete elliptic integral of the first kind
$K'(k)$ $= K(k') =$ complement
k' $= (1 - k^2)^{1/2}$

and the effective dielectric constant ε'_r is given by

$$\varepsilon'_r = \frac{1 + \varepsilon_r}{2}$$

The above expression for Z_o is not very useful for design purposes since it relies on uncommon elliptic functions as opposed to simple calculator functions. Hilberg [8] has derived a remarkable expression for this elliptic integral ratio which is accurate to better than 3 parts per million. Using Hilberg's approximation, the characteristic impedance of CPW can be expressed as follows: For $0 < k \leq 1/\sqrt{2}$

$$Z_o = \frac{30}{\sqrt{\varepsilon'_r}} \ln \left[2 \frac{1 + (1 - k^2)^{1/4}}{1 - (1 - k^2)^{1/4}} \right]$$

and for $1/\sqrt{2} \leq k < 1$,

$$Z_o = \frac{30\pi^2/\sqrt{\varepsilon'_r}}{\ln \left[2 \frac{1 + \sqrt{k}}{1 - \sqrt{k}} \right]}$$

The characteristic impedance is plotted as a function of the slot width to conductor width ratio in Fig. 4.4

4.2.3 Low inductance grounds and crossovers

Microstrip and coplanar waveguide are adequate for interconnections that do not require conductor crossovers or that are not required to contact the backside ground metallization. Often, however, such connections are needed. In particular, with MS, which does not have a topside ground plane, some means of achieving a low-inductance ground is essential.

Two general methods of grounding are available: (1) the 'wrap-around' ground; and (2) the 'via' hole ground. The former requires a topside metallization pattern near the periphery of the chip which can be connected to the chip ground. The via hole technique, on the other hand, allows placement of grounds through the substrate where desired. Holes are chemically etched from the backside through the substrate until the top metallization pattern is reached. These holes are subsequently metallized at the same time as the ground plane to provide continuity between this plane and the desired topside pad. Fig. 4.8 illustrates typical via holes formed by using an anisotropic etch through a 100 μm thick substrate. The hole size decreases from 150 × 200 μm at the backside to 50 × 75 μm at the topside. The estimated inductance of a via hole is 0.5 pH/μm of substrate thickness.

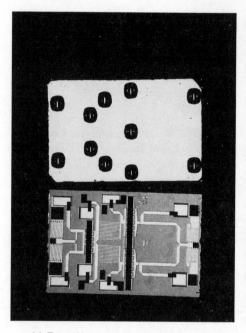

(a) Top and bottom views of MMIC showing via holes

(b) Via hole detail

Fig. 4.8 Via holes are used to achieve low inductance grounds.

Low inductance grounds are especially important in the source lead of power FETs and high f_T millimeter wave devices as described in Section 4.2.5.4. An inductance in the source lead generates a real part (a resistive loss) in the gate circuit, and hence reduces the power gain.

The second interconnect problem arises when it is necessary to connect the individual cells of a power FET without resorting to wire bonds. These interconnects are also required to have a low inductance. Here the so-called 'air-bridge' crossover is useful. This crossover consists of a deposited strap which crosses over one or more conductors with an air gap in between for low capacitive coupling.

An air-bridge source crossover, which interconnects two adjacent source pads of a power FET, is shown in Fig. 4.9. The air gap is approximately 6

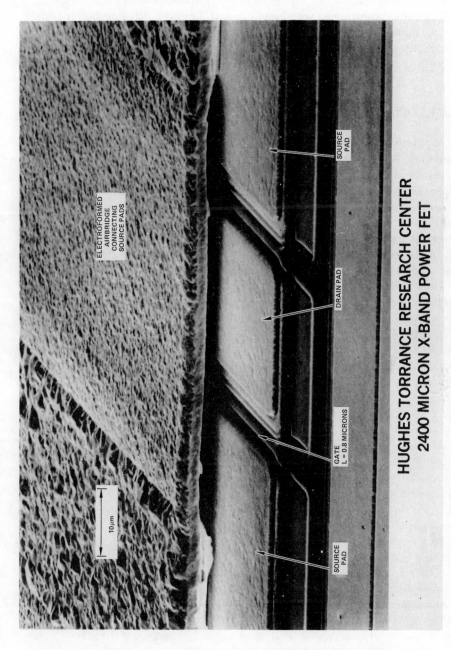

Fig. 4.9 SEM microphotograph of source air-bridge interconnecting the source pads of a power FET.

μm. Clearly shown is the 0.8 μm gate and the larger drain pad underneath the crossover.

It is evident that the air-bridge technology allows one to interconnect all cells without resorting to wire bonding, and therefore satisfies the criterion for a monolithic circuit. Air-bridge interconnects are also useful for microstrip crossovers. A good example is a planar spiral inductor, which requires a contact to the inner terminal.

4.2.4　Lumped elements

Both lumped elements (dimensions < 0.1 wavelength) and distributed elements, that is, elements composed of sections of transmission line, are generally required for monolithic circuit design. Lumped LC elements, often used as matching elements and resistors, are required for biasing networks, gain equalization and resistive terminations. Due to the large capacitance values required, lumped thin-film capacitors are absolutely essential for bias bypass applications. Planar inductors can be extremely useful for matching purposes, especially at the lower end of the microwave spectrum where transmission line elements are very large, physically.

The choice of whether to employ lumped or distributed elements depends on the frequency of operation. Lumped elements are useful through X-band up to, perhaps, 20 GHz. For frequencies beyond this range, distributed elements are preferred. It is difficult to realise a truly lumped element, even at the lower frequencies, because of the parasitic capacitance to ground associated with thin GaAs substrates. In this section we shall review the design principles of planar lumped elements.

4.2.4.1　Capacitors
In monolithic circuits, capacitors are required for dc blocks, RF bypass and matching applications. Some possible geometries for planar capacitors are shown in Fig. 4.10. The first three, which use no dielectric film and depend on electrostatic coupling via the substrate, generally are suitable for applications where low values of capacitances are required (less than 1.0 pF). A distributed version of (a), where the overlap length is $\lambda/4$, is particularly useful at millimeter wave frequencies as a dc block. The last two geometries, the so-called overlay structures which use dielectric films, are suitable for low-impedance (power) circuitry and bypass and blocking applications. The overlay structure is suitable for capacitance values in the 1 to 40 pF range.

The two sources of loss in planar capacitors are conductor losses in the metallization, and losses in the dielectric film. Since the first three configurations illustrated in Fig. 4.10 are edge-coupled capacitors, high charge and current concentrations near the edges tend to limit the Q-factors. At X-band frequencies, the Q-factor is typically in the range of 40 to 80, despite the fact that no dielectric losses are present. The last two geometries

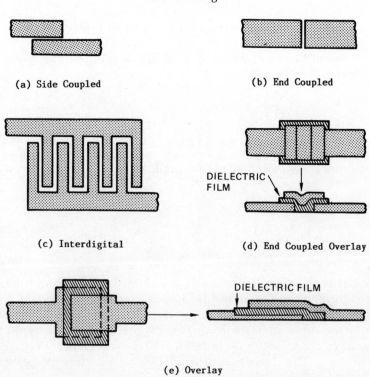

Fig. 4.10 Planar capacitor geometries for MMIC applications.

distribute the current more uniformly over the metal electrodes due to their 'parallel plate' geometry. However, even here, Q-factors of only 50 to 100 are typical at X-band (10 GHz) because of dielectric film losses.

4.2.4.1.1 Overlay capacitors. Overlay or metal-dielectric metal capacitors are typically fabricated with a bottom metal layer of Cr-Au-Cr, a dielectric layer, such as Si_3N_4, 2000 Å thick, and a top metal of Cr-Au. An air-bridge is used to connect to the top metal layer to avoid edge breakdown. At this time, the state-of-the-art in overlay capacitors is such that they can be fabricated with relatively high yield (> 95%) and breakdown voltages in excess of 75 V. The density of the dielectric, the thickness of the layer, and the dielectric constant can be controlled to the extent that a capacitance tolerance of ±10% is obtainable.

Clearly the dielectric film performs a key role for overlay capacitors. Some important properties of dielectric films are (1) dielectric constant, (2) capacitance/area, (3) microwave losses, (4) breakdown field, (5) temperature coefficient, (6) film integrity (pinhole density and stability over time), and (7) method and temperature of deposition. This last requirement is obviously important, because the technology used for film deposition must be compatible with the technology used to fabricate the active devices (FETs). Dielectric films which easily satisfy this criterion are SiO_x and Si_3N_4.

Some useful figures of merit [9] for dielectric films are the capacitance-breakdown voltage product

$$F_{cv} = \left(\frac{C}{A}\right)V_b \tag{7}$$

$$= \varepsilon_r\varepsilon_0 E_b \tag{8}$$

$$= 8 \text{ to } 30 \times 10^3 \text{ pF} \cdot \text{V mm}^{-2} \tag{9}$$

and the capitance-dielectric Q-factor product

$$F_{cq} = \left(\frac{C}{A}\right)Q_d \tag{10}$$

$$= \frac{(C/A)}{\tan\delta_d} = \frac{\varepsilon_r\varepsilon_o}{(t)\tan\delta_d} = 10^3 \text{ to } 10^6 \text{ pF mm}^{-2} \tag{11}$$

where C/A is the capacitance per unit area, V_b is the breakdown voltage, E_b is the corresponding breakdown field, ε_r is the dielectric constant, t is the dielectric thickness and $\tan\delta_d$ is the dielectric loss tangent. Breakdown fields of the order of 1–2 MV cm^{-1} are typical of good dielectric films. Dielectric constants are in the order of 4–20. Loss tangents can range from 10^{-1} to 10^{-3}. The candidate dielectric films and their properties are summarized in Table 4.2.

The overlay structure is shown in greater detail in Fig. 4.11 (a). Taking into account the longitudinal current paths in the metal contacts, one may analyse this device as a lossy transmission line. For maximum Q, the longitudinal and transverse dimensions are small compared with a wave-

Table 4.2. Properties of dielectric films for overlay capacitors

Dielectric material	ε_r	C/A^* (pF/mm^2)	$(C/A).Q_d^*$ (pF/mm^2)	$(C/A).V_b$ (pF V/mm^2)	TCC (ppm/°C)	Comments
SiO_x	4.5–6.8	300	Low ($\sim 10^4$)	Medium ($> 10^4$)	100–500	Evaporated
SiO_2	4–5	200	Medium ($\sim 10^5$)	Medium ($> 10^4$)	50	Evaporated, CVD or sputtered
Si_3N_4	6–7	300	High ($\sim 10^6$)	High ($\sim 10^5$)	25–35	Sputtered or CVD
Ta_2O_5	20–25	1100	Medium ($\sim 10^5$)	High ($> 10^5$)	0–200	Sputtered and anodized
Al_2O_3	6–9	400	High ($\sim 10^6$)	High ($> 10^5$)	300–500	CVD, sputtered and anodized
Polyimide	3–4.5	35	High	—	–500	Organic film, spun and cured

* Dielectric film thickness is 2000 Å, except for polymide which is 10 000 Å.

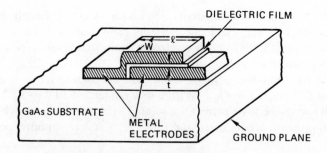

(a) Capacitor geometry

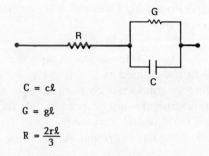

$C = c\ell$

$G = g\ell$

$R = \dfrac{2r\ell}{3}$

(b) Simplified Equivalent Circuit

Fig. 4.11 Overlay capacitor geometry equivalent circuit.

length in the dielectric film. In this case, a good approximation to the capacitor is the simple equivalent circuit shown in Fig. 4.11 (b). The input impedance of this lossy capacitor is given by

$$Z_{in} = \frac{2rl}{3} + \frac{1}{gl + j\omega cl}$$

where $r = R_s/w$ represents the resistance, per unit length of the metal plates and c and g denote the capacitance and conductance per unit length of the dielectric film. The relation between g and c is determined by the loss tangent of the film, $g = \omega c \tan\delta_d$. The series resistance in the plates is determined by the skin resistance if the metal thickness exceeds the skin depth, or the bulk metal resistance if the reverse is true. Usually, the bottom metal layer is evaporated only, and hence is only about 0.5 μm thick, which may be less than the skin depth. The top metal is normally built up to a thickness of several micrometers or more by plating. When the skin effect loss condition prevails, the Q-factor corresponding to these losses is given by the expression

$$Q_c = \frac{3}{2\omega R_s (C/A)^2}$$

where R_s is the surface skin resistivity and l is the electrode length. Note the strong dependence on electrode length. This arises because of the longitudinal current path in each electrode. Note that if one electrode, say the bottom plate, is very thin, Q_c is decreased.

The dielectric Q-factor is $Q_d = 1/\tan\delta$, and the total Q-factor is given by the relation $Q^{-1} = Q_d^{-1} + Q_c^{-1}$. With a loss tangent of 10^{-3} to 10^{-4} (typical for bulk SiO_2 or Si_3N_4), we expect the total Q to be dominated by conductor losses at microwave frequencies and that Qs of several hundred could be achievable.

The first thorough study of overlay capacitors was done by Caulton [10] who determined that the dielectric loss of deposited SiO_2 or Si_3N_4 was much higher than anticipated and that Qs of only 30–50 were common at X-band. This could be improved by heat treating the films. The current state-of-the-art is such that capacitors with Qs of 30–100 can be fabricated on monolithic circuits with good yield.

4.2.4.1.2 Inter-digitated capacitors. Inter-digitated capacitors are composed of a single layer of metallization defined as shown in Fig. 4.10 (c) forming a number of interleaved fingers connected on each end by a shorting bar. Alley [11] has analysed this structure and proposed a low frequency model as shown in Fig. 4.12 when l and W are short compared to a wavelength. C_T is the capacitance of the shorting bar to ground and C_1 is the capacitance of the fingers to ground.

$$C_2 = \left(\frac{\varepsilon_r + 1}{W}\right) l \left[(N - 3) A_1 + A_2\right] \text{ (pF/inch)}$$

where

$$A_1, A_2 = f(X/T)$$

A_1 and A_2 are geometry-dependent constants relating interfinger capacitance to X/T ratio and are given in references [11] and [12].

As the dimensions of the capacitor become large compared to a wavelength, we must consider the distributed nature of the structure [11]. Using coupled transmission line theory, we can represent pairs of coupled fingers at any given frequency by a lumped equivalent model. In Fig. 4.13, these

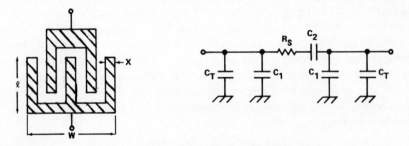

Fig. 4.12 Interdigitated capacitor geometry and equivalent circuit.

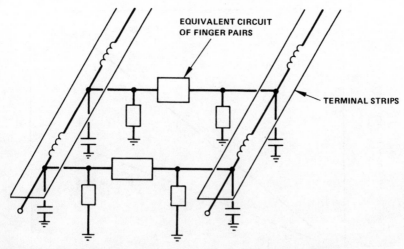

Fig. 4.13 Distributed equivalent circuit model of the interdigitated capacitor.

models then form a periodic discontinuity along the transmission lines formed by the pair of terminal strips. If the fingers are spaced closely together we can write distributed transmission line equations for the terminal strips which are now reactively loaded and coupled by the finger pairs. We now have two different resonant modes in the interdigital capacitor; lengthwise along the fingers or transversely along the terminal strips.

The Q of the interdigital capacitor can be approximated by

$$Q_c = \frac{XN}{WC\frac{4}{3}R_s}$$

for metallization thickness greater than several skin depths, where X is the finger width and R_S is the sheet resistivity. This equation predicts a Q somewhat higher than measured and Fig. 4.14 gives a curve of capacitance versus number of fingers for two finger lengths and their corresponding Qs at X-band. Note that doubling the finger length and holding capacitance fixed, divides Q by four.

Microwave Qs of 75 and 50 have been measured at 10 and 15 GHz respectively for a 0.3 pF interdigital capacitor with 5 μm gaps, 10 μm lines and a metallization thickness of 1.5 μm.

4.2.4.2 Inductors

In monolithic circuits, inductors are used as matching elements and RF chokes for bias insertion. Typical inductance values for monolithic circuits fall in the range from 0.5 to 10 nH with the higher values required at lower microwave frequencies. Fig. 4.15 illustrates various geometries that can be used for thin-film inductors. Aside from the high-impedance line sector, all of the structures depend on mutual coupling between the various line segments to achieve a high inductance in a small area. In any multi-segment

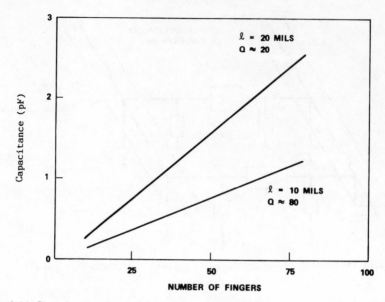

Fig. 4.14 Capacitance as a function of the number of fingers.

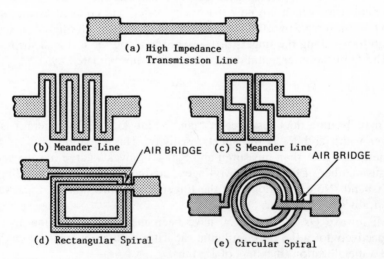

Fig. 4.15 Planar inductor geometries for MMIC applications.

design, one must insure that the total line length is a small fraction of a wavelength, otherwise the conductor cannot be treated as 'lumped'. For inductors with a total line length that is *not* a small fraction of a wavelength, the inductor must be modelled as a distributed element. A typical application of spiral inductors in a monolithic IC is shown in Fig. 4.16. Note the air-bridge crossovers.

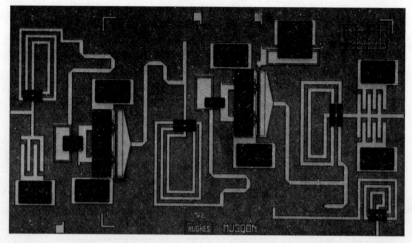

Fig. 4.16 Application of spiral inductors to MMICs. Chip size 1.9 × 1.1 mm.

Terman [13] has developed a simple approximation for the low frequency inductance of a circular spiral inductor.

$$L = \frac{a^2 n^2}{8a + 11c}$$

where L is in nH, and

$$a = (d_o + d_i)/4$$
$$c = (d_o - d_i)/2$$

n is the number of turns, and d_o and d_i are the outer and inner diameters, respectively in mils. The maximim Q, resulting when $d_i/d_o = 0.2$, is given by

$$Q \approx \frac{130 W L^{1/2}}{d_o^{1/2} K} \left(\frac{f(\text{GHz})}{2} \right)^{1/2} \left(\frac{\varrho(\text{Cu})}{\varrho} \right)^{1/2}$$

Podell's [14] approximation for spiral inductors is given as

$$L(\text{nH}) = 0.0124 D n^2$$
$$Q \simeq D \sqrt{f(\text{GHz})}$$
$$Z_o \simeq 28 n \ln(16 T/D)$$

where all the dimensions are in mils, D is the inductor diameter, and T is the metallization thickness.

While the circular spirals exhibit a higher Q, square or rectangular spirals are easier to generate and hence are more useful. The geometry of a rectangular spiral inductor and its equivalent circuit are shown in Fig. 4.17. Based on the method of Greenhouse [15], the inductance of several different size rectangular spirals are plotted in Fig. 4.18 as a function of the number of segments.

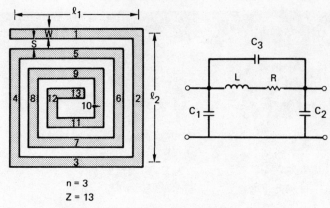

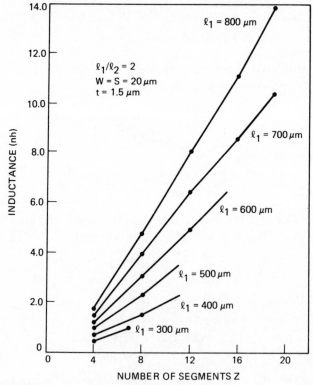

Fig. 4.17 Geometry of rectangular spiral inductor and its equivalent circuit.

Fig. 4.18 Inductor as a function of the number of segments.

These expressions neglect the ground plane and distributed effects. When thin substrates ($h < d$) are used, the presence of the ground plane decreases the inductance by typically 10–20%.

Krafesik [16] has derived a closed form expression for modelling rectangular spiral inductors which includes mutual inductance, ground plane and phase shift effects. A lumped-pi equivalent circuit, shown in Fig. 4.17, with element values that vary with frequency, is used to model the inductor to twice the self-resonant frequency. While this approach is similar to the Greenhouse [15] and Grover [17] approach, it also includes ground plane effects, capacitance calculations, and the effects of phase shift on inductance and capacitance values. When compared to measured results, the calculated results are accurate to 5% up to the self-resonant frequency.

Spiral inductors with a total (unwrapped) line length, l, that is not a small fraction of a wavelength must be modelled as a distributed circuit. Insight into the distributed nature of a spiral inductor can be obtained by considering the generalized transmission line equation.

$$Z_{in} = Z_o \frac{Z_L \cosh \gamma l + Z_o \sinh \gamma l}{Z_o \cosh \gamma l + Z_L \sinh \gamma l}$$

where Z_o is the characteristic impedance of the spiral, Z_L is the line terminating impedance, and γ is the complex propagation constant $\alpha + j\beta$ used to account for line loss. While this equation is strictly true only for uniform TEM transmission lines, it can be applied with reasonable accuracy to less than ideal geometries. For example, a spiral inductor is in reality a non-uniform transmission line with a variable Z_o along its length. However, by assigning an effective Z_o to the spiral, it can be represented quite accurately by the above equation. For a spiral inductor with a total unwrapped length of less than $\lambda/4$, the effective Z_o can be calculated as the even mode impedance of the strip above the ground plane.

This model approximately accounts for the distributed capacitance along the inductor spiral due to the ground plane and the adjacent lines. The inductor is then self-resonant when the total electrical length is $\lambda/4$. The self-resonant frequency is then

$$f_o = \frac{c}{4 \sqrt{\varepsilon'_r} \, l}$$

which c is the velocity of light and ε'_r is the even mode effective dielectric constant of the spiral line.

One application, which takes advantage of the distributed properties of a spiral inductor, is a self-resonant ($l = \lambda/4$) spiral used as an RF choke for bias insertion. Assuming an even mode characteristic impedance of 150 ohms, a minimum reactance of over 900 ohms can be realized over a 20% bandwidth.

The Q of a distributed spiral inductor is illustrated in Fig. 4.19. Due to skin effect, the low frequency Q increases as the square root of frequency. It then reaches a maximum for $f/f_o = 0.371$ ($l = 0.371 \, \lambda/4$) before falling to zero at the self-resonant frequency. This maximum Q value is equal to 79% of the extrapolated low frequency Q.

In practice, inductor Q-factors of typical 50 are observed at X-band frequencies with somewhat higher values at higher frequencies. There appears to be no way to improve the Q-factor significantly, because of the high unfavourable ratio of metal surface area to dielectric volume.

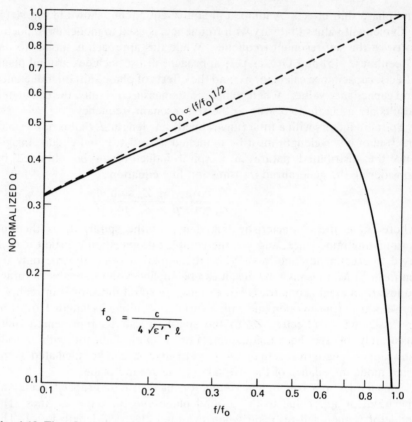

Fig. 4.19 The Q of the spiral inductor decreases due to distributed effects. f_o corresponds to the inductor self-resonant frequency.

Higher Q value inductors can be realised by using short sections of microstrip transmission line. Two sources of loss, conductor and radiation losses, are important for these distributed microstrip inductors. Conductor losses vary inversely with the substrate thickness, and increase as the line impedance increases. The conductor Q-factor for a transmission line is given by

$$Q_c = \frac{27.3}{(\alpha\lambda_g)}$$

where $(\alpha\lambda_g)$ is the line loss expressed in decibels. Since $(\alpha\lambda_g)$ decreases as $f^{-1/2}$, Q_c increases as the square root of frequency, as for thin-film inductors. On the other hand, the radiation Q is given as

$$Q_r = \frac{R}{(fh)^2}$$

where h is the substrate thickness and R is a function of w/h, the dielectric constant of the substrate and the line termination. Note that the radiation Q

decreases as the square of the frequency and the substrate thickness h. Thus any attempt to increase the conductor Q-factor by increasing the frequency and substrate thickness is eventually negated by the decrease of the radiation Q. For frequencies above X-band, open-circuit stub resonators are dominated by radiation losses, unless the substrate is less than 250 μm thick. This radiation also can cause coupling to adjacent circuits. For 100 μm thick substrates, the Q of microstrip inductors is typically 100 at X-band frequencies.

4.2.4.3 Resistors
Resistors are essential monolithic circuit elements required for terminating such components as hybrid couplers, power combiners and splitters. They are also important elements in lossy gain equalization circuits and as negative feedback elements. Some factors to be considered in the design of such monolithic resistors are: (1) the sheet resistivity available; (2) thermal stability or temperature coefficient of the resistive material; (3) the thermal resistance of the resistor; and (4) parasitic elements affecting the frequency response.

A monolithic resistor, shown in Fig. 4.20, can be realised in a variety of forms which fall into three categories: (1) semiconductor films; (2) deposited metal films; and (3) cermets. Resistors based on semiconductors can be fabricated by forming an isolated area of conducting expitaxial film on the substrate, for example, by mesa etching or by isolation implant of the surrounding conducting film. Another way is by implanting a high-resistivity region within the semi-insulating substrate. Metal film resistors are formed by evaporating a metal layer over the substrate and forming the desired pattern by photolithography. Cermet resistors are formed from films consisting of a mixture of metal and a dielectric. However, because of the dielectric, they are expected to exhibit an RC frequency dependence similar to that of carbon resistors, which may be a problem in the microwave band.

Metal films are preferred over semiconducting films because the latter exhibit a nonlinear behaviour at high dc current densities and a rather strong temperature dependence. Not all metal films are suitable for monolithic circuits, since their deposition must be compatible with the IC process. Table 4.3 lists some candidate resistive films.

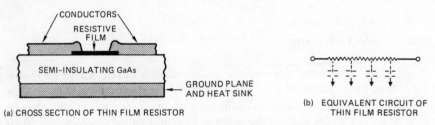

(a) CROSS SECTION OF THIN FILM RESISTOR

(b) EQUIVALENT CIRCUIT OF THIN FILM RESISTOR

Fig. 4.20 Monolithic resistor for MMIC applications.

Table 4.3. Properties of candidate resistive films

Material	Resistivity ($\mu\Omega$-cm)	ICR (ppm/°C)	Method of deposition	Stability	Comments
Chromium	13 (Bulk)	+3000 (Bulk)	Evaporated Sputtered	G–E	Excellent adherence to GaAs
Ti	55–135	+2500	Evaporated Sputtered	G–E	Excellent adherence to GaAs
Ta	180–220	–100 to +500	Sputtered	E	Can be anodized
NiCr	60–600	+200	Evap.(300°C) Sputtered	G–E	Stabilized by slow anneal at 300°C
TaN	280	–180 to –300	Reac. Sput.	G	Cannot be anodized
Ta$_2$N	300	–50 to –110	Reac. Sput.	E	Can be anodized

A problem common to all planar resistors is the parasitic capacitance to ground in the dielectric and the distributed inductance of the film, which makes such resistors exhibit a frequency dependence at high frequencies. If the substrate backside is metallized, the resistive material forms a lossy microstrip line and can be analysed as such.

For low thermal resistance, one should keep the area of the film as large as possible. To minimize discontinuity effects in width, the width of the resistive film load should not differ markedly from the width of the line feeding it. This means that the resistive element should be as long as possible to minimize thermal resistance. This length is specified by the sheet resistivity of the film and is given by the formula

$$l = \frac{wR}{\varrho_s}$$

where w is the width of the film, R the desired load resistance, and ϱ_s the sheet resistance of the film.

If the length of the load is increased while decreasing the sheet resistivity to achieve a low thermal resistance, the load may begin to exhibit the behaviour of a lossy transmission line rather than a lumped resistor. Fig. 4.21 shows how the VSWR increases dramatically at low values of ϱ_s due to the length of the load. Also shown is the thermal resistance. Clearly, a tradeoff is necessary between VSWR and thermal resistance.

4.2.5 GaAs FET devices

A field-effect transistor is a unipolar semiconductor device whose current is controlled by an electric field. This current flows in a doped semiconductor layer called a channel under the influence of an electric field. This electric field is applied to the ends of the channel by ohmic contacts which are called the source and drain electrodes. The controlling element, the gate, is

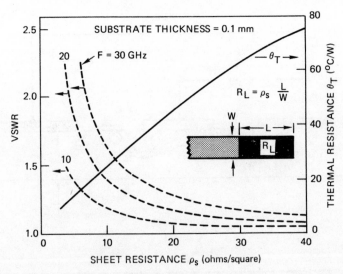

Fig. 4.21 Resistor VSWR and thermal resistance as a function of the GaAs sheet resistance (after Pucel).

positioned between the source and the drain on the channel (active layer). The gate can be realised by several different techniques including a p–n junction (JFET), metal on an oxide layer (MOSFET), metal on an insulating layer other than an oxide (MISFET) and a metal-semiconductor junction (MESFET). While other device structures are being investigated, the MESFET structure is generally employed with GaAs MMICs.

For most microwave applications, gallium arsenide (GaAs) is preferred over silicon, since electrons have six times higher low-field mobility and two times higher maximum drift velocity in GaAs as opposed to silicon. These material properties translate into lower parasitic resistances, a larger transconductance and a lower transit time for electrons in the high field region under the gate. In turn, these device characteristics result in improved microwave performance, such as lower noise figures, higher gain and higher cut-off frequencies. Further, unlike silicon, GaAs has a high resistivity substrate ($\varrho > 10^7 \, \Omega - \text{cm}$) which makes it an ideal medium for integrating other microwave components on the same chip.

4.2.5.1 MESFET geometry

The cross section of a GaAs MESFET (metal-semiconductor field-effect transistor) is shown in Fig. 4.22. The MESFET is a planar device with all three electrodes coplanar on the surface. This fact makes the MESFET ideally suited for planar IC applications. The vertical geometry consists of a thin (0.2–0.6 μm) n-type GaAs layer doped to a level of typically $10^{17} \, \text{cm}^{-3}$ fabricated on a semi-insulating GaAs substrate. This thin active layer can be formed by several different techniques including ion implantation, vapour or liquid epitaxial techniques, metal organic chemical vapour deposition (MOCVD) and molecular beam epitaxy (MBE) to name a few. Typical

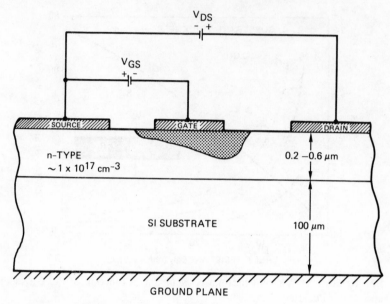

Fig. 4.22 Cross section of a GaAs MESFET.

dopants include sulphur, tin and silicon. A buffer layer of pure GaAs is often included between the active layer and the semi-insulating substrate to act as a barrier against the diffusion of impurities from the substrate to the active layer.

The source and drain contacts are ohmic while the gate contact is a Schottky-barrier contact. A depletion region (an area devoid of free electrons) is formed beneath the gate as a result of the Schottky barrier effect. The dimensions of the gate have a major impact on the performance of the FET. The short dimension of the gate, called the gate length, determines the electron transit time through the device and thereby has a major impact on the device gain and noise performance. In order to minimize the parasitic gate-to-source resistance, the gate is frequently offset toward the source.

A practical device structure will normally consist of several such gate fingers connected in parallel rather than one long structure in the W dimension. The reason for this is due to distributed gate effects as explained below. A power device will normally consist of many such elemental devices connected in parallel in order to increase the current capacity of the resulting device.

The current–voltage characteristics of a MESFET are illustrated in Fig. 4.23. Electrons in the active layer flow from the source to the drain when a positive voltage V_{DS} is applied to the drain. For a fixed gate voltage V_{GS} and small values of V_{DS}, the active layer behaves like a linear resistor – the current increases linearly with the applied voltage. As the drain-source voltage is increased, the electron drift velocity does not increase proportionately and consequently the current starts to saturate. Finally the electron drift velocity reaches its maximum value and the drain current remains constant (saturates) for any further increase in voltage. The voltage at which the drain current saturates is termed V_{SAT} and is typically 2 V. For a fixed positive drain-source voltage and a negative gate voltage, the depletion region under the gate widens and deepens with increasing gate voltage eventually reaching across the active layer to the semi-insulating substrate. Consequently, the channel current will decrease proportionately until it is completely cut off.

The gate voltage which results in the depletion region reaching across the channel is termed the pinch-off voltage. For low values of drain-source voltage (the linear region), this gate action produces a variable channel resistance. In the saturation region, this gate action produces a variable current source. In summary, the negative gate voltage controls the depth of the depletion region under the gate which in turn restricts the flow of electrons from the source to the drain. Consequently, the gate voltage controls the drain current.

For monolithic circuit applications, the FET device is operated in both linear and saturated modes. In the linear region, the channel resistance is controlled by the gate voltage resulting in a voltage contolled variable resistor. This effect can be used, for example, to create an analogue voltage controlled attenuator. In the digital version of this effect, the gate is switched between 0 V and pinch-off causing the channel impedance to change between low and high impedance states, respectively. The impedance of the low resistance state is determined by the channel resistance while impedance of the high resistance state is limited by the source-drain capacitance. By combining several devices, this effect can be used to realise a multiple-throw RF switch. No holding current is required in either state.

In the saturated (active) region, the FET is a near ideal current source and

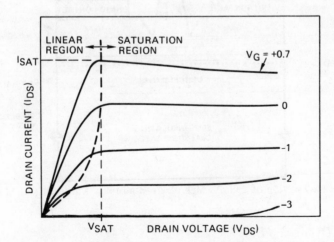

Fig. 4.23 MESFET current–voltage (I–V) characteristics.

consequently, it can provide gain. The GaAs FET is an excellent low noise amplifier as well as a power amplifier. It can also be used as an oscillator and, in either the single or dual-gate configurations, as a mixer.

4.2.5.2 HEMT structure

The HEMT (high electron mobility transistor) device structure, shown in Fig. 4.24, is similar to that of a MESFET, except for its material configuration. Typically, a HEMT structure consists of a 0.5 to 1 μm thick undoped GaAs layer, a 50 to 100 Å thick undoped $Al_xGa_{1-x}As$ ($x \simeq 0.3$) layer, a 500 to 600 Å thick heavily doped $Al_xGa_{1-x}As$ ($x \simeq 0.3$) layer and a 500 Å thick heavily doped GaAs layer. Those layers are grown sequentially. The doping concentration of the heavily doped GaAs and $Al_xGa_{1-x}As$ layers is around 10^{18} cm^{-3}. The two-dimensional electron gas (2 DEG) created underneath the undoped $Al_xGa_{1-x}As$ layer by spilled electrons from the heavily doped $Al_xGa_{1-x}As$ layer exhibits extremely high mobility at low temperature, because scattering due to ionized impurities is minimized. A 2 DEG mobility in excess of 100 000 cm^{2}V^{-1}s^{-1} at 77°K has been achieved. At room temperature, it decreases to around 8000–9000 cm^{2}V^{-1}s^{-1}. Nevertheless, the 2 DEG mobility is still higher than that of the conventional n-type GaAs material at room temperature by a factor of two. These high mobility values of the 2 DEG result in the high transconductance, g_m, of HEMTs. The Schottky barrier gate controls the flow of the 2 DEG under the gate by a depletion region punched through the heavily doped $Al_xGa_{1-x}As$ layer. Chao *et al* [18] have recently demonstrated an intrinsic transconductance as high as 580 mS/mm from HEMTs at room temperature. This value is almost twice the best transconductance obtained with equivalent GaAs MESFETs. Further details of HEMTs and their applications are covered in Chapter 7 of this book.

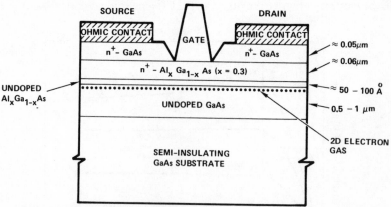

Fig. 4.24 Cross section of a GaAs high electron mobility transistor (HEMT).

4.2.5.3 Dual-gate FET

The structure of a dual-gate FET is shown in Fig. 4.25. It consists of a

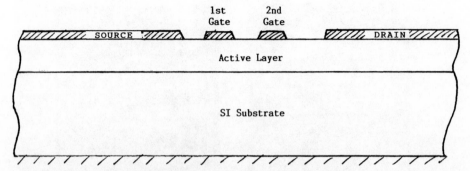

Fig. 4.25 Cross section of a dual-gate MESFET structure.

conventional FET geometry with two gates in the channel instead of one. The dual-gate device can equivalently be considered to be two single-gate FETs connected in series. Hence, the operation and characteristics of the composite (dual-gate) FET can be analysed as two series connected single-gate FETs sharing the same drain-source bias current.

The dual-gate FET is an important circuit element for monolithic ICs. By using the second gate as the control gate, it can function as a voltage controlled amplifier with a variable gain range of typically 30 dB or, in the digital mode, as a switch with an on–off ratio of 30 dB. The dual-gate FET can also be used as a mixer by applying the local oscillator signal to one gate and the RF signal to the other gate. As an amplifier, the second gate is usually RF grounded. This configuration can be analysed by considering the dual-gate FET to be two series connected single-gate FETs as described above, and with the second gate grounded, this geometry is recognized as the cascode connected amplifier configuration. This configuration has certain advantages in terms of stability and gain [19]. In addition to variable gain amplifier applications, the dual-gate FET is an important control element for phase shifter and switch applications [20]. A photograph of a dual-gate device is shown in Fig. 4.26.

4.2.5.4 *Equivalent circuit model*
A rigorous RF equivalent circuit for a MESFET or HEMT device would model the gate structure as a distributed RC network. However, a simple lumped-element circuit is capable of adequately describing the device performance accurately if the gate fingers are kept short (electrical length of less than 10°). The small signal equivalent circuit for the GaAs MESFET is shown in Fig. 4.27 for the common source configuration. While the common gate and common drain configurations are useful for certain circuit functions, the common source configuration is by far the most versatile and therefore the most popular. It provides moderate impedance levels and good gain performance. In the intrinsic model, the depletion region capacitance under the gate is denoted by the gate-source capacitance C_{gs} and its charging resistance in the channel by R_i. C_{dg} models the depletion region

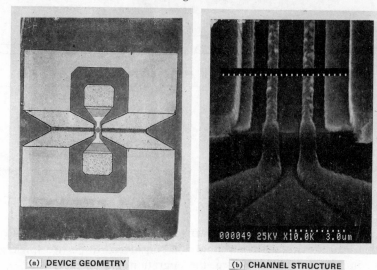

(a) DEVICE GEOMETRY   (b) CHANNEL STRUCTURE

Fig. 4.26 Dual gate MESFET devices.

capacitance between the gate and the drain, and C_{dc} is the capacitance of the high field dipole under the gate. C_{dc} is responsible for the device gain roll-off at a rate faster than the traditional 6 dB per octave at higher frequencies. R_{ds} represents the drain-source channel resistance. The transconductance $g_m = g_{mo}e^{-j\omega\tau}$ relates the channel current to the voltage across the depletion region capacitance, C_{gs}. g_{mo} is independent of frequency and τ is the delay associated with the time required for the electrons of travel under the gate at maximum drift velocity (2×10^7 cm/sec). For a 0.25 µm gate length FET, the gate delay τ is typically 1.5 p$_s$. The extrinsic (parasitic) elements are: R_g the gate-metal resistance, R_s the source channel and contact resistance, R_d the drain channel and contact resistance and C_{ds} the substrate capacitance. The element values of a typical GaAs MESFET with a 0.25 µm gate length and a 60 µm gate width are listed in Table 4.4. These element values were derived from the 2 to 18 GHz S-parameter data plotted in Fig. 4.28.

The microwave and millimeter wave performance of MESFET and HEMT devices are dependent on device geometry and material parameters. In the device geometry, the most critical parameter is the gate length L. Decreasing the gate length decreases the capacitance C_{gs} and increases the transconductance g_m, consequently improving the current-gain bandwidth f_T. For the short-gate-length microwave MESFET and HEMT devices, f_T is proportional to $1/L$ [21]. As a result of this gate length dependence, high frequency operation is achieved by shrinking the gate length to the minimum size that can be realised by a given gate fabrication technology. Current UV and electron-beam lithographic techniques limit the gate length to approximately 0.5 and 0.1 µm, respectively. However, shrinking the gate length also has several negative consequences including increasing the gate resistance and decreasing the drain output resistance, R_{ds} [22]. To

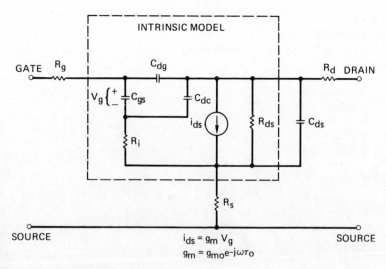

(a) Equivalent Circuit

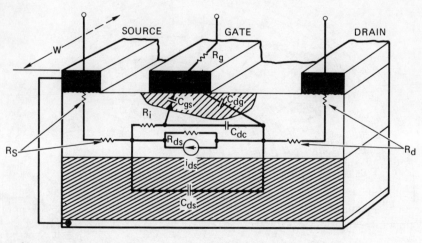

(b) Cross-Sectional View

Fig. 4.27 Small signal equivalent circuit of GaAs MESFET.

counteract this output resistance effect, the channel thickness also must be reduced in direct proportion to the gate length keeping the channel thickness less than or equal to the gate length. This is turn implies a higher doping level. However, the doping level is limited to approximately 4×10^{17} cm^{-3} by breakdown phenomena.

The above considerations yield a millimeter wave device with a very short gate length (~ 0.1 μm) and a very thin (~ 0.1 μm) highly doped ($N = 3$–4×10^{17} cm^{-3}) channel. Such an intrinsic MESFET device is expected to exhibit an f_{max} of well over 200 GHz.

Table 4.4. Equivalent circuit parameters of a low noise GaAs MESFET with a 0.25 μm × 60 μm gate

Intrinsic elements		Extrinsic elements	
g_m	= 16 mmho	C_{ds}	= 0.02 pF
τ_o	= 1.3 ps	R_g	= 3.0 Ω
C_{gs}	= 0.076 pF	R_d	= 4.7 Ω
C_{dg}	= 0.0035 pF	R_s	= 5.7 Ω
C_{dc}	= 0.011 pF		
R_i	= 2.7 Ω		
R_{ds}	= 600 Ω		

dc bias

$$V_{DS} = 3V$$
$$V_{GS} = 0$$
$$I_{DS} = 9 \text{ mA}$$

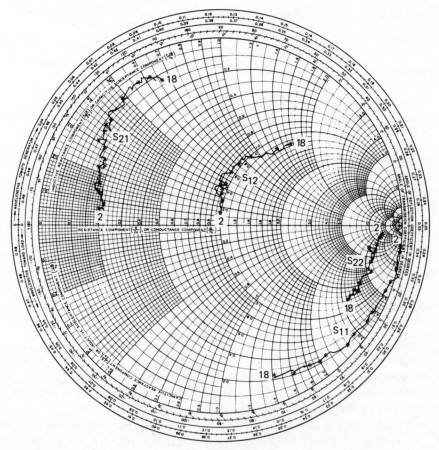

Fig. 4.28 S-parameter data of a 0.25 μm × 60 μm gate MESFET over 2 to 18 GHz. Full scale value for S_{11}, S_{22} and S_{21} is 1 and full scale value for S_{12} is 0.1.

As described above, while the intrinsic device parameters notably the ratio of the device g_m to the gate-source capacitance (the f_T) are critical, the extrinsic elements (R_g, R_s, L_s) are often equally important in determining the device high freqency performance. Insight into the key elements controlling high frequency performance can be obtained by examining the expression for the device maximum available gain (MAG). A rigorous expression including all the parasitic elements would be hopelessly complicated preventing insight into the key parameters and their interaction. An approximate expression for MAG including the parasitic elements is given by [23].

$$\text{MAG} = \left(\frac{f_T}{f}\right)^2 \frac{1}{4G_d(R_i+R_s+R_g+\pi f_T L_s) + 4\pi f_T C_{gd}(R_i+R_s+R_g+2\pi f_t L_s)} \quad (12)$$

where f_T, the frequency at unity current gain is

$$f_T = g_m/2\pi C_{gs} = \frac{v_s}{\pi L_g} \quad (13)$$

This equation indicates that the device gain decreases at the traditional rate of 6 dB/octave as the frequency increases. The frequency at which the MAG is equal to unity is termed f_{max} and represents the upper limit for device operation. f_{max} denotes the boundary between an active device and a passive circuit. Above that frequency the FET is purely a passive device and therefore can provide no gain. f_{max} can be expressed as

$$f_{max} = f_T[4G_d(R_i + R_s + R_g + \pi f_T L_s) + 4\pi f_T C_{gd}(R_i + R_s + R_g + 2\pi f_T L_s)]^{-\frac{1}{2}}$$

This equation illustrates the important role the parasitic elements (R_g, R_s and L_s) play in determining the upper frequency limit of device operation. For a given gate length, the gate resistance can be minimized by employing short gate fingers and/or a gate with a mushroom-shaped cross section as shown in Fig. 4.29 (a). The source resistance is usually minimized by a combination of design techniques including: (1) off-setting the gate in the channel toward the source as shown in Fig. 4.29 (b), (2) recessing the gate in the channel, and (3) employing an N^+ (typically 2×10^{18} cm^{-3}) contact layer to improve the source ohmic contact resistance.

The source inductance plays an increasingly important role in determining the device gain at higher frequencies. For millimeter wave devices with a high f_T, the effect of the source inductance is magnified by the $\pi f_T L_s$ and $2\pi f_T L_s$ terms which reduce the gain to a greater extent than with a lower f_T device. The gain loss of a 0.25 μm gate device at 60 GHz is plotted in Fig. 4.30 as a function of the source inductance. A similar source inductance problem occurs with power devices due to their large gate periphery and resultant low impedance level. A solution to this inductance problem for power applications involves employing a push–pull circuit configuration as described in Section 4.4.

For monolithic IC applications, the source inductance is usually determined by the inductance of via holes which connect the top side source

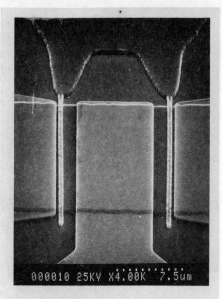

(a) High Profile Gate for minimum gate resistance.

(b) Off-set gate for minimum source resistance.

Fig. 4.29 Techniques for minimizing parasitic gate and source resistances.

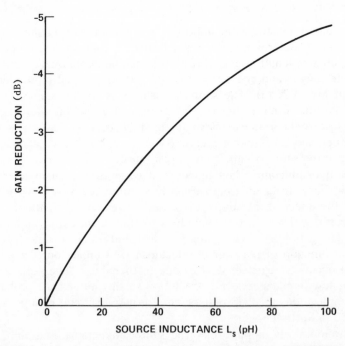

Fig. 4.30 Gain reduction at 60 GHz for a 0.25 μm × 60 μm gate MESFET.

metallization to ground on the backside of the chip. The inductance of a via hole through a 100 μm thick GaAs substrate is typically 50 pH. With this method of grounding, the source inductance can be minimized by thinning the GaAs substrate and/or employing multiple via holes in parallel to ground. However, there are other practical considerations such as element Q and substrate thickness which limit the minimum achievable inductance. At millimeter wave frequencies, a better solution to this inductance problem involves the utilization of the coplanar waveguide (CPW) configuration as described previously. Due to its coplanar geometry (ground plane on the same surface with the centre conductor), it is possible to utilize the CPW ground plane directly as the FET source thereby essentially eliminating the source inductance.

An important parasitic not accounted for in equation (12) is the channel feedback capacitance, C_{dc}. This capacitance is associated with the dipole domain that forms in the high field region under the gate. This capacitance creates internal positive feedback in the FET which produces a peak in the unilateral gain response at a frequency between f_T and f_{max} [24]. Above the gain peak, the device gain rolls off at a rate of 12 dB per octave instead of the traditional 6 dB per octave. While it does not have a significant impact on f_{max}, it does have a substantial effect on the unilateral gain at lower frequencies. This effect can create a substantial error in f_{max} if it is estimated by extrapolating the low frequency gain to 0 dB at a rate of 6 dB per octave.

The maximum unilateral gain and MAG are plotted in Fig. 4.31 for the device parameter previously presented in Table 4.4. As shown in Fig. 4.31, f_{max} is approximately 94 GHz, while equation (12) predicts a value of 112 GHz.

4.2.5.5 *Low noise device considerations*
Principally, the GaAs MESFET is a low-noise device because only the majority carriers participate in its operation. However, in practical GaAs MESFETs, extrinsic resistances are unavoidably added to the intrinsic device. At normal operating temperatures, these parasitic resistances dominate the noise properties of the device.

The study of noise in field-effect transistors was initiated by van der Ziel ([25], [26]) based on Shockley's gradual-channel approximation [27]. Later the noise behaviour of GaAs MESFETs in the common-source configuration has been investigated by a number of workers, as summarized by Liechti in his excellent review article [28]. Pucel *et al.* [29] performed an exhaustive noise analysis of pratical FET devices with parasitic elements and derived an expression for the device minimum noise figure in terms of g_m, C_{gs}, R_g and R_s.

While the relationship between the noise parameter and the equivalent circuit elements has been given in rigorous and complicated forms, it is desirable to express the noise properties in terms of a simple analytical express. For frequencies up to at least f_T, Fukui [30] has found empirically that the minimum noise figure can be expressed as

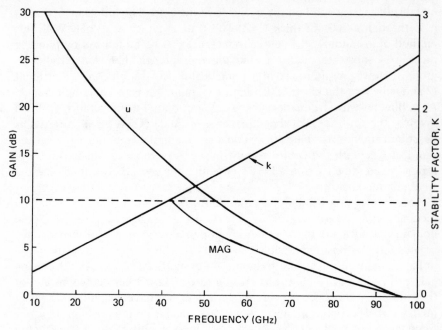

Fig. 4.31 Maximum unilateral gain, U, maximum available gain, MAG, and the stability factor, k, as a function of frequency for the 0.25 μm × 60 μm device parameters listed in Table 4.4.

$$F_{min} = 1 + kfL \sqrt{g_{mo} (R_g + R_s)} \qquad (15)$$

where

k is empirically derived noise coefficient of approximately 0.27 for MESFETs and 0.17 for HEMT devices

f is frequency (GHz)

L is gate length (μm)

g_{mo} is transconductance at $V_g = 0$ (mhos)

R_g is gate metallization resistance (ohms)

R_s is total source series resistance, (ohms)

In this expression, the parameters g_{mo}, R_g, and R_s are fixed independent of the bias point. While g_m is in general a function of the bias point, for this expression it is evaluated at $V_g = 0$ and the operating drain voltage (typically 3 V). Hence, by making a series of measurements on the device by the method outlined in [31], it is possible to predict the minimum device noise figure.

This expression indicates the importance of the gate length L and the parasitic elements, R_g and R_s, to the minimum noise figure. While decreasing the gate length obviously decreases the noise figure, it also increases the gate resistance due to the reduction in the gate metal cross-section. In order to compensate for this, the unit gate width (finger length) also has to be reduced in order to maintain the same gate metallization resistance. This often results in a complicated device structure due to the subsequent requirement for paralleling many gate fingers in order to maintain a favourable device impedance level. Another solution to this problem involves increasing the gate cross sectional area (forming a mushroom-like cross-setion) as described above.

Equation (15) relates the device minimum noise figure to elementary device model circuit elements, g_m, R_g, R_s, C_{gs}. However, Fukui [30] has shown empirically that it is also possible to relate the minimum device noise figure to geometical and material parameters of the device.

$$F_0 = 1 + fK \left[\frac{NL^5}{a} \right]^{1/6} \left[\frac{17z^2}{hL_g} \ (1+s) \ + \ \frac{2.1}{a_1^{0.5} N_1^{0.66}} \ + \ \frac{1.1L_2}{a_2 N_2^{0.82}} + \frac{1.1L_3}{a_3 N_3^{0.82}} \right]^{1/2} (16)$$

where the device parameters are defined in Fig. 4.32. K, the noise coefficient, is approximately 0.04 for MESFET devices, and the skin effect terms $s = 0.08 \ \sqrt{(fhL_g)}$. The first term inside the bracket represents the gate resistance including skin effect, while the last three terms are the source resistance. It should be noted that the minimum noise figure as given in equation (16) is independent of the device total gate width, but varies with the unit gate width (finger length) through the R_g term.

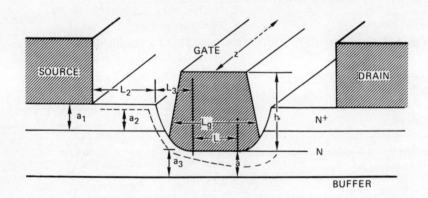

a   effective thickness of the active channel, in μm;
N   effective free-carrier concentration in the active channel, in 10^{16} cm^{-3};
z   unit gate width in mm;
h   average gate metallization height, in μm;
L_g average gate metallization length, in μm;
a_1 effective channel thickness under the source electrode, in μm;
N_1 effective free-carrier concentration in the channel under the source electrode in 10^{16} cm^{-3};
L_2, L_3 effective length of each sectional channel between the source and gate electrodes, in μm;
a_2, a_3 effective thickness of the sectional channel, in μm;
N_2, N_3 effective free-carrier concentration of the sectional channel, in 10^{16} cm^{-3}.

Fig. 4.32 Parameter definition for Fukui's noise equation.

Based on equation (16) above, the noise performance of 0.5 μm and a 0.25 μm gate length FET devices are plotted in Fig. 4.33. A summary of their parameters are contained in Table 4.5.

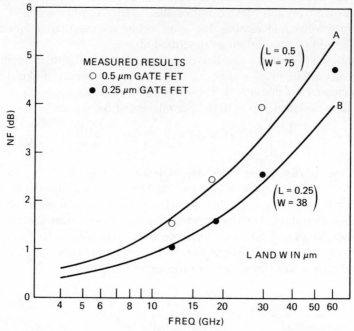

Fig. 4.33 Noise performance of 0.25 μm and 0.5 μm gate length MESFETs.

Table 4.5. Device and material design parameters for the 0.5 μm and 0.25 μm gate devices

Parameter	0.5 μm gate (A)	0.25 μm gate (B)
L (μm)	0.5	0.25
W (μm)	75.0	38.0
a (μm)	0.13	0.10
a_1 (μm)	0.36	0.32
a_2 (μm)	0.28	0.24
L_2 (μm)	0.5	0.5
h (μm)*	0.4	0.15
R_c (10^{-6} Ωcm²)	3.0	3.0
N (10^{17} cm⁻³)	2.5	2.5
K	0.034	0.034

* Effective height, as the gate cross section is not a rectangle

4.2.5.6 *Power FET design considerations*

Since the breakdown voltage is limited by fundamental material parameters (doping level), high power GaAs FET device operation can be realised by effectively connecting many small-signal devices in parallel. Critical to the performance of the power device is the manner in which the sub-FET elements are put in parallel. The basic building block of the power device is the unit gate width or finger length. Gain, output power and efficiency are all a function of the basic device geometry. Since the RF output power per mm of gate periphery is limited, the total gate width of the power FET basically determines the output power capability. Generally, an inter-digitated device geometry is employed with the an array of gate fingers separating drain-source electrodes as shown in Fig. 4.34. The fundamental parameters of the FET device are active layer doping and thickness, gate recess, gate length, unit gate width (finger length) and source-drain spacing and the gate-to-gate spacing. The former influences the power gain and the latter determines the thermal resistance.

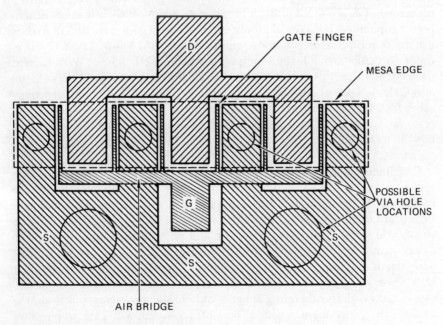

Fig. 4.34 Power MESFET geometry.

Channel doping level and thickness. The optimum doping level and channel thickness are a function of the desired frequency of operation. Experimental data from many laboratories at frequencies ranging from 4 to 16 GHz indicate that the MESFET is capable of a power density of approximately one watt of output power per mm of gate width [32] decreasing with doping level. However, for gain considerations the higher frequencies require a higher doping level. Based on experiments in several laboratories, the optimum doping level appears to vary from 8×10^{16} at 8

GHz to 1.6×10^{17} at 15 GHz to 2×10^{17} at 20 GHz. The optimum value of I_{DSS} is constant at approximately 350 mA per mm of gate width independent of the doping level and frequency. This implies that the channel thickness must decrease as $1/N$.

Gate recess. A recessed gate geometry improves the devices output power by reducing the parasitic source-gate resistance R_s. This in turn reduces the saturation voltage and increases the transconductance. For a channel doping level of 1.7×10^{17}, the optimum gate recess depth is in the 400 Å to 800 Å range. In general, it has been determined empirically that the recess depth should be 25% to 50% of the channel thickness [33].

Gate length. The gate length is the major parameter determining the device gain and thereby the operating frequency range, since for a given device the gain decreases at a rate of 6 dB per octave. In general, it is desirable to utilize the largest gate length consistent with the required gain at the operating frequency. Devices with shorter gates will of course have a higher gain but are more difficult to fabricate and have slightly lower output power capability. For typical power applications, a gain level of 5 to 6 dB is usually required. This generally implies a device MAG of 11 to 12 dB in order to achieve the highest output power and efficiency. With current device structures, this in turn requires a gate length of approximately 1.2 μm at 8 GHz, 0.8 μm at 10 GHz, and 0.5 μm at 15 GHz. Since there are many other factors that effect the device gain such as the source inductance, gate resistance and source resistance, it is often possible to extend a given gate length technology to higher frequencies by optimizing other device parameters.

Gate finger width. A power FET geometry generally consists of many gate fingers connected in parallel to a gate summing bar. Each gate finger is in reality an RC transmission line excited at one end. If the individual gate fingers are made too wide, the microwave signal propagating down the gate stripes will suffer excessive attenuation and phase shift. Consequently, the device gain and output power will suffer. This is particularly true for short gate length structures due to their small cross sectional area and resultant high resistance per unit length. However, in order to avoid a long device aspect ratio with its inherent geometry and phasing problems, it is desirable to make the gate fingers as wide as possible consistent with gate propagation losses.

Measurements confirmed by several labs indicate gate finger widths of up to 150 μm at X-band frequencies and 75 to 100 μm for Ku-band applications [34]. Based on a distributed model for the gate fingers, the calculated gain loss at 20 GHz due to the gate attenuation is shown in Fig. 4.35. The parameters of the device are summarized in Table 4.6.

4.3 AMPLIFIER DESIGN PRINCIPLES

A microwave amplifier usually consists of a cascade of several active devices with interstage and input/output matching networks as shown in Fig. 4.36.

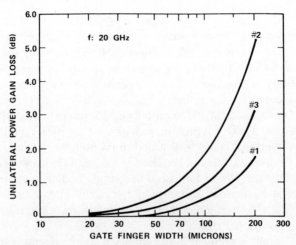

Fig. 4.35 Gain loss due to gate finger attenuation.

Table 4.6. Device parameters for data plotted in Fig. 4.35

	No. 1	No. 2	No. 3
Carrier concentration N (cm^{-3})	1×10^{17}	1.5×10^{17}	1.5×10^{17}
Channel thickness a (μm)	0.25	0.2	0.2
Gate length L_g (μm)	1.0	0.5	0.75
Gate height L_h (μm)	0.8	0.5	0.75
R (Ω/μm)	0.05	0.16	0.11
C (pF/μm)	1.0×10^{-3}	6.0×10^{-4}	7.5×10^{-4}
L (pH/μm)	1.0	1.8	1.5
r ($\Omega\mu$m)	1.5×10^3	1.0×10^3	1.0×10^3

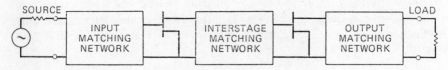

Fig. 4.36 Microwave amplifier configuration.

Depending on the application, the first several stages are designed to minimize noise contributions while the last several stages are optimized to provide maximum power, maximum efficiency or minimum distortion. The role of the input/output networks is to transform the system impedance (usually 50 ohms) to an impedance at the device which allows the FET to provide minimum noise on the input and maximum power or minimum distortion on the output. The inter-stage network transforms the output impedance of one stage to match the input of the next while providing the required 6 dB per octave gain compensation over the band of interest. The passive inter-stage and input/output networks also contain the bias and isolation networks. Some type of isolation is often required between stages to prevent interaction which can cause gain ripple and possibly oscillations. To ensure stability, inter-stage isolation is generally required between stages that are not unconditionally stable.

One type of isolation network commonly employed in hybrid MIC circuits is the quadrature 3 dB coupler in a balanced amplifier configuration. However, this approach is not well suited to monolithic applications due to the excessive GaAs substrate are required. Two amplifiers and two couplers are required per stage. When compared to a single-ended configuration, the balanced approach requires more than twice the substrate area. As a consequence, the single-ended configuration is usually employed in mono-lithic circuit applications.

The FET is usually employed in the common source circuit configuration due to its high gain and moderate input/output impedance levels. While other configurations are useful for special applications, the common source configuration is by far the most popular.

4.3.1 Stability and gain

An important parameter affecting the overall amplifier design is the device stability factor k. In terms of the device S-parameters, it can be expressed as

$$k = \frac{1 - |S_{11}|^2 - |S_{22}|^2 + |S_{11}S_{22} - S_{12}S_{21}|^2}{2|S_{21}| \, |S_{12}|}$$

The first step in an amplifier design is to examine the stability factor for all frequencies where the gain is greater than unity and hence oscillations are possible. The significance of the stability factor k is that for $k > 1$, the device is unconditionally stable and no combination of load and source reflection coefficients can cause oscillations. For this case, unique source and load reflection coefficients [35] exist which result in the maximum available gain MAG given by

$$\text{MAG} = \left[\frac{S_{21}}{S_{12}} \right] [k - \sqrt{k^2 - 1}]$$

On the other hand, if $k < 1$ certain combinations of load and source reflection coefficients can cause oscillations. For this case, there are several alternative design approaches. First of all, design techniques exist for designing 'conditionally' stable amplifiers. In such designs care must be taken to avoid certain regions of instability in the source and load reflection coefficient planes. This design method is adequately described in several references ([35]–[37]), and therefore, it will not be repeated here. Another approach, suitable for monolithic circuit applications, employs resistive matching elements to make $k > 1$. This method is particularly advantageous in that the resistive matching network can also be used for gain equalization. It is also an effective means of increasing the device stability at lower frequencies where it tends to be a problem. This second method will be described in greater detail later. A third method involves negative feedback with either purely reactive or lossy networks to achieve $k > 1$. Reactive feedback is particularly attractive for low noise applications, where it is possible to achieve a simultaneous noise and gain match [38].

The following steps summarize the amplifier design procedure:

(1) Determine the device S-parameters either by measurement or by model. Note that the hybrid circuit parasitics (such as bond wires) are not present in the monolithic medium. However, other parasitic elements such as the source inductance must still be included in the model.
(2) Calculate the device stability factor k and MAG or MSG versus frequency over a wide bandwidth. Stability problems often occur at frequencies below the desired band of operation.
(3) For $k > 1$, generate a unilateral model for the device and design the input/output networks as described below. Gain equalization may be included in the networks.
(4) For $k < 1$, either follow the procedure outlined in reference [36], or use lossy matching or feedback to create a device with $k > 1$. Then follow the procedure in (3).
(5) After arriving at the initial input/output networks, enter the actual device parameters and the matching networks in a commercial circuit analysis program (such as Supercompact or Touchstone) and compute the response over the desired band of frequencies. Include the bias networks in this simulation. Check the stability over a broad frequency range – particularly below the design band. Optimize the circuit.
(6) Layout the circuit on a CAD system and identify the parasitic elements (microstrip junctions and bends, via hole inductance, etc.). Include these parasitics in the analysis program and re-optimize the circuit to include them.

4.3.2 Bias networks

The design of the bias circuits for monolithic ICs is as important as the design of the matching networks. A good RF design becomes useless if the amplifier oscillates due to an improper bias network design. The bias circuit determines the device operating point (power or low noise), amplifier stability particularly at lower frequencies, temperature stability and often gain.

Depending on the application – low noise, high gain, class A power, class AB or B high efficiency – an optimum dc operating point exists. Table 4.7 summarizes some typical bias conditions. Low noise amplifiers operate at a relatively low drain-source voltage V_{DS} and current I_{DS}, typically $I_{DS} =$

Table 4.7. FET bias conditions

Application	Drain voltage V_{DS} (V)	Drain current I_{DS}
Low noise	2–3	$0.15 I_{DSS}$
High gain	4	I_{DSS}
High power (Class A)	8–10	$0.5 I_{DSS}$
High efficiency (Class AB or B)	6–8	0–$0.3 I_{DSS}$

$0.15I_{DS}$ and $V_{DS} = 3$ V. For higher gain, the bias point is adjusted upward to a higher I_{DS} level, often all the way to I_{DSS} for maximum gain. For power applications, the bias point must be shifted to a higher voltage (typically 8 to 10 V) and an I_{DS} level of approximately $0.5I_{DSS}$. For high efficiency Class AB or B operation, I_{DS} and V_{DS} must both be reduced from their high power values.

A selection of commonly employed bias networks is shown in Fig. 4.37. These circuits can be used for low noise, high gain, high power, and high efficiency applications. The bias circuit in Fig. 4.37 (a) uses separate supply voltages for the gate and the drain bias. In order to avoid unsafe operating conditions, the gate voltage V_G must be applied first and removed last. However, in spite of this complication, this network is the most popular particularly for power applications since it results in the lowest source inductance and hence the highest gain. The bias circuits in Figs 4.37 (b) and 4.37 (c) also use two bias supplies, and since the gate voltage must be applied before the drain, V_S and V_G must be applied first in Figs 4.37 (b) and 4.37 (c), respectively.

The bias circuits in Figs 4.37 (d) and 4.37 (e) require only one power supply, and as a result they are quite popular. This type of network energizes the gate, drain and source simultaneously, and thereby avoids the turn-on and turn-off problems of the two-supply networks. The bias point can be adjusted by varying the value of the source resistor. Due to the power dissipated in the source resistor, the efficiency of the amplifier is reduced. As a result, this approach is not usually employed with power amplifiers.

Due to the source resistor, this single supply approach has good protection from turn-on transients. Transient protection can also be achieved with the other two-supply circuits by employing long RC time constants in the positive supply and short RC time constants in the negative supply.

Since the source is not grounded directly in Figs 4.37 (b)–(e), a low inductance source bypass capacitor is required. The presence of source inductance can cause significant gain reduction particularly at higher frequencies and cause instabilities resulting in oscillations. Bias circuit oscillations can also arise due to insufficient bypass capacitance at the RF choke. These oscillations are usually in the video frequency range up to,

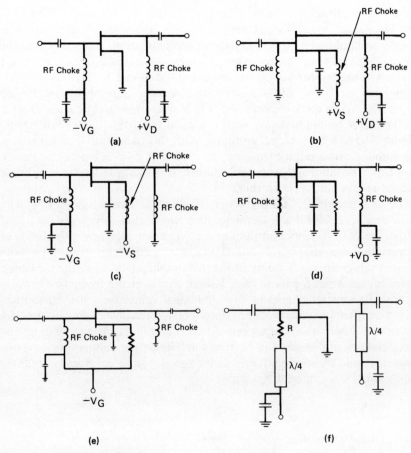

Fig. 4.37 FET bias circuits.

perhaps, several hundred megahertz. In addition to an on-chip bypass capacitor of typically 10 pF, a 100 to 1000 pF chip capacitor is usually required immediately adjacent to the GaAs chip. Another stabilization technique, which is more effective for instabilities near the operating band, utilizes resistance in series with the gate biasing circuit as shown in Fig. 4.37 (f).

4.3.3 Unilateral model

In Section 4.2.5 the equivalent circuit model for the GaAs FET was presented. In practice the model element values are derived from measured S-parameters by using computer optimization routines to 'fit' the model response to the measured results. This modelling method has been quite successful in predicting the small signal frequency response of GaAs FET devices at frequencies up to 60 GHz [39]. However, due to its complexity, this 14-element model is not very useful for amplifier design purposes. Usually a simplified model, a unilateral approximation, is employed in the circuit design process. The final network is then optimized with the complete 14-element model substituted for the unilateral model.

For current GaAs FET devices, the reverse transmission scattering parameter, S_{12}, is quite small ($S_{12} < 0.1$) at microwave frequencies. Hence, the unilateral assumption ($S_{12} = 0$) is a good approximation providing the stability factor k is greater than unity. With this unilateral approximation, the amplifier design problem is simplified into two independent design problems. The input and output matching networks are addressed separately without interaction between them.

The unilateral model for a GaAs FET device is shown in Fig. 4.38. The shunt and series feedback elements have been eliminated. As described below, the Q of the input/output equivalent circuits determine the limits of the resultant amplifier bandwidth. For a small signal gain stage, the model element values are obtained by fitting this simplified model to the measured device input/output S-parameters, S_{11} and S_{22}, over the frequency band of interest. In contrast, the input circuit element values for a low noise stage are derived from a noise model as described in Section 4.5. In the case of a power amplifier, the output circuit is determined from large signal considerations as described in Section 4.4. The output parameters for a low noise stage and the input parameters for a power stage are usually obtained from a small signal model.

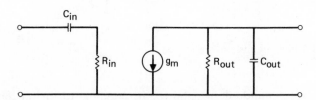

Fig. 4.38 MESFET unilateral model approximation.

With this unilateral model, the multistage amplifier design problem can be reduced to the three simple independent matching problems shown in Fig. 4.39. In each case the problem can be stated in terms of realising some desired driving-point impedance behaviour over the frequency band of interest.

While the form of these matching networks, and the basis on which the desired impedance characteristic is determined will certainly differ with the application, the method by which the networks are obtained is the same, regardless of the application.

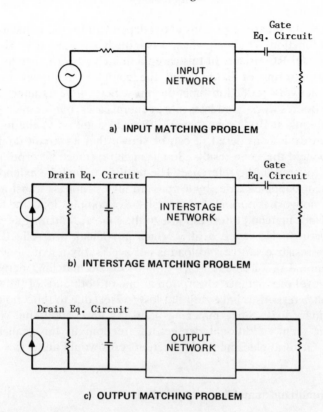

Fig. 4.39 Three basic matching problems of a multistage amplifier.

4.3.4 Q-bandwidth limits on impedance match

Before attempting to design a broadband matching network, the perform-ance limits must first be determined. The inexperienced circuit designer often attempts to obtain matching networks blindly by numerical optimiza-tion. If the desired performance level is not achievable, which is often the case, considerable computer and engineering time is wasted without any useful results.

The relative reactive to resistive (susceptive to conductive) behaviour of the load immittance sets the limits on achievable broadband performance. This behaviour is sometimes described in terms of a parameter called load-Q. Bode [40] showed that the integral of return-loss *vs* frequency is bound by a constant, and this constant is inversely proportional to load-Q. The area under this curve represents the value of the integral. If load-Q is decreased, the available area is correspondingly increased and the same level of matching performance is achievable over a wider band. If the load-Q is unchanged, a reduction in desired bandwidth will allow an improvement in matching performance. This constant is dependent on the behaviour of the reflection function. The load which was initially considered by Bode was a simple parallel RC circuit. In this case the match performance limit can be described in terms of load-Q and the complex frequency location of matching network reflection function zeros. Several years later, Fano [41] extended Bode's work to address more general cases, but for most purposes, a detailed look at Bode's work will suffice. In addition to the parallel RC case which Bode considered, it can be shown that its circuit dual, the RL case, also yield the same results. So all single reactance absorption lowpass cases are covered in the reference. The results can then be extended to the two element bandpass cases by lowpass to high bandpass transformation.

An RC network is normally used to absorb a complex load and to provide an impedance matched filter response to the source. Matching networks are filter structures. However, a filter which provides a low reflection match between a resistive source and load is not necessarily a matching network. The additional requirement which is imposed on matching networks is a required level of reactance absorption at one or both ends of the structure. Typical filter responses have zero flat-loss (offset) due to reflection zeros on the imaginary axis. Since matching networks have additional constraints placed on them, additional degrees of freedom in the realization are required. General placement of reflection zeros would allow this freedom.

4.3.5 Broadband matching network design

The design of lumped LC impedance matching networks is now considered. Applicability of the approach to be discussed is broad, since most practical applications can be reduced to the problem of obtaining a desired impedance behaviour at one or both sides of a two-port network.

It is important to note that there are two fundamentally different ways of addressing the LC matching network problem. One is the classic filter synthesis approach, which requires that a load model be available for absorption into the matching network through the synthesis process. The other approach operates directly on the interface impedance requirements without a model.

The classic filter synthesis approach to broadband impedance matching network design usually involves four activities:

(1) Approximation – a functional representation of the desired response,
(2) Realisation   – the network synthesis step which satisfies the approximation and absorbs the load model,
(3) Mapping   – a frequency domain transformation to the desired passband,
(4) Load model   – a one-port which is formed by an LC two-port which is resistively terminated.

Although most design requirements are for passbands which do not extend down to dc, the lowpass representation is very useful. Since frequency domain mappings can be used to extend approximations to other bands, it is not necessary to address the approximation problem separately for each type of network to be considered. In some cases, even the completed network realisations can be transformed directly. The lowpass (LP) to bandpass (BP) mapping is one such case.

It is clear that an optimum matching network will only provide good matching inside the desired pass-band. This is a consequence of the Bode analysis, since the area under the return-loss frequency response is fixed. One of several classic approximations can be applied here to achieve an abrupt transition between pass-band and stop-band. Usually the phase transfer characteristic is not a primary consideration. For cases where phase linearity is important, additional bandwidth margin can be included in the design or another approximation such as the Bessel or Gaussian can be used. Of potential interest here are the Butterworth, Chebyshev, and Elliptic responses. Each of these has been previously used for matching network design. However, the Chebyshev (equal ripple) response is by far the most popular. This is because it offers superior performance to the Butterworth, and is more easily realised than the Elliptic forms. Consequently, the sensitivity to element variations also falls between the Butterworth and Elliptic cases. Further details on network synthesis and analysis for filters and matching networks can be found in references [42] to [54].

4.4 POWER AMPLIFIER DESIGN

In the proceeding discussion, the general problem of designing impedance matching networks was considered. Systematic matching procedures were developed for the input, output and interstage matching networks. In this section the design of monolithic power amplifiers is presented, and these general matching principles are applied to the problem of power amplifier design.

Power amplifier design is considerably more complex than small-signal linear amplifier design due to the following factors:

(1) In order to realise high output power levels, a large gate periphery device is required. This large device results in low input/output impedance levels (typically an input impedance of 1 Ω/watt of output power). This in turn produces broadband matching difficulties and excessive matching circuit losses.

(2) The gain performance of a power amplifier is often prescribed for a range of power levels – not just the maximum output power level. Consequently, the design must be examined over the required power range and an adequate compromise reached. However, device data as a function of power level is usually quite limited, which makes it difficult to predict the subsequent gain variation.

(3) The FET device is operated in the non-linear region which results in the S-parameters being a function of the power level. Current methods of S-parameter characterization at high power levels are not adequate. With a linear synthesis procedure, gain ripple is attributed to the mismatch loss. However, with a power amplifier, due to nonlinar effects, output mismatch can create a greater power loss than linear theory can predict. For example, linear theory predicts that a 2:1 VSWR will produce a gain loss of 0.51 dB. However, with power amplifiers, a 2:1 VSWR on the output can often produce a power loss of 1 dB or greater.

4.4.1 Bias point

In contrast to low noise amplifier design, power amplifiers are generally designed to provide maximum power to a load at high efficiency. Generally speaking a design optimized for maximum output power will not provide the highest efficiency and *vice versa*. The device bias conditions and optimum load are quite different for maximum output power as opposed to maximum efficiency.

The bias point of the device has a major impact on the device output power and efficiency. Considerable insight on the effect of the bias point can be obtained by considering the device current–voltage (I–V) characteristic in conjunction with various load lines. Fig. 4.40 illustrates the drain family of characteristic curves. For drain voltages below V_{SAT}, the device is in the linear region where the drain current is approximately proportional to the drain voltage. For voltages greater than V_{SAT}, the device is in the saturated region where the device drain current is independent of the drain voltage. The saturation region is limited on the low side by V_{SAT} and on the high side by the avalanche breakdown voltage V_A. Maximum output power will occur when the load line allows maximum voltage and current excursions.

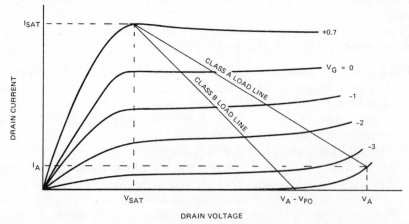

Fig. 4.40 Drain current–voltage characteristics with class A and class B load lines.

The load line for class A and class B operations are also shown in the figure. For class A operation, the FET device is biased at typically $1/2I_{SAT}$, and for class B the device is biased at or near pinch-off. While class A provides the highest power, class B operation generally results in the highest efficiency. For class B operation, the gate voltage swing is $2 (V_b + V_{po})$ where V_b is the built-in voltage and V_{po} is the pinch-off voltage. In class A operation the gate bias voltage is typically $\frac{1}{2} (V_{po} - V_b)$ and the total swing will then be $V_b + V_{po}$. Due to the higher voltage swing, the input power is up to 6 dB higher in class B operation than in class A for the same output power. This means that the gain of a class B stage is approximately 6 dB lower than a class A biased stage.

4.4.1.1 Class A operation

As shown in Fig. 4.40, in the saturated class A mode, the FET device is switched along the load line between the points (V_{SAT}, I_{SAT}) and (V_A, I_A). The resulting quiescent bias point for this condition is $V_D = (V_A + V_{SAT})/2$ and $I_D = (I_{SAT} + I_A)/2$. The avalanche voltage V_A is generally not well defined and hence it is often desirable to operate at a fixed bias voltage determined by other considerations such as reliability. The corresponding output power is given by

$$P_A = \frac{1}{8} (I_{SAT} - I_A) (V_A - V_{SAT}) \tag{17}$$

From the operating point, the input dc power is

$$P_{DC} = \frac{(V_A + V_{SAT}) (I_{SAT} + I_A)}{4} \tag{18}$$

and the resulting drain efficiency is

$$\eta_D = \frac{1}{2} \left(\frac{I_{SAT} - I_A}{I_{SAT} + I_A}\right) \cdot \left(\frac{V_A - V_{SAT}}{V_A + V_{SAT}}\right) \tag{19}$$

The power added efficiency is further reduced by the gain as $\eta_{add} = \eta_D (1 - 1/G)$ where G is the amplifier power gain. The efficiency of a class A amplifier is reduced from the classical value of 50% due to a reduced voltage and current swing. The voltage swing is reduced by V_{SAT} and the current swing by I_A. V_{SAT} is usually in the 1.5–2 V range while V_A varies from 15 to 30 V depending on the channel geometry and doping level. In general, the impact of V_{SAT} is to limit the maximum class A efficiency to approximately 45%. While the saturation voltage has an effect on efficiency, the dominant mechanism appears to be the avalanche current [55]. If this is true, reducing the drain voltage should reduce avalanche current and thereby increase the efficiency. On the other hand, if the saturation voltage is the limiting mechanism, one would expect that reducing the drain voltage would only decrease the efficiency. Fig. 4.41 illustrates the effect of drain voltage on efficiency for one cell (0.8 × 1000 μm) Hughes 4080 device. Clearly, the drain efficiency increases with decreasing drain voltage lending credibility to the avalanche current argument.

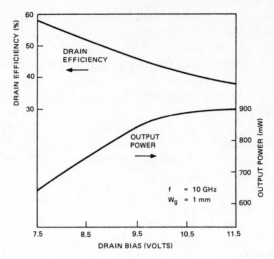

Fig. 4.41 Drain efficiency and output power as a function of the drain bias voltage.

4.4.1.2   Class B operation

While class A operation provides the highest power, class B or more accurately, class AB usually results in the highest efficiency. True class B operation is rarely used with FET devices due to g_m compression near pinch-off and the resultant low device gain. This is over and above the ~6 dB gain loss, compared to class A, due to the greater input voltage swing requirement. The output power is lower with class B due to the increased (negative) gate voltage, which reduces the available voltage swing at the drain. The maximum voltage between the gate and the drain is limited by the gate-drain breakdown voltage V_B. With $V_A = V_B - 2 V_{PO}$ and $I_A \sim 0$, the maximum power output can then be expressed as

$$P_B = I_{MAX} (V_B - 2 V_{PO} - V_{SAT})/8 \qquad (20)$$

Comparing this to the class A output power (Equation (17)) with $V_A = V_B - V_{PO}$

$$P_A = I_{MAX} (V_B - V_{PO} - V_{SAT})/8 \qquad (21)$$

The difference, $I_{MAX} (V_{PO}/8)$, can be a substantial quantity.

The maximum drain efficiency for class B operation is given by

$$\eta_D = \frac{\pi}{4} \left(\frac{I_{SAT} - I_A}{I_{SAT} + I_A} \right) \cdot \left(\frac{V_A - V_{SAT}}{V_A + V_{SAT}} \right) \qquad (22)$$

Again the drain efficiency is reduced from the theoretical limit (78.5%) for a classical class B amplifier by the finite values of I_A and V_{SAT}. The effect of the bias point on efficiency is illustrated in Fig. 4.42 for a Hughes 4080 device operating at 10 GHz. As shown in the figure, the gate voltage is varied from –2 V, corresponding to class A operation, to –5 V where class B operation is obtained. Due to the reduction in gain, the power added

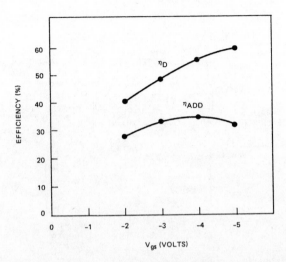

Fig. 4.42 Drain and power-added efficiency as a function of the gate voltage.

efficiency reaches a maximum for a gate voltage of approximately –4 V. However, the drain efficiency continues to increase with negative gate voltage all the way to pinch-off (class B operation).

4.4.2 Large signal characterization

A number of methods are currently being used to determine the output impedance under large signal conditions. One can of course use the load-pull method where the device is tuned with variable tuners to the desired performance level. The optimum load can then be determined by measuring the output network containing the tuners. This method is time-consuming since it has to be done on a narrow band basis separately for each frequency and requires a well controlled repeatable circuit environment. Despite these shortcomings, this method usually produces reliable results. It is currently being used at frequencies up to 18 GHz with relatively small tuner losses (< 0.5 dB) and errors. For production applications, where a large number of measurements are necessary, a fully automated tuning set has been reported by Cusack *et al.* [56].

An improved method by Takayama [57], shown in Fig. 4.43, generates the tuner electronically by feeding some of the input power back into the output. This incident signal can be varied in phase and amplitude and thus arbitrary load impedances can be generated electronically. The advantage is the elimination of mechanical tuners which do not allow orthogonal tuning and which create reproducibility problems. However, this method has a basic problem of accurately determining the output power for output reflection coefficients close to unity, as it is apt to be for large unmatched

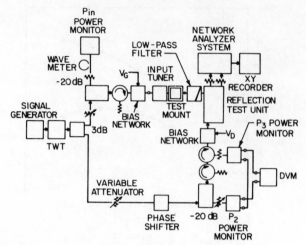

Fig. 4.43 Electronic load-pull system (after Takayama).

devices. This method lends itself particularly well to computer control, since it does allow orthogonal tuning.

4.4.3 Push–pull configuration

The matching networks for broadband power amplifiers must provide both reactance absorption at the device side and impedance transformation between the 50 Ω system impedance and the optimum device load. A design which requires less impedance transformation is clearly more attractive for monolithic integration.

Large power FETs are comprised of many smaller FETs connected in parallel. The result is a reduction in impedance level by a factor of $1/N$, for N paralleled FETs. For example, the total input resistance of a 4 mm gate width FET is of the order of 0.6 ohms. Instead of parallel combining FET segments or cells, pairs of segments can be effectively operated in series by employing the push-pull configuration. This approach, illustrated schematically in Fig. 4.44, has been applied to monolithic power amplifiers.

The advantages of the push-pull configuration over that of single-ended operation are summarized as:

(1) Four times higher impedance level for a given output power level. Hence, circuit losses are reduced and efficiency is increased.
(2) The common lead impedance is not in the signal path. So, gain and stability are improved.
(3) Even harmonics are suppresed from the output.
(4) Second harmonic reactive tuning for improved efficiency can be easily implemented in the 'common lead'. Fundamental currents in the common lead cancel; hence, second harmonic tuning does not affect fundamental tuning and *vice versa*.

In comparison to a single-ended approach, the push-pull configuration inherently provides lower losses and higher gain. Both effects have a beneficial effect on power-added efficiency. Large power FETs require very low impedance interfaces. As described above, the input resistance of a 4 mm (conventional parallel) FET is of the order of 0.6 ohms. Correspondingly, the output parallel loading level is approximately 12 ohms. For a given total gate width, the push-pull combined impedances are greater by a factor of four. Hence, this approach yields the impedance levels of a 1 mm device from a device four times that size. The significance of this is clear when monolithic circuit losses are considered. Power can be delivered more easily and efficiently into 2.4 ohms than into 0.6 ohms.

The source inductance has a major effect on the device gain. Large conventional parallel-combined power FETs have a source inductance of the order of 0.05 nH for each mm of gate width. This requires many via holes through the substrate to minimize the source inductance to an acceptable level. Alternatively, the push-pull configuration offers much lower source inductance in the signal path. This is due to the fact that the signal currents flow only in the odd mode, while most of the source inductance resides in the even mode (common mode). In comparison to the single-ended approach, the potential gain improvement ranges from 1 dB in X-band to more than 2 dB at Ku-band frequencies. In addition, the reduced source inductance results in less sensitivity to C_{gs} variation, thereby improving the amplifier yield.

The push-pull configuration has often been credited with greater

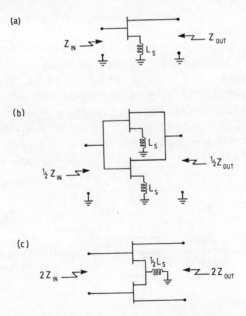

Fig. 4.44 Push-pull offers a 4:1 impedance advantage over parallel combined cells.

bandwidth potential. Strictly speaking, this is not true. Because of the 4:1 impedance advantage with push-pull, the need for impedance transformation is reduced. In principle, however, two real terminations can be 'matched' over arbitrary bandwidths. The fundamental bandwidth limiting parameter is load Q. Push-pull operation does not change the Qs. As real parts are raised, so are series and shunt reactances. The push-pull configuration, because of the 4:1 impedance advantage, does offer reduced circuit losses with practical networks. As the impedance transformation ratio is reduced, the monolithic integratability of the circuit is improved. This is the significant point here. However, if the reactance level (load Q) is too great for the desired bandwidth, push-pull cannot help.

4.5 LOW NOISE FET AMPLIFIER DESIGN

The noise performance of a low-noise FET amplifier is determined by both the device noise parameters and the input/output matching networks. While the performance limits are set by the FET device, it is the job of the matching networks to extract that potential performance over the frequency band of interest. As such, the matching networks determine the amplifier bandwidth, control the gain magnitude and flatness, and minimize the noise figure over the desired bandwidth. In this section, the interaction of the device noise parameters with the input/output matching networks is examined.

4.5.1 Noise match

A low-noise FET amplifier stage is shown schematically in Fig. 4.45. For the following discussion, this stage is considered to be the first stage of a multistage amplifier. The role of the input and the output matching networks are quite different. The input matching network M_1 is required to transform the system input impedance (usually 50 ohms) to an impedance at the device input which minimizes the device noise contributions, while the output network must extract the maximum device gain, i.e., network M_2

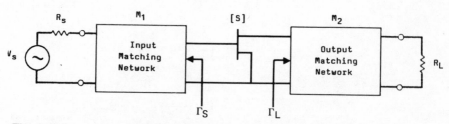

Fig. 4.45 Low noise amplifier stage.

must conjugately match the resulting device output impedance. The effect of the input reflection coefficient on the stage noise figure can be expressed as

$$F(\Gamma_s) = F_{min} + 4r_n \frac{|\Gamma_s - \Gamma_o|^2}{(1 - |\Gamma_s|^2) |1 + \Gamma_o|^2} \tag{23}$$

where

F_{min} = minimum noise figure
r_n = normalized equivalent noise resistance = $R_n/50$
Γ_s = the source reflection coefficient
Γ_o = the optimum source reflection coefficient for F_{min}

This equation expresses the relationship between the stage noise figure and the input reflection coefficient, Γ_s. When $\Gamma_s = \Gamma_o$, the minimum noise figure F_{min} results. The three parameters F_{min}, R_n, and Γ_o are referred to as the device noise parameters and are functions of the FET device and its bias point. The noise figure and the associated gain are plotted, as a function of the drain bias current, in Fig. 4.46 for a typical 0.5 μm gate MESFET device. Note that the minimum noise figure and the maximum gain occur at different bias points. This fact creates a tradeoff in the bias point as described below. R_n is a measure of the sensitivity of the device to the source impedance. For a broadband design, a small value of R_n is desired.

Equation (23) is plotted in Fig. 4.47 for a typical 0.5 μm MESFET device. The optimum source reflection coefficient and circles of constant noise figure are plotted on the input reflection coefficient, Γ_s, plane. Also indicated in the figure is the reflection coefficient for maximum gain (conjugate match). In general, for the common source configuration, the reflection coefficient for minimum noise is not the same as the reflection coefficient for maximum gain. This implies that when the device is matched for minimum noise, the amplifier input reflection coefficient will not be zero. This input mismatch is characteristic of low-noise, common source, FET amplifiers matched for minimum noise. With hybrid MIC amplifiers, a 'balanced' amplifier configuration is usually employed to eliminate the input reflection. However, the balanced configuration is not practical for monolithic circuit applications due to its size requirements. For MMIC designs, other techniques such as series inductive feedback [58] must be used to reduce the input VSWR.

While the input network provides the optimum impedance for minimum noise figure, the role of the output network, M_2, is to extract the maximum device gain. In order to accomplish this, F_2 must conjugately match the output impedance of the FET over the band of interest. The required load reflection coefficient can be expressed in terms of the input reflection coefficient and the device S-parameters as

$$\Gamma_L = (S'_{22})^* \tag{24}$$

where

$$S'_{22} = S_{22} + \frac{S_{12}S_{21}\Gamma_s}{1 - S_{11}\Gamma_s} \tag{25}$$

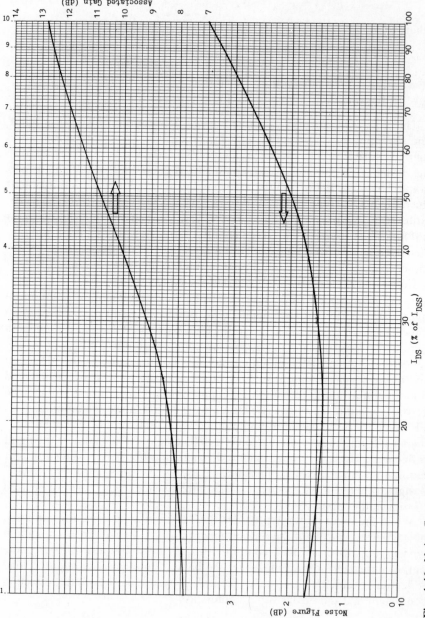

Fig. 4.46 Noise figure and associated gain as a function of the drain bias current for a typical 0.5 μm MESFET device at 12 GHz.

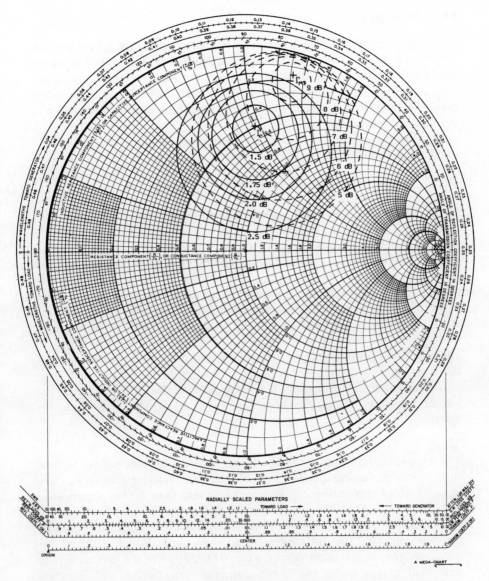

Fig. 4.47 Noise figure circles (solid curves) and available gain circles (dashed curves) for a typical 0.5 μm MESFET device at 12 GHz.

The resulting device gain, termed the available power gain, is given by

$$G_A = \frac{|S_{21}|^2 (1 - |\Gamma_s|^2)}{|1 - S_{11}\Gamma_s|^2 (1 - |S'_{22}|^2)} \tag{26}$$

and is a function only of the input reflection coefficient and the device S-parameters.

4.5.2 Stage gain

While the stage noise figure is obviously important, the stage gain also plays an important role in the noise figure of a multistage amplifier. This is due to the fact that stages other than the first also contribute to the noise in the amplifier. While the noise figure of the overall amplifier is determined primarily by the first stage, the second and third stages often make a significant contribution.

This can be seen by examining the expression for the noise figure of a multi-stage amplifier.

$$F_T = F_1 + \frac{F_2 - 1}{G_1} + \frac{F_3 - 1}{G_1 G_2} + \ldots \tag{27}$$

In this equation, F_1, F_2 and F_3 are the noise figures of the first three stages and G_1, G_2 and G_3 are the corresponding available gains. While the first stage noise figure adds directly to the total noise figure, the noise contribution of the second stage is reduced by the gain of the first and the noise contribution of the third stage is reduced by the gain of the first and second stages. For example, in a multistage amplifier with a stage noise figure of 2 dB and an available gain of 6 dB, the stages following the first stage contribute approximately 0.5 dB to the total multi-stage amplifier noise figure. In general, it is important to design and bias the stages to minimize the overall amplifier noise figure and not just the noise figure of the first stage.

An important figure of merit for a low noise device, which includes the impact of gain, is the cascaded noise figure (often called the noise measure) given by

$$F_c = 1 + \frac{F - 1}{1 - 1/G} = \frac{F - 1/G}{1 - 1/G} \tag{28}$$

This expression represents the noise figure of an infinite cascade of identical stages each one having a noise figure of F and an available gain G. For low noise amplifer applications, F_c is more important than F. The significance of this figure of merit is that it represents the lower limit of noise performance for a multistage amplifier. While in practice the first several stages are operated for minimum F_c, the last several stages are biased and matched for maximum gain or maximum power. Due to the impact of the gain on F_c, the individual stages of a multistage amplifier are biased and matched (designed) to minimize the cascade noise figure and not simply the noise figure. MESFETs are typically biased at $0.1-0.2$ I_{DSS} while HEMTs are usually operated in the enhancement mode (positive gate voltage) for minimum cascaded noise figure.

4.5.3 Noise parameters

The device noise parameters F_{min}, R_n and Γ_o and their frequency dependence are critical to the sucessful design of low noise FET amplifiers.

F_{min} can be determined from a knowledge of the device geometry as described in Section 4.2.5, or it can be determined empirically, in a hybrid MIC medium, by 'tuning up' an amplifier stage for minimum noise figure. While such a procedure has several sources of potential error, at frequencies below 18 GHz, the test fixture and the circuit losses can be measured with sufficient accuracy to produce reasonable results. This device characterization procedure has even been extended to 60 GHz [59].

However, R_n and Γ_o are more difficult to obtain. In principle these three parameters can be determined with the aid of equation (23) and at least seven independent measurements of the noise figure with seven different but known source reflection coefficients. This data provides a system of over-constrained equations with four unknowns (Γ_o is complex). The noise parameters are then obtained by using curve-fitting techniques. This operation must be repeated at each frequency of interest. A variation of this method has been implemented with a varactor tuned source [60].

Podell [61] has proposed a simplified noise model for the FET which predicts the device noise parameters based on a knowledge of F_{min} at one frequency and the device equivalent circuit. The basic procedure is summarized in Fig. 4.48. The FET input circuit is represented as a simple series RC circuit with R equal to the sum of R_g, R_i and R_s and $C = C_{gs}$. This information is obtained from the equivalent circuit model or simply from S_{11} data. The optimum source admittance (for minimum noise) is derived from this series RC network by first transforming this series circuit to its parallel equivalent.

$$g_1 = \frac{1}{(Q_1^2 + 1) R_1}$$

$$b_1 = \frac{\omega C_1}{1 + 1/Q_1^2}$$

where

$$Q_1 = \frac{1}{\omega C_1 R_1}$$

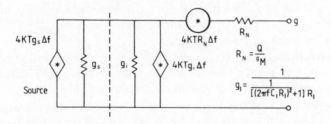

Fig. 4.48 Podell's noise model.

The optimum source admittance, $y_o = g_o + jb_o$, is found to be related to this parallel equivalent circuit by

$$g_o \quad = g_1 \sqrt{1 + 1/A}$$

$$b_o \quad = -b_1 \quad = -\frac{\omega C_1}{1 + 1/Q_1^2} \approx -\omega C_1$$

$$R_n \quad = A/g_1$$

where R_n, the equivalent noise resistance, is assumed to be dependent only on the device bias conditions, and A is related to the minimum noise figure as

$$A = \frac{(F_{min} - 1)^2}{4 F_{min}}$$

Note that while the optimum source admittance conjugately matches the imaginary part of the device input admittance, it presents a real part g_o which is larger than g_1 by the factor $\sqrt{1 + 1/A}$. This implies that the matching circuit Q for minimum noise figure is also lower by the same factor. The implication of a lower Q is of course greater bandwidth potential.

This model predicts a noise figure which to first order increases with frequency as $F = 1 + Bf$. This is in basic agreement with Fukui's model presented in Section 4.2.5.

This simplified model produces several other interesting conclusions including predicting the degree of mismatch under noise match conditions and an expression relating the associated gain to MAG.

$$|\Gamma| \quad = \frac{1}{F_{min}}$$

$$\frac{G_A}{MAG} \quad = \frac{F^2_{min} - 1}{F^2_{min}}$$

Note that for very low noise figures ($F_{min} \to 1$), the input reflection coefficient approaches unity and the ratio of the associated gain to MAG approaches zero. However, these expressions are only approximations since they neglect, among other things, source inductance and drain-to-gate feedback. It is well known that the source inductance increases the real part of the input impedance by a factor of approximately $g_m L_s/C_{gs}$. This is a negative feedback effect which has no affect on the device noise measure ([62], [63]). In fact it has been shown that with a combination of both source (series) and drain-gate (shunt) feedback it is possible to simultaneously noise and power match a device at one frequency [62].

The significance of this theory is that the device noise parameters can be predicted over a broad range of frequencies based on only a knowledge of F_{min} at one frequency and the device input equivalent circuit.

4.5.4 Design procedure

The matching network considerations discussed in Section 4.3.5 can be used to determine the required complexity of the input matching network for a low noise stage. With the unilateral device approximation, the input impedance to be matched for optimum noise figure or noise measure can be accurately modelled over a large frequency range by a constant value series or parallel RC circuit. The output impedance is accurately modelled as a parallel RC circuit which could also be optimally matched for minimum reflection coefficient over the band. If this were done, the gain of the stage would slope at approximately 6 dB per octave due to the inherent gain roll-off of the FET. To achieve the desired flat gain characteristic, the output matching network must be designed to compensate for this by absorbing or reflecting an increasing fraction of power as frequency decreases.

The input matching circuit can be developed as follows:

(1) The required complexity of the matching network is determined from the Q of the input impedance for noise matching. A knowledge of the maximum mismatch which can be tolerated while meeting the noise specification is required. This information is derived from the device noise parameters.
(2) A network topology is selected and the element values are calculated.
(3) The network element values are optimized using the complete device equivalent circuit and known parasitic elements. The circuit layout is accomplished with the aid of a CAD system.

When the input matching circuit for the FET device is matched to minimize noise measure, the associated available gain from the FET will have a gain-frequency slope of 5 to 6 dB per octave due to the intrinsic device roll-off. This is the response which would result if the FET output were conjugately matched at each frequency in the band. To achieve flat gain response over the desired band, the output matching network must compensate for this roll-off. The output network must be designed to provide the necessary impedance matching to achieve the maximum associated gain at the upper end of the desired frequency band. It must also provide a frequency dependent gain equilization by mismatch or resistive attenuation to compensate for the FET gain slope at lower frequencies. The choice of reactive or resistive compensation is dependent on the application. In a high gain, single-ended amplifier, where a minimum requirement for additional isolation is desired, the resistive compensation is preferred. Since power is absorbed rather than reflected, problems associated with inter-stage interaction are reduced.

4.5.5 Active matching

The design of GaAs monolithic circuits is generally a compromise between performance, chip size, power consumption and overall yield. All these factors must be considered during the design phase. Furthermore, the sensitivity of the circuit performance to processing variations, material parameters, design accuracy and repeatability is critical to the success of the effort. By employing active devices as impedance transformation elements (termed 'active matching'), it is possible to develop a design which is more tolerant of component variations and results in a smaller chip size. This in turn results in higher yields and ultimately lower production costs.

The advantages and disadvantages of active matching are summarized in Table 4.8. The smaller chip size results from the use of active components for matching rather than large passive LC matching networks. The active components can include a common gate configuration for the input matching, a common drain configuration for the output matching and the common source configuration with or without feedback for gain. Active matching is in general less sensitive to component and process variations due to the fact that gain is achieved by employing active devices rather than critical reactive matching. This also results in lower Q requirements for the components which in turn implies smaller component size and a thinner metallization pattern.

Table 4.8. Active matching considerations

Advantages	Disadvantages
(1) Smaller chip size	(1) Non-optimum performance
(2) Less sensitive to parameter variations	(2) Requires submicron gate length FETs
(3) Higher overall yield	(3) Higher power consumption
(4) Lower cost	

Three excellent examples of active matching designs are Rockwell's 0.7 to 9 GHz amplifier [64], Plessey's 2 to 4 GHz amplifier [65], and Hewlett Packard's 5 MHz to 3 GHz instrumentation amplifier [66]. The Rockwell amplifier demonstrated a gain of 7 ± 1 dB over the 0.7 to 9 GHz band with a chip size of 2.5 by 2.5 mm. The Plessey amplifier achieved a gain of 19 dB over the 2 to 4 GHz frequency range with a chip size of 2.2 by 3.0 mm (86.6 $\times$ 118.1 mils). This represents a 60% reduction in chip area compared to a passively matched amplifier with similar (only 9 dB gain) performance characteristics. The HP amplifier demonstrated a gain of 26 dB with a chip size of only 0.44 by 0.65 mm (17.3 $\times$ 25.6 mils). These three examples illustrate the ability of active matching techniques to reduce the chip size.

One of the disadvantages of active matching is that it results in non-optimum performance from the amplifier. A common gate stage used for input matching will result in a higher noise figure and lower associated gain than a corresponding common source stage with LC matching. However, the common gate stage will, in general, have a better VSWR over a broader bandwidth.

Another disadvantge is that active matching requires higher gain devices. This in turn implies a shorter gate length in order to achieve the higher f_T. The higher gain is required to compensate for the lack of reactive matching.

An additional disadvantage of active matching is higher power consumption. Since resistive elements, as opposed to inductors, are used for bias insertion, power is dissipated in these elements. While this power dissipation is offset somewhat by the smaller bias currents required with the smaller gate periphery actively matched devices, the passively matched approach probably has an advantage in power dissipation.

The three possible FET configurations are summarized in Table 4.9. They are generated by alternately grounding one of the three terminals (source, gate and drain) of the FET. The voltage gain and the transducer power gain expressions are dc values, and hence are valid only at low frequencies. The common source configuration exhibits the highest gain while the common gate configuration can provide a broadband input match and the common drain configuration can provide a broadband output match.

Table 4.9. Properties of the three FET configuration

Device parameter	Common source	Common gate	Common drain
Input admittance	Low	g_m	~ 0
Output admittance	Low	~ 0	g_m
Voltage gain, A_v	$-g_m R_L$	$g_m R_L$	$\dfrac{g_m R_L}{1 + g_m R_L}$
Transducer power gain, G_T	$G_o{}^*$	$\dfrac{G_o{}^*}{(1 + R_S g_m)^2}$	$\dfrac{G_o{}^*}{(1 + R_L g_m)^2}$
Typical G_T (calculated)	12 dB	6 dB	2 dB
		($R_L = 200\ \Omega$)	($R_S = 100\ \Omega$)
Measured G_T at 8 GHz			
(a) 50 Ω source and load	9 dB	-0.6 dB	1.2 dB
(b) New source or load (calculated from data)	—	4.4 dB	2.5 dB
		($R_L = 200\ \Omega$)	($R_S = 100\ \Omega$)

$^*\ G_o = 4 R_S R_L g_m{}^2$

Common source configuration. The conventional common source (CS) configuration has low input and output admittances and high gain in a 50 Ω system. The gain values for the CS configuration in Table 4.9 are representative of a 300 μm gate width device. The input and output admittances are capacitive with a series RC circuit representing the input and a parallel RC circuit representing the output. The CS configuration also exhibits excellent noise characteristics. The high gain, low noise and moderate impedance level of the CS configuration account for its popularity.

Common gate configuration. The characteristics of the common-gate (CG) configuration, as an active matching element, has been analysed by Niclas [67]. The CG configuration exhibits a broadband resistive input, a very high output impedance and low gain in a 50 Ω system. The input admittance is approximately equal to the device g_m and this approximation holds over a wide bandwidth. The output admittance has virtually no real part and can be represented by a small capacitor. While the CG confituration exhibits little gain in a 50 Ω system, due to its high output impedance, it can provide significant gain when operated with a higher resistance (200 Ω) load. The calculated and measured gain values in Table 4.9 are based on a 150 μm gate configuration and are not as good as a CS configuration. The CG configuration can provide a wieband input match with a reasonably good noise figure.

Common drain configuration. The common drain (CD) configuration is potentially useful as an output stage. It has a capacitive input admittance with a near zero real part and an output admittance approximately equal to g_m. While the CD configuration has good output matching capability, it exhibits low gain. With a source impedance of 100 Ω and a load of 50 Ω, the theoretical low frequency gain is only 3 dB. The device input capacitance limits the source impedance to approximately 100 Ω without reactive matching. No improvement in gain is realized by further raising the source impedance. Due to the low gain of this configuration, it is of limited value for monolithic circuit applications.

4.6 DISTRIBUTED AMPLIFIER DESIGN

The distributed amplifier (DA) approach, first proposed by Percival [68] in 1937, has enjoyed a renaissance in the past several years due to the emergence of the GaAs FET and monolithic circuits ([69]−[71]). The topology of the DA is particularly suited to MMICs because its passive circuit predominantly consists of inductors which can be realised in the form of short lengths of microstrip line. Applied originally to electron tubes, this approach has the unique capability of adding device transconductance without adding device parasitic capacitance. This is accomplished by incorporating the parasitic shunt capacitances of the FET device into an artificial low-pass transmission line. The bandwidth is now, ideally, limited only by the cut-off frequency of the periodic structure.

The design of the DA involves a careful choice of the FET device parameters, the number of active devices or sections, the impedance of the microstrip segments and the cut-off frequency of the artificial transmission lines. These parameters are examined in the following discussion.

4.6.1 Distributed amplifer configuration

The distributed amplifier approach, shown schematically in Fig. 4.49,

OUTPUT TRANSMISSION LINE

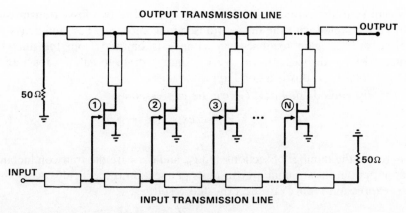

Fig. 4.49 Distributed amplifier approach.

consists of an input and output transmission line coupled by FET devices spaced uniformly on the lines. The input transmission line is periodically loaded by the FET gate-source capacitance and is terminated in its characteristic impedance. Similarly, the output transmission line is loaded by the FET drain-source capacitance but requires additional capacitive loading to equalize the characteristic impedances and velocities of the two lines. This is accomplished by employing a transmission line (TL) in series with the drain. The series TL is used for gain peaking at the high end of the band to compensate for losses on the input TL.

The device input/output capacitance is incorporated into the transmission line per-unit-length capacitance and hence is no longer a frequency limiting parasitic. This DA approach allows, in effect, the FET device g_m to be paralleled without the corresponding paralleling of the device input/output capacitance. The net result is increased bandwidth.

4.6.2 Amplifier circuit analysis

The equivalent circuit of the DA circuit, shown in Fig. 4.50, consists of an input transmission line formed with the gate-source circuit and an output

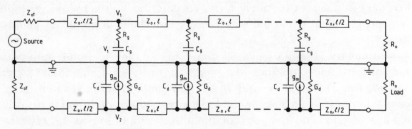

Fig. 4.50 Distributed amplifier equivalent circuit.

transmission line formed with the drain circuit. The two lossy transmission lines are coupled with the device transconductance, g_m. For simplicity, the device drain-to-gate feedback capacitance is omitted from the analysis. Since FET devices have very small drain-to-gate capacitances ($C_{gd} < 0.03\ C_{gs}$), this is a good approximation.

The distributed amplifier, G_T can be expressed as

$$G_T = |g_m|^2\ Z_o^2\ \frac{n^2}{4} \exp{}^- (\alpha_1 + \alpha_2)^n \tag{29}$$

where n is the number of sections and g_m and $\alpha_{1,2}$ are the transconductance and input/output attenuation constants per section respectively. As $\omega \to 0$, this expression reduces to the expected result.

$$G_{T\ (\omega \to 0)} = \left(\frac{n\ g_{mo}}{\dfrac{2}{Z_o} + nG_d}\right)^2 \tag{30}$$

The above equations illustrate an important characteristic of the distributed amplifier. The transducer gain increases as the square of the number of sections, n, for small values of n. However, due to the finite losses on the input/output lines, the exponential term eventually dominates resulting in a critical number of sections for maximum gain. Beyond this point, further increases in length (n) actually result in a gain reduction. The maximum gain occurs when the exponent in the gain equation (equation (29)) is equal to 2, or

$$n_{max} = \frac{2}{\alpha_1 + \alpha_2} \tag{31}$$

and the corresponding expression for the maximum gain is

$$G_{Tmax} = \left(\frac{|g_m|Z_o\ e^{-1}}{\alpha_1 + \alpha_2}\right)^2 \tag{32}$$

In addition to predicting the optimum length of the distributed amplifier circuit, the gain equation can also be used to predict the frequency response of the distributed amplifier. In equation (29), the only frequency dependent terms are $|g_m|$, α_1 and α_2. For submicron gate FET devices $|g_m|$ is only a mild function of frequency. For example, with a gate RC time constant of 1 ps, $|g_m|$ is down only 0.4 dB at 50 GHz. α_2 is only slightly a function of frequency. α_1, the input transmission line attenuation coefficient, is the dominant frequency dependent term. As described in the following section, α_1 is proportional to ω^2 and is responsible for the high frequency gain roll-off.

In summary, the presence of finite losses on the input and output transmission lines modifies the classical distributed amplifier n^2 gain dependence. These losses result in an optimum number of sections (length) for maximum gain. Increasing the length beyond this optimum point results in a loss of gain. The losses, predominantly the input line losses, effectively determine the frequency response of the distributed amplifier.

4.6.3 Transmission line parameters

As described in the previous section, the propagation constants of the input/output transmission lines play a major role in determining the maximum gain and frequency response of the DA. In this section, the characteristic impedance and complex propagation constant of a periodically loaded transmission line are presented.

Each section of the distributed amplifier can be represented by the lumped-distributed network shown in Fig. 4.51. This section is repeated n times forming an input/output transmission line. The periodic shunt admittance, Y_1, is separated by a length, l, of transmission line with a characteristic impedance Z_o. For the input line Y_1 consists of the gate series RC equivalent circuit, and for the output line Y_1 consists of the drain parallel RC circuit.

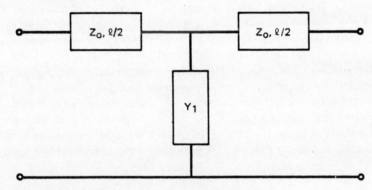

Fig. 4.51 Lumped-distributed transmission line section.

The resulting expessions for the characteristic impedance and the propagation constant of the overall network (section) are

$$Z_{OT} = Z_o \left[\frac{2 + Y_1 Z_o \tanh \gamma l/2}{2 + Y_1 Z_o \coth \gamma l/2} \right]^{1/2} \tag{33}$$

$$\gamma_T = \cosh^{-1} \left(\frac{\cosh \gamma l + \dfrac{Y_1}{2Y_o} \sinh \gamma l}{} \right) \tag{34}$$

where the T subscript has been added to distinguish the parameters of the section from those of the TL elements.

These expressions are exact including the effect of line losses. However, to provide insight into the dominant processes, several simplifying assumptions are necessary. First, the TL segments are assumed to be lossless. Secondly, Y_1 is assumed to be a lossy but 'high–Q' ($\text{Im}[Y_1] \gg \text{Re}[Y_1]$) capacitor, C_g. Under these assumptions Equation (33) reduces to

$$Z_{OT} = Z_o \left[\frac{1 - \dfrac{\phi \, C_g}{2 \, Cl} \tan \dfrac{\phi}{2}}{1 + \dfrac{\phi \, C_g}{2 \, Cl} \cot \dfrac{\phi}{2}} \right]^{1/2} \tag{35}$$

where ϕ is the electrical length of the TL segments. As indicated by the above equation, Z_{OT} is a function of frequency becoming imaginary for sufficiently large ω. This expression can be used to determine the cut-off characteristics of the input/output transmission line by solving for the frequency for which $Z_{OT} = 0$. Beyond this frequency, operation is in the 'stop band' and the transmission line attenuates the signal. Due to distributed effects, the cutoff frequency will always be higher than the frequency computed using the classical expression for a lumped element transmission line ($f_c = 1/(\pi \sqrt{LC})$).

Based on the above equation, the characteristic impedance of a lumped-distributed transmission line is plotted as a function of frequency in Fig. 4.52 with the ratio C_g/Cl as a parameter. Cl is the transmission line capacitance per section. The independent variable, frequency, is normalized and is shown in the figure as the transmission line electrical length ϕ. For small values of ϕ, the characteristic impedance is nearly constant as one would expect. As the transmission line approaches a quarter wavelength, Z_{OT} drops rapidly passing through zero at $\phi \sim 100°$ for $C_g/Cl = 1$. Note also that smaller values of C_g/Cl result in a higher cut-off frequency. This is due to the fact that smaller values of C_g/Cl represent less capacitive loading and hence more nearly approximates a uniform transmission line.

The characteristic impedance of the TL segments can be determined by considering the impact of the C_g/Cl ratio and practical considerations. If C_g/Cl is made very small, the high frequency response is enhanced, but the device gate periphery per section is reduced which reduces the gain and

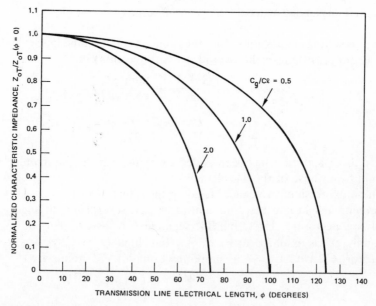

Fig. 4.52 Characteristic impedance of the lumped-distributed section.

power per section thereby requiring more sections. On the other hand, increasing the C_g/Cl ratio degrades the high frequency response and raises the required Z_O of the TL segments. A good compromise appears to be a C_g/Cl ratio of 1.

Using the same low loss assumptions the complex propagation constant expression, equation (34), can be reduced to a more manageable form. The real part of the propagation constant is of particular interest since, as described in the previous section, it determines, to a large measure, the amplifier gain and bandwidth. Assuming a gate equivalent circuit consisting of a series RC network, the attenuation constant for the input transmission line is approximately given by

$$\alpha_1 \quad \sim \quad \frac{\omega^2 \tau_2 C_g}{2Y_{OT}} \tag{36}$$

where $\tau_2 = R_g C_g$ is the gate RC time constant.

Note the ω^2 dependence of α_1. This is the dominant frequency determining term in the gain expression.

Similarly, for a drain equivalent circuit consisting of a parallel RC network, the attenuation constant for the output transmission line is given by

$$\alpha_2 \quad = \quad \frac{G_d}{2Y_{OT}} \tag{37}$$

4.6.4 Design example

As an example, the design of a K-band (18 to 26.5 GHz) distributed amplifier will be presented. Based on equation (35) and $C_g/Cl = 1$, the line length ϕ is selected such that the maximum VSWR on the lumped-distributed transmission line is 1.1:1 at the band edges. With $Z_{OT}(\phi_1) = 1.1 \times 50\ \Omega$ and $Z_{OT}(\phi_2) = 50/1.1\ \Omega$, the maximum value of ϕ that satisfies this requirement is $\phi_2 = 73°$ and $\phi_1 = 49.6°$. This results in a cut-off frequency of 35.6 GHz. In contrast, the lumped element expression $(f_c = 1/[(\pi \sqrt{(LC)}])$ yields a value of 29.3 GHz. From Fig. 4.52, the value of $Z_{OT}(\phi = 0)$ is then 60.4 Ω and the resulting values for Z_o and C_g are as follows:

$$Z_o \quad = \left(\frac{C_g}{Cl} + 1\right)^{1/2} Z_{OT}\ (\phi = 0)$$

$$= \sqrt{2}\ (60.4) = 85.4\ \Omega$$

$$C_g \quad = \quad Cl \quad = \quad Y_o\ \frac{l}{v} = \frac{\phi\ \text{(Degrees)}}{0.36\ Z_o\ f(\text{GHz})}\text{pF}$$

$$= \frac{73}{0.36(85.4)(26.5)} = 0.090\ \text{pF}$$

Based on a typical gate capacitance per unit periphery of 1.32 ff/μm for the Hughes 0.5 μm gate length FETs, the required gate width per section is

$$L = 0.09/0.00132 = 68 \ \mu m$$

From equation (36), and a typical gate time constant, $\tau_2 = 1.3$ ps, the attenuation constant on the input line increases as ω^2 and at 26.5 GHz is given as

$$\alpha_1 = \frac{(2 \ \pi \times 26.5 \times 10^9)^2 \ (1.3 \times 10^{-12}) \ (0.09 \times 10^{-12})}{2 \times 0.022}$$

$$= 0.074 \ \text{nepers/section} = 0.64 \ \text{dB/section}$$

On the output line, with $G_d = 0.8$ mS, the attenuation is frequency independent and is given by

$$\alpha_2 = \frac{0.0008}{2 \times 0.022}$$

$$= 0.02 \ \text{nepers/sections} = 0.17 \ \text{dB/section}$$

Based on a total attenuation of 0.094 nepers/section (0.82 dB/section) and equation (29), n was selected to be 7. The resulting gain is calculated to be 6.3 dB at dc decreasing to 3.7 dB at 26.5 GHz. This gain roll-off, due to the gate line attenuation, can be compensated on the drain line by distributing the drain compensation capacitance over a series transmission line keeping the low frequency capacitance, $Y_o l/v$, constant. This can be accomplished semi-empirically with the aid of a computer network analysis routine.

A photograph of a 7-section distributed amplifier chip, based on this design, is shown in Fig. 4.53. The circuit is fabricated on a 100 μm thick GaAs substrate with dimensions of 1.1 × 3.2 mm. Via holes are etched from the back of the chip to form the grounded source contacts. The device geometry consists of an interdigital structure with two 0.5 × 34 μm gate fingers connected to the gate line. The resistive terminations and the bias insertion networks are not included on-chip.

The gain performance of a typical amplifier chip is shown in Fig. 4.54, and the corresponding noise figure is shown in Fig. 4.55. The bias conditions are summarized as $V_G = 0$, $V_D = 2.5$ V and $I_D = 45$ mA. Over the 2 to 26.5 GHz band, the gain is 6.0 ± 0.3 dB. The gain decreases to 4 dB at 32 GHz before falling rapidly above 32 GHz. As shown in Fig. 4.55, the noise figure increases gradually from 4.2 dB at 8 GHz to 5.4 dB at 26.5 GHz. The input/output VSWR is typically better than 1.5:1 with a worst case of 1.7:1 at 22 GHz. The output power is typically 13 dBm at the 1 dB gain compression point.

A similar DA design with a 1/4 μm gate structure has demonstrated a 2 to 40 GHz bandwidth [72].

The above results demonstrate the viability of the DA approach to produce broadband (greater than a decade) low noise amplification. The

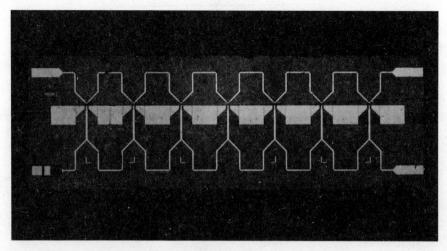

Fig. 4.53 7-section distributed amplifier chip.

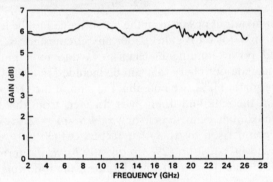

Fig. 4.54 Gain and noise figure of the 7-section distributed amplifier.

noise performance is also respectable for a broadband amplifier. In fact, for large *n* (5 or more sections), Aitchison [73] has shown that the minimum noise figure of a DA approaches that of a reactively matched narrow band amplifier.

4.6.5 Power considerations

The output power of a DA is limited by the following mechanisms [74].

(1) RF limiting on the gate line due to forward gate conduction.
(2) The total gate periphery is limited by gain and other considerations.
(3) FET gate-drain breakdown voltage.
(4) Optimum load line requirements.

The first power-limiting mechanism is due to the finite RF voltage swing that can be allowed on the input gate line. This swing is limited on the positive RF cycle by the forward conduction of the gate diode and on the negative cycle by the pinch-off voltage of the device. Hence, assuming a 50 Ω input

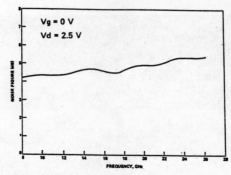

Fig. 4.55 7-section distributed amplifier noise figure.

impedance and FET devices with a pinch-off voltage of –4 V biased at a drain current $I_{DSS}/2$, the maximum input RF power to the amplifier is limited by

$$P_{in,\ max} = \frac{(4 + 0.5)^2}{8 \times 50} = 0.051 \text{ W}$$

Thus, maximum output power from the amplifier cannot be larger than Gain x $P_{in,max}$ (200 mW for 6 dB gain) under any circumstances.

The second power-limiting mechanism is due to a limitation on the maximum total gate periphery that can be included in a single-stage design. Referring to equation [29], we note that the gain of the DA is limited by the attenuation on the gate and drain lines. In fact, from this simplified gain expression, maximum gain occurs for $(\alpha_1 + \alpha_2)n = 2$. α_2 can usually be neglected in comparison to α_1. Other factors, which also reduce gain but are not included in equation (29), frequently force the term to be chosen less than two. Hence, from gain considerations the following inequality has to be satisfied:

$$\alpha_1 n < 2$$

From equation (36), expressing the gate line attenuation constant α_1 in terms of the FET input parameters R_g and C_{gs}, we find

$$R_g\ \omega^2\ C_g^2 Z_O\ n < 4$$

In this equation, R_g varies inversely and C_{gs} varies directly with gate periphery for a given FET geometry. Hence, in terms of gate periphery w per FET, this equation becomes

$$nw\ \omega^2 \leq \text{constant.}$$

Thus, for a specified maximum frequency of operation and for a given FET geometry, there is an upper limit to the maximum total gate periphery, nw, that can be employed. This maximum periphery determines the gain and consequently the output power of a DA. For example, with the design presented in the previous section, the total gate periphery was 476 μm resulting in a gain of approximately 6 dB, and the maximum frequency of operation was approximately 30 GHz.

The third power-limiting mechanism is the FET gate-drain breakdown voltage. The drain terminals must be able to sustain the RF voltage swings on the output transmission line. This voltage is given by

$$V_{max} (P\text{--}P) = V_B - V_{PO} - V_{SAT}$$

where V_B is the gate-to-drain breakdown voltage, V_{PO} is the pinch-off voltage and V_{SAT} is drain saturation voltage.

Using this expression, for a given output impedance R_L, the maximum saturated output power is

$$P_{max} = (V_B - V_{PO} - V_{SAT})^2/(8R_L).$$

With a breakdown voltage of 15 V and a 50 Ω output load, approximately 200 mW of output power can be expected. Note that reducing the output impedance is not a good solution for increasing the output power since it also results in a corresponding gain reduction.

The fourth power-limiting mechanism is related to the optimum ac load line requirements. For a DA, the loadline that each individual FET sees is predetermined by the drain line characteristic impedance. The only remaining design flexibility is the gate periphery of the unit FET. The total gate periphery is already determined from gate attenuation and gain considerations as described above. For example, we have established that the total periphery allowed is around 500 μm for the 30 GHz DA. For a 7-section design, each FET has a periphery of approximately 70 μm. The optimum load line for such a device is 500 Ω, representing a significant mismatch to the 50 Ω output impedance. Such an impedance mismatch has a significant effect on the output power and consequently on the efficiency of the DA.

Several approaches have been developed to address the above DA power limitations. One example is the work of Ayasli [74] where a power combining approach has been utilized to achieve a power level of 250 mW over the 2 to 20 GHz band. Another example is the capacitively coupled DA scheme ([75], [76]) which has produced power levels up to 800 mW [77].

4.7 CONCLUSIONS

Over the past several years, GaAs monolithic technology has made great technical and commercial advances. On the commercial front, dozens of new start-up companies have been formed over the past two years to exploit GaAs MMIC technology. These and established companies are now producing MMICs selling in small quantities for $10 per chip for the simplest functions to $400–500 per chip for 1/2 watt, 6 to 18 GHz amplifiers. At the time of this writing (second quarter 1987), several start-up companies are reporting shipments of 10 000 chips per month. Further price reductions are inevitable. Based on the history of silicon ICs, we can expect to see an order of magnitude decrease in prices in the next several years. This, of course, is predicated on the development of a major market ($> 10^6$ chips/year) for

MMICs. Current price projections indicate that 1 W, 6 to 18 GHz ICs will be selling for less than $5 each in small quantities (< 100) by 1992.

On the technical front MMICs are coming of age. Design techniques material and processing technology are maturing resulting in high performance, reproducible and hence affordable ICs. Hughes recently produced over 3000 X-band power (1.6 W) MMICs with yields approaching 45% [78].

As the MMIC technology matures, it is pushing into millimeter-wave frequencies. Distributed amplifiers operating at frequencies up to 40 GHz [79], monolithic receivers at 30 GHz [80], and monolithic power amplifiers at 44 GHz [81] have been reported. Monolithic oscillators operating at 69 GHz [82] and 115 GHz [83] have also been reported.

In addition to new levels of performance, the level of integration is also increasing. Whole subsystems are being integrated on a single chip. Podell [84] has recently demonstrated a 4 GHz receiver front-end (LNA, mixer, LO and IF amplifier) on a single chip of GaAs with dimensions of 34 × 39 mils.

An even higher level of integration has recently been demonstrated by Raytheon and Texas Instruments with their X-band single-chip transmit/ receive (T/R) module. The Raytheon chip consists of a 5-bit phase shifter, 3-stage low noise amplifier, 4-stage power amplifier and two T/R switches all on a single chip with an area of 24 mm^2 [85]. The TI chip contains a 4-bit phase shifter, a 4-stage power amplifier, a 3-stage low noise amplifer and two T/R switches resulting in a total chip area of 58.5 mm^2 [86]. Using selective ion implantation, dc yields as high as 32% have been reported with these large chips. The achievement of these single chip T/R modules marks a milestone in the integration level and function density of MMICs.

However, some industry leaders have been to question this 'radar on a chip' philosophy. The resulting large chips are generally low in yield, complicated to process and difficult to handle and mount. With a chip area of 58.5 mm as described above, only 60 or so chips are potentially available from a 3 inch wafer. With a yield of 25%, only 15 chips are produced per wafer.

In contrast, many applications are emerging that use a combination of monolithic and hybrid circuits extracting the best from both worlds. A good example of such an application is Plessey's S-band T/R module [87]. It contains five MMICs and a thick-film alumina substrate used as a 'mother board' providing the interconnection function, impedance matching, and the module control circuit. This combination of monolithic and hybrid circuitry allows an optimal mix of both technologies: MMICs are used where a large density of active devices are required, such as an amplifier, and hybrid circuit technology is employed when the required component is physically large or the required circuit Q is high, for example, a quadrature coupler, power combiner or a narrow band filter. Often the hermetic package itself can serve as the hybrid circuit providing additional matching, bias networks and filtering as required. This multichip hybrid/monolithic approach rather than a single mega-chip approach also has the potential of dramatically improving the individual chip yield and thereby the system cost providing the cost of interconnecting the chips is less than the increased cost of the mega-chip.

This chapter has focused primarily on amplifier design principles. However, the concepts presented can also be applied to monolithic circuits in general. The passive and active monolithic circuit elements (transmission lines, resistors, inductors, capacitors and single and dual-gate FET devices), presented in Section 4.2, are the fundamental 'building blocks' for any monolithic circuit and hence their application is not limited to amplifier design. In addition, the amplifier design principles presented in Section 4.3, can be applied to the design of oscillators, switches, phase shifters and mixers to name a few. Although each of these applications have unique problems and require special design considerations, many of the amplifier design principles are useful for monolithic circuit design in general. In particular, the matching network techniques presented here are relevant to all monolithic circuit applications requiring a broadband impedance match to an active or passive element (complex load). This represents almost all microwave applications with the exception of 'direct coupled' circuits. For example, broadband oscillator design requires that a prescribed impedance function be synthesized over a given frequency band. Further, switches and phase shifters, which incorporate active devices, must be matched over broad bandwidths to 50 ohms. Both cases can be readily accomplished by employing the broadband matching concepts presented in Section 4.3.

In summary, monolithic microwave circuits are finally a reality. The factors most contributing to the realization of MMICs are:

(1) the emergence of the Shottky-barrier field-effect transistor (MESFET),
(2) the low-loss microwave properties of semi-insulating GaAs,
(3) the development of GaAs ion implantation techniques,
(4) the development of large diameter (3 inch) crystal growth techniques,
(5) the emergence of potential system applications for monolithic microwave circuits.

ACKNOWLEDGEMENT

The authors wish to thank the members of the Torrance Research Center of Hughes Aircraft Company for their contributions to this work and Gwen Ziegert for her timeless efforts in typing and preparing this manuscript.

REFERENCES

1. Hyltin, T. M. (1965) 'Microstrip transmission on semiconductor substrates' *IEEE Trans. Microwave Theory and Tech.* **MTT–13** 777–781.
2. Mehal, E. and Wacker, R. W. (1968) 'GaAs integrated microwave circuits' *IEEE Trans. Microwave Theory and Tech.* **MTT–16** 451–454.
3. Pengelly, R. S. and Turner, J. A. (1976) 'Monolithic broadband GaAs FET amplifiers' *Electron Lett.* **12** 251–252.

4. Howes, M. J. and Morgan, D. V. (1985) *Gallium Arsenide Materials, Devices and Circuits* John Wiley and Sons, New York.
5. Hammerstad, E. O. and Berrddal, F. (1975) *Microstrip Handbook* Electronics Research Laboratory, University of Trondheim, Norway.
6. Wen, C. P. (1969) 'Coplanar waveguide: a surface strip transmission line suitable for nonreciprocal gyromagnetic device applications' *IEEE Trans. Microwave Theory and Tech.* **MTT–18** 1087–1090.
7. Rowe, D. A. and Lao, B. Y. (1983) 'Numerical analysis of shielded coplanar waveguides' *IEEE Trans. Microwave Theory and Tech.* **MTT–31** 911–915.
8. Hillberg, W. (1969) 'From approximation to exact relations for characteristic impedances' *IEEE Trans. Microwave Theory and Tech.* **MTT–17** 259–265.
9. Pucel, R. A. (1981) 'Design considerations for monolithic microwave circuits' *IEEE Trans. Microwave Theory and Tech.* **MTT–29** 513–534.
10. Sobol, H. and Caulton, M. (1974) 'The technology of microwave integrated circuits' *Advances in Microwaves* **8** New York, Academic Press.
11. Alley, G. D. (1970) 'Interdigital capacitors and their application to lumped-element microwave integrated circuits' *IEEE Trans. Microwave Theory and Tech* **MTT–18** 1028–1033.
12. Smith, J. I. (1969) 'The even and odd mode capacitance parameters for coupled lines in suspended substrate' *1969 G–MTT Symp. Digest* 324–328.
13. Terman, F. E. (1943) *Radio Engineer's Handbook* New York, McGraw-Hill.
14. Podell, A. F. (1980) 'GaAs MICs: expensive, exotic but exciting' *Microwaves* 56–61.
15. Greenhouse, H. M. (1974) 'Design of planar rectangular microelectronic inductors' *IEEE Transactions on Parts, Hydbrids, Packaging* **PHP–10** 101–109.
16. Krafesik, D. M. and Dawson, D. E. (1986) 'A closed-form expression for representing the distributed nature of the spiral inductor' *IEEE Monolithic Circuits Symposium Digest* 87–92.
17. Grover, F. W. (1946) *Inductance calculations* Van Nostrand, Princeton, N.J. Reprinted by Dover Publications (1962) 17–47.
18. Smith, P. M., Chao, P. C., Duh, K. H. G., Lester, L. F., Lee, B. R. and Ballingall, J. M. (1987) 'Advances in HEMT technology and applications' *1987 IEEE MTT–S Int'l Microwave Symposium Digest* 749–752.
19. Liechti, C. A. (1975) 'Performance of dual-gate GaAs MESFETs as gain-controlled, low-noise amplifiers and high-speed modulators' *IEEE Trans. Microwave Theory and Tech.* **MTT–23** 461–469.
20. Chen, Y. K., Hwang, Y. C., Naster, R. J., Ragonese, L. J. and Wang, R. F. (1985) 'A GaAs multi-band digitally-controlled 0°–360° phase shifter' *IEEE GaAs IC Symposium Digest* 125–128.
21. Hower, P. and Bechtel, G. (1973) 'Current saturation and small-signal characteristics of GaAs field-effect transistors' *IEEE Trans. Electron Devices* **ED–20** 213–220.
22. Reiser, M. and Wolf, P. (1972) 'Computer study of submicrometre FET's' *Electron. Lett.* **8** 254–256.
23. Ohkawa, S., Suyama, K. and Ishikawa, H. (1975) 'Low noise GaAs field-effect transistors' *Fujitsu Sci, Tech. J.* **11** 151–173.
24. Steer, M. B. and Trew, R. J. (1986) 'High frequency limits of millimeter-wave transistors' *IEEE Electron Device Letters* **EDL–7** 640–642.
25. van der Ziel, A. (1962) 'Thermal noise in field-effect transistors' *Proc. IRE,* **50** 1808–1812.

26. (1963) 'Gate noise in field effect transistors at moderately high frequencies' *Proc. IEEE* **51** 462–467.

27. Shockley, W. (1952) 'A unipolar field-effect transistor' *Proc. IRE* **40** 1365–1376.

28. Liechti, C. A. (1976) 'Microwave field-effect transistors – 1976' *IEEE Trans. Microwave Theory and Tech.* **MTT–24** 279–300.

29. Pucel, R. A. Haus, H. A. and Statz, H. (1975) 'Signal and noise properties of gallium arsenide microwave field-effect transistors' *Advances in Electronics and Electron Physics* **38** New York: Academic Press 195–265.

30. Fukui, H. (1979) 'Optimal noise figure of microwave GaAs MESFET's' *IEEE Trans. Electron Devices* **ED–26** 1032–1037.

31. Fukui, H. (1979) 'Determination of the basic device parameters of a GaAs MESFET' *Bell Syst. Tech. J.* **58** 771–797.

32. DiLorenzo, J. V. and Wisseman, W. R. (1979) 'GaAs Power MESFET's: design, fabrication, and performance' *IEEE Trans. Microwave Theory and Tech.* **MTT–27** 367–378.

33. Macksey, H. M., Doerbeck, F. H. and Vail, R. C. (1980) 'Optimization of GaAs power MESFET device and material parameters for 15 GHz operation' *IEEE Trans. Electron Devices* **ED–27** 467–471.

34. Frensley, W. R. and Macksey, H. M. (1979) 'Effect of gate strip width on the gain of GaAs MESFETs' *Proc. 7th Biennial Cornell Electrical Engineering Conf.* Ithaca, Cornell University, 445–452.

35. Vendelin, G. D. (1982) *Design of amplifiers and oscillators by the S–parameter method* John Wiley and Sons, New York.

36. Gonzalez, G. (1984) *Microwave transistor amplifiers analysis and design* Prentice-Hall, Inc., Englewood Cliffs, N.J., USA.

37. Ha, T. T. (1981) *Solid-state microwave amplifier design* John Wiley & Sons, New York, N.Y., USA.

38. Engberg, J. 'Simultaneous input power match and noise optimization using feedback' *Proc. 1974 European Microwave Conf.* 385–389.

39. Yau, W., Watkins, E. T., Wang, S. K., Wang, K. and Klatskin, B. (1987) 'A four-stage X-band MOCVD HEMT amplifier' *1987 IEEE MTT–S Digest* 1015–1018.

40. Bode. (1945) *Network analysis and feedback amplifier design* Van Nostrand, New York.

41. Fano, (1948) 'Theoretical limitations on the broadband matching of arbitrary impedances' *MIT Technical Report No. 41*.

42. Matthaei, G. L. (1956) 'Synthesis of Tchebycheff impedance-matching networks, filters, and interstages' *IRE Trans. CT* 163–172.

43. Weinberg, (1962) *Network analysis and synthesis* McGraw-Hill, New York.

44. Levy, (1964) 'Explicit formulas for Chebyshev impedance-matching networks, filters, and interstages' *Proc. IEE* (London) **III** 1099–1106.

45. Green, E. (1954) 'Amplitude-frequency characteristics of ladder networks' Marconi's Wireless Telegraph Co., Essex, England.

46. Takahasi, H. (1951) 'On the Ladder-type filter network with Tchebyshev response' *J. Inst. Elect. Comm. Engrs. Japan* **34**(2).

47. Zverev, (1967) *Handbook of filter synthesis* Wiley, New York.

48. Apel, T. R. (1983) 'Bandpass matching networks can be simplified by maximizing available transformation' *MSN* (Dec.) 105–117.

49. Christian and Eisenmann 'Broadband matching by lowpass transformations'. *Proc. Fourth Annual Allerton Conf. on Circuit and System Theory.*

50. Matthaei, G. L. (1964) 'Tables of Chebyshev impedance-transforming networks of low-pass filter form' *IEEE Proc.* (Aug.) 939–963.
51. Cottee and Joines (1979) 'Synthesis of lumped and distributed networks for impedance matching of complex loads' *IEEE CAS–26* (May) 316–329.
52. Cristal, E. G. (1965) 'Tables of maximally flat impedance transforming networks of low-pass-filter form' *IEEE Trans. Microwave Theory and Tech.* **MTT–13** 693–695.
53. Mellor, (1986) 'Improved computer-aided synthesis tools for the design of matching networks for wide band microwave amplifiers' *IEEE Trans. Microwave Theory and Tech.* **MTT–34** 1276–1281.
54. Apel, T. R. (1984) 'One-port impedance models prove useful for broadband RF power amplifier design' *MSN* (Oct.) 96–105.
55. Wemple, S. H., Steinberger, M. L. and Schlosser, W. D. (1980) 'Relationship between power added efficiency and gate-drain avalanche in GaAs MESFETs' *Electron. Lett.* **16** 459.
56. Cusack, J. M., Perlow, S. M. and Perlman, B. S. (1974) 'Automatic load contour mapping for microwave power transistors' *IEEE Trans. Microwave Theory and Tech.* **MTT–22** 1146.
57. Takayama, Y. (1976) 'A new load-pull characterization method for microwave power transistors'. *1976 IEEE MTT–S International Microwave Symp. Dig.* 218–220.
58. Anastassiou, A. and Strutt, M. (1974) 'Effect of source lead inductance on the noise figure of a GaAs FET' *Proc. IEEE*, **62** 406–408. Also, for corrections, see Iversen, S. (1975) *Proc. IEEE*, **63** 983–984.
59. Watkins, E. T., Schellenberg, J. M., Hackett, L. H., Yamasaki, Y. and Feng, M. (1983) 'A 60 GHz GaAs FET amplifier' *1983 IEEE MTT–S Digest* (June) 145–147.
60. Lane, R. Q. (1978) 'Device noise and gain parameters in 10 seconds' *Microwaves* (Aug.) 53–57.
61. Podell, A. F. (1981) 'A functional GaAs FET noise model' *IEEE Trans. Electron Devices* **ED–28** 511–517.
62. Engberg, J. 'Simultaneous input power match and noise optimization using feedback' *Proc. 1974 European Microwave Conf.* 385–389.
63. Haus, H. and Adler, R. (1959) *Circuit theory of linear noisy networks* Technology Press of MIT and Wiley, New York 53.
64. Petersen, W. C., Decker, D. R., Gupta, A. K., Dully, J. and Ch'en, D. R. (1981) 'A monolithic GaAs 0.1 to 10 GHz amplifier' *1981 IEEE MTT–S Int. Microwave Sym. Digest* (June) 354–355.
65. Pengelly, R. S., Suffolk, J. R., Cockrill, J. R. and Turner, J. A. (1981) 'A comparison between actively and passively matched S-band GaAs onolithic FET amplifiers' *1981 IEEE MTT–S Int. Microwave Sym. Digest* (June), 367–369.
66. Estreich, D. B. (1982) 'A wide band monolithic GaAs IC amplifier' *ISSCC Digest of Technical Paper* (Feb) 194–195.
67. Niclas, K. B. (1985) 'Active matching with common-gate MESFET's' *IEEE Trans. Microwave Theory and Tech.* **MTT–33** 492–499.
68. Percival, W. S. (1937) 'Thermionic valve circuits' British Patent 460562.
69. Ayasli, Y., Reynolds, L. D., Vorhaus, J. L. and Hanes, L. (1982) 'Monolithic 2–20 GHz traveling-wave amplifier' *Electron. Lett.* **18**(14) 596–598.
70. Strid, E. W. and Gleason, K. R. (1982) 'A DC–12 GHz monolithic GaAs FET

distributed amplifer *IEEE Trans. Microwave Theory and Tech.* **MTT–30** 969–975.

71. Niclas, K. B., Wilser, W. T., Kritzer, T. R. and Pereira, P. R. (1983) 'On theory and performance of solid state microwave distributed amplifiers' *IEEE Trans. Microwave Theory and Tech.* **MTT–31** 447–456.

72. Pauley, R. G., Asher, P. G., Schellenberg, J. M. and Yamasaki, H. (1985) 'A 2 to 40 GHz monolithic distributed amplifier' *Technical Digest* 1985 GaAs IC Symposium, Monterey, California.

73. Aitchison, C. S. (1985) 'The intrinsic noise figure of the MESFET distributed amplifier' *IEEE Trans. Microwave Theory and Tech.* **MTT–33** 460–466.

74. Ayasli, Y., Reynolds, L. D., Mozzi, R. L. and Hanes, L. K. (1984) '2–20 GHz GaAs travelling-wave power amplifiers' *IEEE Trans. Microwave Theory and Tech.* **MTT–32** 290–295.

75. Kim, B. and Tserng, T. Q. (1984) '0.5W 2–21 GHz monolithic GaAs distributed amplifier' *Electron. Lett.* **20**(7) 288.

76. Ayasli, Y. A., Miller, S. W., Mozzi, R., and Hanes, L. K. (1984) 'Capacitively coupled travelling-wave power amplifier' *IEEE 1984 Microwave Millimeter-Wave Monolithic Circuits Symp. Dig.* (May) 52–54.

77. Kim, B., Tserng, H. Q. and Shih, H. D. (1985) 'High power distributed amplifier using MBE synthesized material' *IEEE 1985 Microwave Millimeter-Wave Monolithic Circuits Symp. Dig.* (June) 35–37.

78. Wang, S. K., Wang, D. C., Chang, C. D., Siracusa, M. and Liu, L. C. T. (1986) 'Producibility of GaAs monolithic microwave integrated circuits' *Microwave J.* (June) 121–133.

79. Pauley, R. G., Asher, P. G., Schellenberg, J. M. and Yamasaki, H. (1985) 'A 2 to 40 GHz monolithic distributed amplifier' *Technical Digest 1985 GaAs IC Symposium*, Monterey, California.

80. Liu, L. C. T., Liu, C. S., Kessler, J. E. and Wang, S. K. (1986) 'A 30 GHz monolithic receiver' *1986 Microwave and Millimeter Wave Monolithic Circuits Symp*.

81. Kim, B., Tserng, H. Q. and Shih, H. D. (1986) '44 GHz monolithic GaAs FET amplifier' *IEEE Electron Device Letters* **EDL–7** 95–97.

82. Maki, D. W., Schellenberg, J. M., Yamasaki, H. and Liu, L. C. T. (1984) 'A 69 GHz monolithic FET oscillator' *Dig. IEEE 1984 Microwave and Millimeter-Wave Monolithic Circuits Symposium* 62–66.

83. Tserng, H. W. and Kim, B. (1985) 'A 115 GHz monolithic GaAs FET oscillator' *Technical Dig. 1985 IEEE GaAs IC Symposium* 11–13.

84. Podell, A. F. and Nelson, W. W. (1986) 'High volume, low cost, MMIC receiver front end' *IEEE 1986 Microwave and Millimeter-Wave Monolithic Circuits Symp. Dig.* 57–59.

85. Jerinic, G., Durschlag, M., Mozzi, R., Kazian, T., Smith, C., Dormail, J., Wendler, J. and Tajima, Y. (1986) 'X-band single chip T/R module development' *Government Microcircuit Applications Conf. Dig.* 343–345.

86. Witkowski, L. C., Brehm, G. E., Coats, R. R., Heston, D. D., Hudgens, R. D., Lehmann, R. E., Macksey, H. M., Tserng, H. Q. and Wisseman, W. R. (1986) 'A GaAs single chip transmit/receive radar module' *Government Microcircuit Applications Conf. Dig.* 339–342.

87. Green, C. R., Lane, A. A., Tombs, P. N., Shukla, R., Suffolk, J. R., Sparrow, J. A. and Cooper, P. D. (1987) 'A 2 watt GaAs TX/RX module with integral control circuitry for S–band phased array radars' *1987 IEEE MTT–S Int. Microwave Sym. Dig.* 933–936.

Chapter 5

Monolithic Microwave Integrated Circuit Technologies

James A. TURNER

5.1 INTRODUCTION

The development of process technologies for microwave and millimetre wave integrated circuits poses a number of wide ranging challenges for the process engineer. The diversity of frequencies from about 1 GHz to 100 GHz means that a standard technology cannot be adopted for the whole frequency spectrum as the circuit topology required at each end of this frequency range differ widely. At low frequencies, ~1 GHz, traditional distributed (microstrip) style matching elements consume too much surface area of GaAs and so RF techniques using lumped inductors, capacitors and resistors are used. It is only at frequencies in excess of 6−7 GHz that fully distributed circuits are small enough to enable utilization of this type of impedance matching.

If we examine the component and technology requirements for circuits in the microwave and millimetre wave frequency band we can draw up a list such as that given in Table 5.1. The process engineer must first develop technologies for manufacturing all the components given in this table and then devise ways of integrating them into a reproducible, reliable IC process consuming as little GaAs surface area as possible. Throughout this chapter

Table 5.1. Components and technologies for MMICs

FETs (0.5–1.0μm)	$\vee$
Resistors	$\vee$
Capacitors	$\vee$
Inductors	$\vee$
Transmission lines	$\vee$
Multi-level metal	$\vee$
Air bridges	$\vee$
Via hole technology	$\vee$
Material uniformity	$\vee$
No. of gates per chip	≤ 20

the active device in all the circuits described is a gallium arsenide field effect transistor, GaAs MESFET, which has for the past 12 years or so been the 'work horse' for all microwave analogue integrated circuits [1].

The circuit manufacturing procedures can be described by four basic processing technologies:

(1) GaAs etching.
(2) Metal deposition and removal.
(3) Dielectric deposition and removal.
(4) Photo and electron beam lithography.

These are used to generate all the components required in the IC and will be described in Section 5.4.

Let us first examine the basic manufacturing stages in use today for the components listed in Table 5.1.

5.2 ACTIVE DEVICE: THE GaAs MESFET

The MESFET is the vital component of the circuit and this section describes the fabrication technologies used in its manufacture. Fig. 5.1 shows the basic device structure.

The important aspects of the MESFET process technology are:

(1) Control of initial starting material – substrate, buffer layer, active n layer and n+ contact layer.
(2) Isolation of the areas of active material in which the MESFET is fabricated.
(3) Deposition of low contact resistance ohmic source and drain metallizations.
(4) Controlled etching of channel region of the MESFET to define operating current.
(5) Deposition of low resistivity gate metal with dimensions below one micron.
(6) Control of separation between gate electrode and ohmic source contact.

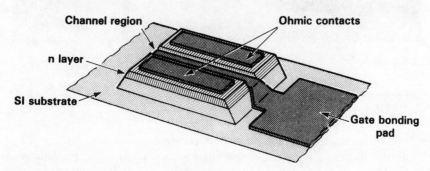

Fig. 5.1 Schematic of a mesa etched MESFET.

One of the major factors affecting MESFET performance is the quality of the material in which the device is made. Currently three major approaches to material supply are being researched. These are:

(1) Vapour phase epitaxy [2].
(2) Molecular beam epitaxy [3].
(3) Ion implantation [4].

These are fully described in chapter 1 of this book and will not be covered here in detail.

The choice of material technology depends to some extent on the particular circuit application but there is no doubt that at present the industry is polarising towards the use of ion implantation mainly on the grounds of cost and parameter control. Fig. 5.2 shows a photograph of an ion implantation machine capable of implanting 350 GaAs wafers per hour. This compares to less than ten wafers per hour from the most sophisticated vapour phase and molecular beam epitaxy equipments. Table 5.2 compares the three material technologies in terms of their useful and less desirable attributes.

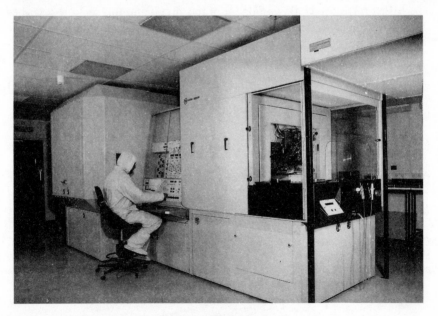

Fig. 5.2 High throughput ion implantation machine.

5.2.1 Ohmic contacts

A basic understanding of the metal-semiconductor junction is necessary to determine the metal systems to use for low contact resistance contacts to GaAs. This is however beyond the scope of this section but can be obtained

Table 5.2. Comparison of material processes

Material technology	Profile control	Electrical quality	Cost per layer
Vapour phase epitaxy (VPE)	Least control of the three techniques	Best quality material in terms of electrical mobility – used for high quality discrete MESFETs	High – growth rate reasonable but scaling up difficult
Molecular beam epitaxy (MBE)	Monolayer accuracy and very versatile	Good – only slightly worse than VPE	Highest – very slow growth rate – not readily adapted for high throughput
Ion implantation (II)	Few percent variation across whole wafer	Acceptable for IC fabrication	Low – true 'mass production' process

from the literature [5]. Suffice to say that an n-type dopant is required which will dope the surface of the GaAs to such an extent as to cause the dominant conduction mechanism to be tunnelling.

A number of dopants have been used, such as silicon, germanium, tin, selenium and tellurium, but the industry has concentrated on an alloy of gold and germanium with additives such as nickel or indium. This is applied as a vacuum deposited film ~5000 Å thick in the proportions of the gold germanium eutectic (88% Au, 12% Ge by weight). The addition of metals such as nickel serve to prevent the gold germanium eutectic alloy 'balling up' during the subsequent heat treatment to form the desired low resistance contact [5]. Generally an overlay of gold is required as the surface of the alloy is not suitable for wire bonding and can also be a difficult one on which to make a reproducible probe contact.

The specific areas of ohmic metallisation are defined by a 'lift-off' process which gives isolated areas of metal with good edge definition. This procedure precedes the alloying step required to produce low resistance contacts.

The conditions for the alloying process tend to be laboratory 'secrets' and depend on the exact metal deposition conditions and the furnacing arrangements (ambient gas type, flow conditions, furnace bore etc). The temperatures and times of alloying are determined experimentally from plots such as given in Fig. 5.3. The condition for minimum contact resistance being chosen as the process requirement. There are many techniques for measuring the contact resistance of the alloyed metal to the semiconductor surface, the best known and perhaps the most accurate is shown schematically in Fig. 5.4 [6]. In this technique the metal contacts are defined with increasing separation on the GaAs surface. Measurement of the

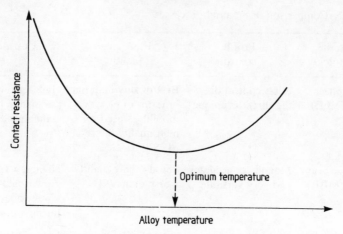

Fig. 5.3 Determination of optimum alloy temperature for ohmic contacts.

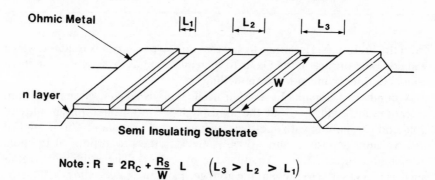

Note : $R = 2R_C + \dfrac{R_S}{W} L \quad \left(L_3 > L_2 > L_1 \right)$

Fig. 5.4 Schematic of contact resistance test structure.

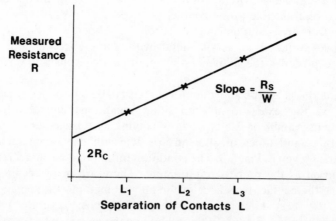

Fig. 5.5 Determination of contact resistance and material resistance from test structure.

resistance between adjacent contacts is used to determine the contact resistance of the metal to semiconductor interface. This is achieved by plotting measured resistance versus contact separation (Fig. 5.5). The intercept on the resistance axis giving a value that is twice the contact resistance. This contact resistance value is related to the width of the metal pattern and so it is convenient to arrange this to be the same as the width of source and drain contact pads on the MESFETS within the circuit. In this way the resistance value determined will relate diectly to the MESFET and can be used to determine the resistances within the device that have a pronounced effect on its microwave performance.

This method can predict contact resistance accurately provided:

(a) Probe resistance is taken into account.
(b) The metallization itself has negligible resistance.
(c) The metal edge definition is good.
(d) The test pattern is isolated from the surrounding GaAs material thereby confining the current flow to that between the defined contact areas.

Recently thermal pulse alloying, TPA, (rather than furnace alloying) has been shown to give lower contact resistance and a faster throughput of wafers. In this process the GaAs wafer is rapidly heated by quartz lamps to a temperature at which the contact resistance is a minimum. The time that the wafer is at the maximum temperature can be as short as 5 seconds.

5.2.2 Gate electrode

The purpose of the gate electrode is to form a rectifying junction to the GaAs such that the resistance of the material between the source and drain contacts can be modulated by an applied voltage.

Two types of junction are used:

(a) p−n diffused or implanted junction
(b) Schottky barrier.

The p−n junction is not common in MMIC technology and is confined at present to digital designs and circuits in which radiation hardness is of prime importance. The most commonly used gate junction is the Schottky barrier or metal semiconductor diode. Here a suitable metal is evaporated onto the surface of the semiconductor to form a rectifying contact. As for the ohmic contact prior knowledge of semiconductor physics is required for the correct choice of metal to be made [5]. Almost all metals will form a Schottky barrier to GaAs but the ultimate choice must (a) adhere well to the surface and (b) be thermally stable. Some metals such as gold and gallium diffuse rapidly in GaAs and this destroys the rectifying characteristics of the diode. However aluminium, chromium, tungsten and titanium meet the two criteria and have been used widely as the gate electrode. Initially aluminium was the most common metal as it is easy to deposit and is thermally stable to above 550°C. However the advent of the MMIC dictated that the gate electrode should be contacted for interconnections within the circuit with some other metal, preferably gold. Aluminium was not ideal in this situation as the gold/aluminium interface was poor from a reliability standpoint — gold aluminium intermetallics are formed that can give high resistance contacts and in some instances voids and undesirable alloy formations in the metallization.

These features are clearly seen in Fig. 5.6 which shows various interactions of a gold bond wire to an aluminium bonding pad on a GaAs MESFET after subjecting to a temperature of 300°C. Fig. 5.6(a) shows how the intermetallic gold/aluminium 'corrosion' can encompass large areas of the bonding pad. Fig. 5.6(b) shows this intermetallic encroaching into the channel region of the MESFET partially destroying the fine metal pattern forming the gate region of the device. These effects are also seen in an integrated circuit environment when aluminium metal films overlaid with gold metallization are temperature stressed. They can however be overcome by developing a technology that prevents a direct contact between gold and aluminium by the use of a barrier layer but this can add considerably to the process complexity. Aluminium has, for integrated circuits at least, given way to other Schottky barrier materials that are free from undesirable intermetallic interactions.

A more acceptable gate metallization for MMICs is a combination of titanium/platinum/gold. In this metal scheme titanium is the Schottky barrier material and is chosen for its good diode properties and because it adheres extremely well to the GaAs surface. A thick gold capping layer is used to improve the electrical conductivity of the gate electrode; however intimate contact of titanium and gold is considered a reliability hazard as gold diffuses quickly through titanium at quite modest temperatures degrading the properties of the gate diode. To overcome this situation a thin interface layer of platinum is deposited onto the titanium film immediately prior to the deposition of the gold cap layer. Platinum has been shown to be a good diffusion barrier and prevents the gold from penetrating into the underlying GaAs material. The importance of a low resistance gate metallisation is illustrated in Fig. 5.7. Here it can be seen that the noise figure (and also the power gain) is strongly dependent on the resistance of the gate stripe [7]. Typical layer thicknesses in the Ti/Pt/Au gate metal system are 500 Å, 1000 Å and 5000 Å respectively. Thin layers of titanium and platinum are used because of their high evaporation temperature which if prolonged can damage the protecting photoresist film used in the gate 'lift-off' process.

As the width of the metal stripe (the gate length in FET parlance) determines the operating frequency of the transistor its accurate definition is vitally important. Again as for the ohmic contacts a lift-off technology is adopted but in this case the gate length is determined by the smallest feature size resolvable by the particular lithographic technique used. Conventionally, contact lithography is used in GaAs and by using mid-ultra violet radiation

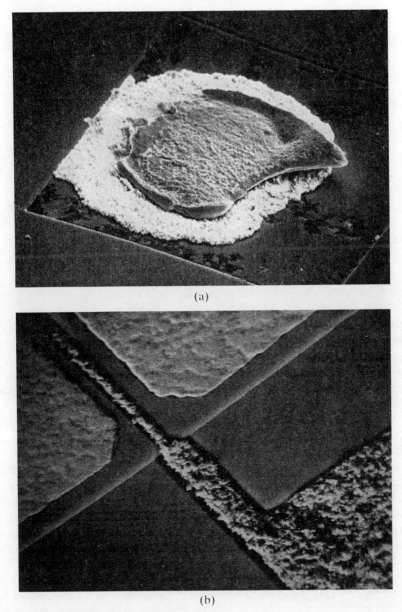

(a)

(b)

Fig. 5.6 Gold-aluminium interactions.

at around 300 nM, line widths of 0.5 microns can be reproducibly fabricated. More recently electron beam lithography has been used to resolve gate dimensions down to below 0.25 microns. However in both technologies the resist profile is an important feature in ensuring that the lift off process gives a clean, well controlled edge to the gate metal stripe.

The deposition and subsequent removal of the gate metal is generally

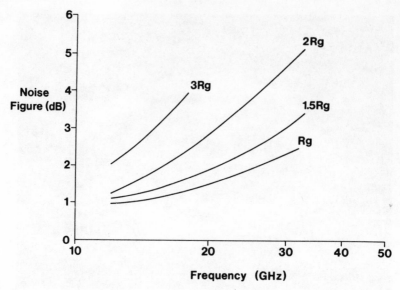

Fig. 5.7 The effect of gate metal resistance on MESFET noise figure.

preceded by a recessing of the GaAs in the area in which the gate is defined. This process assists the lift off process and enables the circuit fabrication engineer to 'tailor' the transistor operating current to suit the particular application. Fig. 5.8 shows this gate recess technique and shows how the recess can be etched and the gate metal deposited using the same photoresist layer.

As only 500–2500 Å of GaAs is removed during the recessing process the etch used, generally a wet etch, must have the properties of controlled slow etch rate and uniform material removal.

Most etches for GaAs use the principle of oxidising the GaAs and then etching away the gallium oxide. The solutions used for this purpose are based on either acid/peroxide or alkaline/peroxide mixtures, both mixtures, when diluted with water, have etch rates suitable for the recess process.

Table 5.3 gives a few examples of GaAs etches [8], some of which are used for recess etching and some for the mesa etch process described in Section 5.2.3.

Table 5.3. Some etches used for mesa and recess etching

Etchant	Composition	Approximate etch rate microns/min
$H_2SO_4/H_2O_2/H_2O$	1:1:1	5.0
Br/CH_3OH	4:100	6.0
Br/CH_3OH	1:100	1.0
$KOH\frac{1}{3}H_2O_2/H_2O$	1:1:10	0.5
$NH_4OH/H_2O_2/H_2O$	1:1:5	2.0

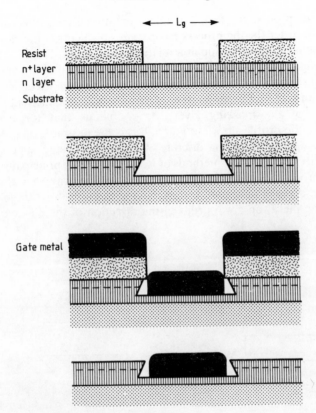

Fig. 5.8 Gate metal 'lift off' process.

5.2.3 Device isolation

In an MMIC the active areas of the FET must be isolated from the rest of the circuit. This is achieved in any one of three different ways. These techniques are necessary to:

(a) restrict the current flow in the device to the gate area of the FET,
(b) to reduce parasitic resistances and capacitances,
(c) to isolate one FET from another,
(d) to expose the underlying semi-insulating material so that the matching elements etc. can be defined on low loss material.

The isolation required can be achieved by

(a) mesa etching,
(b) selective implantation,
(c) implantation of boron, oxygen or protons.

Each one of these techniques produces sufficient isolation between adjacent devices to allow the fabrication of high performance MESFETs.

The mesa etch process is affected by the orientation effects of certain GaAs etches and in order for the process to operate satisfactorily the slope of the mesa edge must be positive and at a relatively low angle. This is to facilitate metal coverage of the step such that contact from the top of the mesa to the surrounding semiconductor is continuous. Mesa edge shapes depend on the etch used and Fig. 5.9 shows the general shape obtained at different orientations of the masking layer. It is obvious that for continuous metallization the profile shown in Fig. 5.9(a) is the desirable one. Although this process is still used for discrete devices it is being superseded in integrated circuits by planar methods of isolation. This is principally because the mesa process places severe restrictions on the way that the circuit designer can arrange his circuit layout, having to ensure that the metallization from the mesa top always runs in the direction of the positive slope.

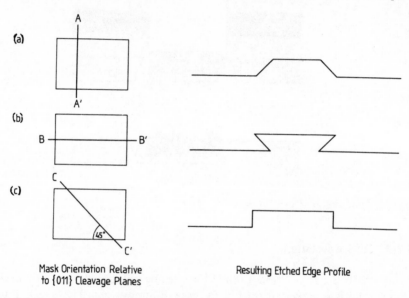

Mask Orientation Relative   Resulting Etched Edge Profile
to {011} Cleavage Planes

Fig. 5.9 Etch profiles for different crystal orientations.

The planar isolation methods currently in use are more complex than the mesa process but do allow considerably more flexibility in circuit design. In the selective implantation process the dopant species is implanted through a masking medium that has been removed from the areas in which doping is required. This creates isolated areas of doped material after a high temperature anneal of the implanted wafer. This process is depicted in Fig. 5.10 (a) where the masking medium can be a dielectric or photo-resist film.

Isolation using boron, oxygen or protons is carried out in a similar way to selective implantation. However isolation is achieved by damaging the lattice causing the resistivity to increase in areas unprotected by the mask, Fig. 5. 10 (b). In this process no annealing is carried out, indeed any high temperature process will cause the damaged areas to revert to their original

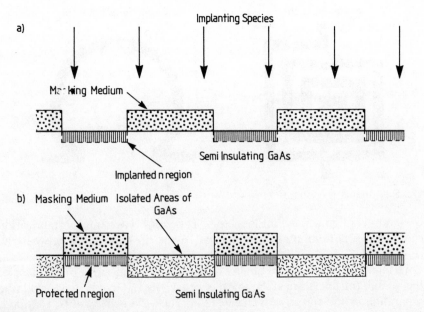

Fig. 5.10 Ion implantation isolation techniques.

electrically conducting state. Table 5.4 summarizes the usefulness of the three processes described above.

Table 5.4. Isolation processes

	Advantages	Disadvantages
Mesa etching	Simple process	Orientation-dependent. Produces non planar surface, introducing limitations to circuit layout
Selective implantation	Produces planar surface	Additional complexity to process
Proton isolation	Produces planar surface	Stability of isolation may be affected by temperature. Additional complexity to process

5.3 CIRCUIT BUILDING BLOCKS

5.3.1 Inductors

Inductors are used particularly in lower frequency circuits as tuning elements between transistors ([9], [10]). Fig. 5.11 shows schematically two

Spiral inductor   **Single loop inductor**

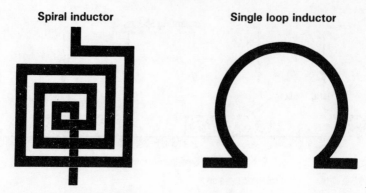

Fig. 5.11 Inductor configurations.

common configurations of inductor. They pose two major fabrication challenges — definition of the metallization and the connection between the centre of the spiral and the rest of the circuit.

To maintain a high circuit quality factor (Q) at low microwave frequencies the metal thickness must be of the order of 3 microns thick and the separation of the spirals is typically 12 microns. To achieve this, sputtered films of a gold based alloy are deposited on to the semi-insulating substrate, annealed and then the spiral pattern defined by ion beam milling. The milling process removes metal not protected by a photo-resist layer at the rate of a few microns per hour.

An alternative technology for achieving a thick metallisation is to plate up the spiral using a similar process to that described for the air-bridge process in Section 5.3.4. One disadvantage of this process is that the edge definition of the plated spiral is somewhat worse than the milled metallization, however it is easier to build up the metal thickness to values considerably greater than 3 microns which can be an advantage in some lower frequency circuits.

The second challenge, that of connecting the centre of the spiral to the rest of the circuit, can be met either by the use of the air-bridge technology or by a metal underpass connection.

The air-bridge technology, similar to that described later in Section 5.3.4, offers a very good microwave solution — it has a low parasitic capacitance associated with it but is prone to mechanical damage. The span is necessarily large and any downward pressure on the bridge will cause it to be distorted and to short to the spirals.

A mechanically more rigid solution is one in which the connection is made to run between two dielectric layers underneath the spiral. This metal is thus isolated from the spiral and from any of the underlying circuit metallization. Fig. 5.12 shows a cross-section through a completed inductor with an underpass connection.

Fig. 5.13 shows a scanning electron microscope picture of a completed spiral inductor with an underpass connection and clearly shows the good definition achieved by ion milling the metallization of the spiral.

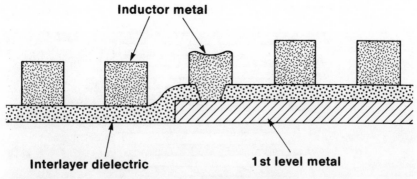

Fig. 5.12 Cross section through an 'underpassed' inductor.

Fig. 5.13 Scanning electron microscope picture of an 'underpassed' spiral inductor.

5.3.2 Capacitors

Capacitors are used within the MMIC for tuning elements, for interstage isolation and for bypass. Two types are currently in use:

(a) Interdigitated for low capacitance (less than 1 pF) [11]
(b) Parallel plate, metal insulator metal for 1–20 pF [10].

The interdigitated capacitor shown in Fig. 5.14 utilizes the capacitative coupling between adjacent fingers. From a process standpoint its fabrication is similar to that of the spiral inductor, namely the deposition of a thick metal film followed by ion beam milling with protecting resist defining the finger areas.

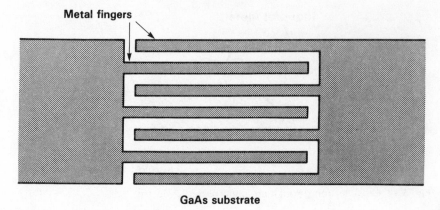

Metal fingers

GaAs substrate

Fig. 5.14 An interdigitated capacitor.

The metal-insulator-metal capacitor (MIM) is formed by two metallised areas separated by a dielectric material. Fig. 5.15 shows this type of capacitor. The dielectric used is chosen to give a reasonable capacitance/unit area, is easy to deposit, and stable in use. Table 5.5 gives three dielectrics in common use, the most common being Si_3N_4 or SiO_2.

The dielectric thicknesses used have a major outcome on the final yield of the circuit. Generally there are many more capacitors in an MMIC than

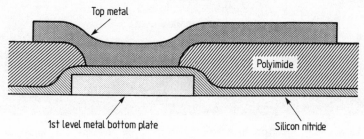

Top metal

Polyimide

1st level metal bottom plate

Silicon nitride

Fig. 5.15 Section through an MIM capacitor.

Table 5.5. Capacitor dielectrics

Material	Dielectric constant	C/A pf/mm^2	Q	Deposition technique
Silicon nitride	6.5	480	Good	Plasma assisted CVD
Silicon dioxide	4.0	340	Good to very good	Low pressure CVD
Polyimide	3–4.5	30	Good to very good	Spun and cured film

MESFETs and so the capacitor yield could dominate the yield of the circuit. Generally dielectric thicknesses of 1000–2000 Å are chosen which are sufficient to minimize the pinholes in the film that cause d.c. short circuits while maintaining a reasonable capacitance per unit area.

5.3.3 Resistors

Resistors are used in the MMIC for determining the bias applied to the transistors, for feedback networks and for terminations in power combiners and Lange couplers [10]. Two families of resistor are commonly used, those made from the GaAs active layers and those produced in deposited resistive films.

GaAs resistors. Perhaps the simplest way to produce a resistor is to use the GaAs material in which the active MESFET is fabricated. In this approach the resistor is fabricated in a similar manner to the FET itself apart from the fact that, in the resistor structure, the gate electrode is omitted. Two types of GaAs resistor are shown in Fig. 5.16. Fig. 5.16 (a) shows a mesa resistor where isolation is achieved by etching. In Fig. 5.16 (b) isolation is achieved by selectively implanting resistive areas in the semi-insulating GaAs substrate. In both resistor types calculation of the resistance is straightforward, being a function of the sheet resistance of the material and the dimensions of the resistor 'stripe'. It is also important to consider the resistance of the metal contacts to the GaAs as this can add many ohms to the total resistance of the structure.

The use of GaAs resistors is not without its problems as an ungated resistor can act, under certain bias conditions, as a planar transverse electron oscillator (Gunn oscillator). It is therefore important to ensure that

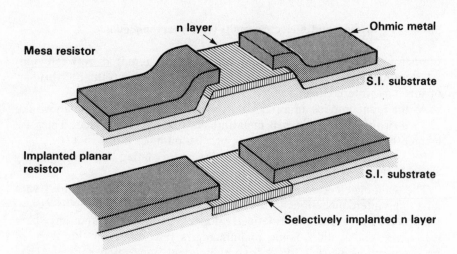

Fig. 5.16 Resistor technologies using GaAs active layer.

the bias voltage across the terminals of the resistor is below the threshold voltage for the onset of Gunn oscillations. It is also important to note that at voltages close to the onset of oscillation the I–V characteristic of the GaAs resistor is non linear and so the circuit designer has to exercise care in the layout of the resistor to ensure that it will operate in the 'low field' linear condition.

Thin film resistors. In Table 5.6 is listed a number of metal compositions with their electrical properties that have and are being used for resistors in MMICs. Cermets are potentially a very good resistor material as changes in composition of the silicon monoxide, chromium mix can alter the sheet resistivity of the deposited layer. However control of the composition is not easy and more work is necessary on the deposition technology if cermets are to be widely accepted as a resistor material.

Generally, whichever material from Table 5.6 is chosen additional process steps are required in order to incorporate it into an MMIC process — consequently wherever possible GaAs resistors are used.

Table 5.6. Some resistor materials used in MMICs

Material	TCR ppm/°C	Ω/sq	Manufacturing technique
TaN	−100	90 typically	Reactive spluttering
CrSiO (cermet)	−300 to +100	50–500	Sputtering from composite target
Bulk GaAs	+3200	300 for 10^{17} material	Epitaxial or ion implanted
NiCr	200	90 typically	Sputtering

5.3.4 Air-bridges and dielectric multi-level interconnections

In many sections of an MMIC a connection is required between non-adjacent metallized areas. This can be accomplished using either air-bridges or dielectric interconnections.

As the name implies an air-bridge is a bridge of metal running above the surface of the circuit between metal areas to be interconnected. They are formed by a photo-resist and plating process outlined in Fig. 5.17.

By careful control of the resist thicknesses and plating conditions, a very strong well-defined bridge can be formed.

The scanning electron microscope picture in Fig. 5.18 shows an FET with air-bridge interconnections to the source contact areas. Properly executed, the process can produce well-defined bridges between metal lands anywhere across the GaAs slice. Some manufacturers rely solely on this form of interconnection using it for passive component integration as well as for interconnecting the MESFET.

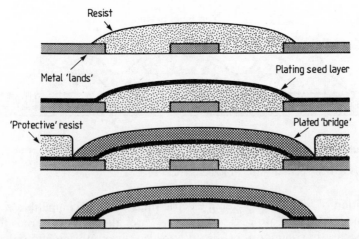

Fig. 5.17 Air bridge plating process.

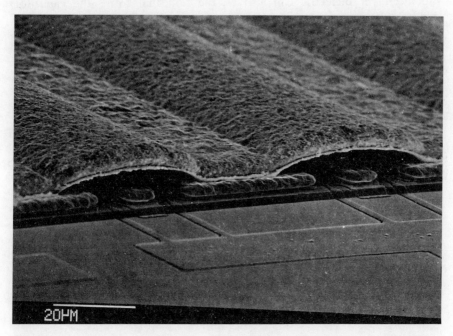

Fig. 5.18 Scanning electron microscope picture of an air-bridged power MESFET.

An alternative interconnection process is to make use of multi-levels of dielectric films. In this process a metal film can be made to run from contact area to contact area between two levels of dielectric. This is shown in Fig. 5.19. One advantge of this approach is the fact that it is essentially a planar process and overcomes the possibility that air-bridges can be damaged during die separation and mounting. It does, of course, involve the

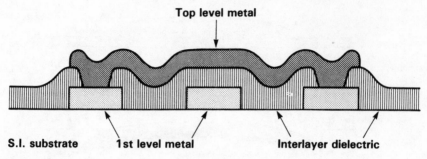

Fig. 5.19 Cross section through a dielectric 'cross over'.

deposition of 'pin hole' — free dielectric films that are compatible with the remainder of the IC process technology. Three dielectric materials for this purpose are in common use, silicon dioxide, silicon nitride and polyimide, the same as used for the capacitor dielectric. In fact for fabrication simplicity the interlayer dielectric can also be the film that forms the capacitor dielectric.

In many multi-level schemes polyimide is a preferred dielectric material as it has the ability to 'planarize' the surface of the MMIC making subsequent metallization over sharp underlying irregularities a relatively easy task.

5.3.5   Via hole technology

As circuits become more complex it is desirable to remove the constraint of r.f. grounding the circuit around its periphery and processes are being developed to directly ground individual components. One way that this can be achieved is by a 'via hole' process in which holes are made from the back of the wafer to contact grounding pads on the front side [12]. This is illustrated in Fig. 5.20. Once the holes have been made it can be filled with the appropriate metallization which makes a direct contact to the front side metal areas requiring to be grounded. The process of mounting the chip onto a suitable substrate automatically grounds the components on the front side of the wafer.

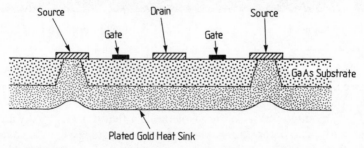

Fig. 5.20 Via hole process for individual component grounding.

In principle the process is straightforward — in practice the etching of the via hole, particularly in a thick (~200 μm) wafer can cause many technological problems. In thin substrates, such as those used for very high frequency or power MMICs, a wet etching process can be used. However the isotropic nature of most wet etches precludes this technique for the thicker substrates. In these cases dry (anisotropic) etch processes are used such as reactive ion etching. This process will be described in a later section of this chapter.

5.3.6 Component and technology integration

Defining the components that were described in the preceding sections of this chapter is not sufficient for the realisation of a full integrated circuit process. Its development is an iterative exercise where the requirements of the technologies to produce the components are combined with those needed to interconnect them. Generally many compromises have to be made before a manufacturable process can be finally realised involving not only aspects of the technology, but also satisfying the requirements of circuit design, assembly and packaging. The interaction necessary between circuit processing engineers, circuit designers and users cannot be overstressed and all successful processes are the result of constant consultation between *all* interested parties.

One example of a GaAs integrated circuit process is shown schematically in Fig. 5.21. In this figure it can be seen how the many layers of the circuit are built up sequentially starting from the initial defining of the MESFET and ending up with the passivation of the complete circuit. This is representative of the many process technologies in use worldwide; however there are many variations. It is probably true to say there are as many IC processes as there are laboratories fabricating them, each one subtly different from the other. Fig. 5.21 shows final metal interconnections being made using interlevel dielectric isolation, another and perhaps more popular approach is to use air-bridge interconnections for this process.

5.4. IC FABRICATION PROCESSES

In this section we describe the fabrication techniques that are used to produce the structures described in the preceding sections of this chapter.

Once the active n-layer has been formed, the whole integrated circuit can be processed using three basic fabrication techniques. These are:

(a) Lithography
(b) Deposition — metal and dielectric
(c) Etching — metal, dielectric and semiconductor.

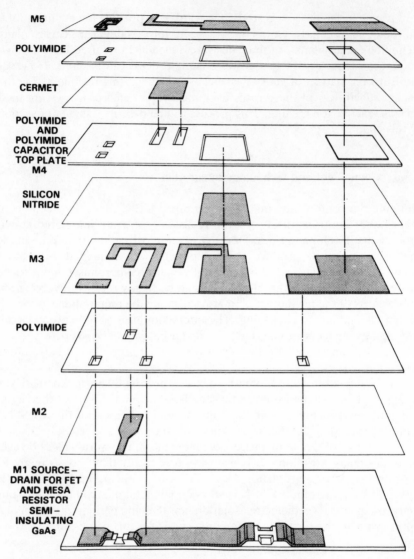

M5

POLYIMIDE

CERMET

POLYIMIDE AND POLYIMIDE CAPACITOR TOP PLATE M4

SILICON NITRIDE

M3

POLYIMIDE

M2

M1 SOURCE – DRAIN FOR FET AND MESA RESISTOR SEMI – INSULATING GaAs

Fig. 5.21 Schematic of GaAs MMIC process.

Various sequences of these three techniques are needed to allow complete circuits to be fabricated [13].

5.4.1. Lithography

Lithography forms a vital part of any process technology, for all process stages are preceded by a lithography step that defines the pattern for etching or for metal deposition. Basically the lithography process involves defining a pattern in a thin photo-sensitive film (photo-resist) by a process of optical exposure through a photographic mask and subsequent differential removal by immersion in a resist developer solution. Fig. 5.22 shows two different methods of photolithography; by direct contact and by optical stepper. In the direct contact method (Fig. 22 (a)) a mask containing the pattern to be defined is placed in hard contact with the photo-sensitive resist film that has previously been spun on to the GaAs wafer as a uniform thickness film. The resist not protected by the dense areas in the mask is irradiated with ultra violet light. The irradiated film of resist is placed in a developer solution and, if the resist is a 'positive' one, those resist areas exposed by the radiation will be removed to expose the substrate surface. If the resist is a negative one then those resist areas not exposed will be removed in the developer. The resist film left after development can then be used as an etch resistance film or as a film for the life-off process described previously in Fig. 5.8.

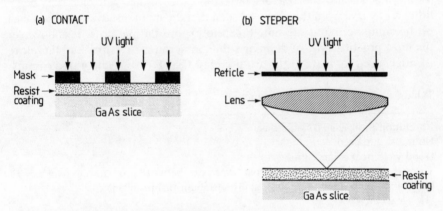

Fig. 5.22 Contact lithography and DSW configurations.

The optical stepper process, Fig. 22 (b), is similar to the in-contact one in terms of exposure and development of the resist except that the mask is held some distance away from the resist coated sample and is exposed by projecting an image of the reticle (mask) on to the surface of the GaAs slice.

Each process has advantages and disadvantages over the other and these are listed in Table 5.7 on page 310.

As far as producing high performance MESFETs is concerned the preferred method of lithography is contact printing for at present it gives the shortest gate lengths.

However neither approach is capable of defining dimensions less than 0.5 micron. For sub-0.5 micron geometries electron beam lithography is used. Here a very narrow beam of electrons is used to expose electron sensitive resists. This type of lithography overcomes the problems of present optical processes and enables geometries down to 0.1 microns or less to be defined.

Table 5.7. Comparison of contact lithography and DSW

Photolith method	Advantages	Disadvantages
Contact lithography	Available at 0.5 μm resolution	Mask procurement difficult Mask wear due to contact with resist Require very flat masks and wafers
Direct step on wafer	No mask wear 5:1 or 10:1 projection system so mask procurement less difficult	Resolution guaranteed only to 1 micron Process conditions need to be very carefully controlled

Table 5.8 shows the factors affecting the choice of electron beam lithography and from this it can be seen that the main disadvantages are cost of the equipment and the time taken to expose the patterns. It is however the most practical way of defining sub-0.5 micron gate stripes and therefore is widely used for state of the art discrete MESFETs and integrated circuits.

Table 5.8. Electron beam lithography.

Resolution to at least 0.2 microns
No masks required
Good yield over a large area
Throughput time limited as each pattern has to be written separately (can mix E-beam and photolithography to partially surmount this problem)
High machine cost

Another real advantage of the electron beam process is its high yield. Fig. 5.23 compares the yield of 0.5 micron gate lines of different gate width and clearly shows a yield advantage at wide gate dimensions. For this reason many companies are using the electron beam technology for gate dimensions up to 1 micron despite its throughput limitations.

5.4.2 Deposition processes

The fabrication of an MMIC requires the deposition of both metal and dielectric films and this is achieved using some of the different techniques explained in the following sections of this chapter.

Metal deposition techniques. Metal films are applied to GaAs wafers either by evaporation or sputtering dependent on the required final metal thickness. In the evaporation process the metal is heated to a temperature

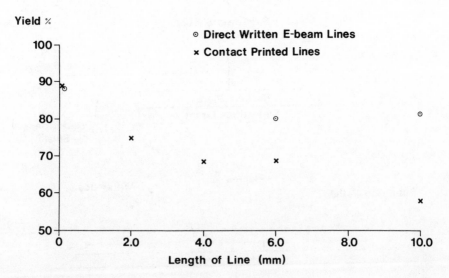

Fig. 5.23 Yield of 0.5 micron gate lengths of different widths for contact lithography and E-beam direct write.

sufficient to cause vaporization. This vaporized metal then deposits on to the GaAs slice situated inside the vacuum chamber held at a pressure of around 10^{-7} torr. Vaporization of the metal is achieved by either heating the metal in an electrically heated coil or by directing an electron beam on to the charge of the metal to be deposited. This method of metal deposition is a somewhat directive one and is not recommended for metallizing over steep edges but is good for lift-off.

The deposition of thick films (≥ 1 micron) and of alloys is best carried out by sputtering. A schematic of a sputtering equipment is shown in Fig. 5.24. In this process high energy ions are directed on to a metal target knocking metal atoms from the target which redeposit on to the GaAs slice. This is a low temperature process and relatively non-directive and hence good for step coverage. Step coverage is enhanced by biassing the substrate as this causes deposited metal atoms to be redeposited on to steep steps. It is particularly good for alloy deposition for the nature of the process causes the deposited film to retain the composition of the target.

Plasma-assisted chemical vapour deposition (PECVD). None of the above processes can effectively be used for depositing dielectric films. The standard process for dielectric deposition is chemical vapour deposition in which the relevant gases are reacted to form the nitride or oxide which deposits onto the substrate at a high temperature. Such processes cannot however be used in GaAs technologies as the temperatures required will adversely affect prior process steps (e.g. ohmic contacts) and may cause the GaAs to decompose. Consequently PECVD processing is used. Here the temperature of the reaction required to form the dielectric film is lowered by the formation of a plasma in which the energetic electrons increase the

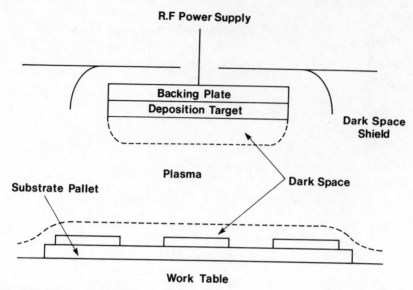

Fig. 5.24 Schematic of metal sputtering equipment.

reaction temperature allowing substrate temperatures below 300°C to be maintained. Silicon nitride is by far the most popular dielectric film used in GaAs processing and this is formed by reacting silane and ammonia in a closed system under a plasma [14].

$$3SiH_4 + 4NH_3 \rightarrow Si_3N_4 + 12H_2$$

One of the disadvantages of the PECVD processes is that the deposited films tend to be amorphous and are not particularly dense. However by optimizing the deposition conditions, films of sufficient quality for IC fabrication can be grown.

Table 5.9. Comparison of deposition techniques

	Main advantages	Main disadvantages
Filament evaporation	Easy process Relatively low cost equipment	Poor for refractory metals Wafers can reach high temperatures
Electron beam evaporation	Good for refractory metals and lift-off processes	Expensive equipment
Plasma assisted chemical vapour deposition	Low temperature process	Produces amorphous films with high etch rate
Sputtering	Good for thick metal, step coverage and alloys	Cannot be easily used for lift-off processes

5.4.3 Etching processes

Once the metal and dielectric films have been deposited they need to be selectively removed to form the various patterns that build up the integrated circuit. This is carried out by a combination of lithography and etching. There are many etching processes available to the GaAs processing engineers. The major ones are listed below and will be described in some detail.

(a) Ion beam milling [15]
(b) Reactive ion etching and plasma etching [16]
(c) Wet chemical etching.

Of these processes only (c) is not a 'dry' process and the move away from wet chemical etching is becoming an important aspect of GaAs technology. In many instances the dry process technology is being adopted because of its controllability and cost effectiveness. In others it is proving to be the only practical method of removing many metals and dielectrics.

Ion beam milling. This is a form of electronic 'sand blasting'. A schematic of the milling equipment is shown in Fig. 5.25. An energetic ion beam is directed onto the material to be etched and metal removal is purely by the impact of these ions on to its surface. Inert gases are used in this process and no chemical reaction takes place. There is provision within the milling equipment to rotate and tilt the substrate material. This is important and

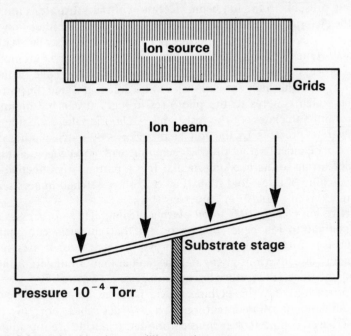

Fig. 5.25 Schematic of ion beam milling equipment.

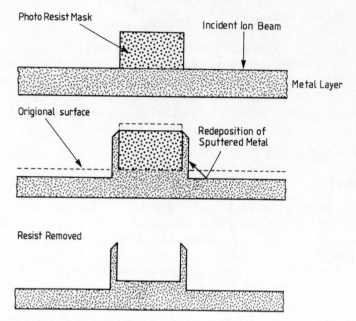

Fig. 5.26 Edge build up on ion beam milled structures.

experience has shown that etch rate is a function of the angle that the substrate presents to the ion beam. Rotation of the substrate is intended to unify the etch rate across the wafer. Ion beam milling is a rather slow process with etch rates around 1−2 microns per hour − it is also not without its problems. One problem that inevitably arises when first attempting the milling process is the build up of etched material at the edge of the metal structure. How this happens is shown in Fig. 5.26 and is due to sputtering of material which adheres to the photo-resist edge. It can be minimized by optimizing the thickness of the resist and by choosing the correct angle that the substrate presents to the ion beam. Correctly carried out ion beam milling can be an important process technique with good edge definition and no undercutting of the masking media. It is a particularly effective process for the etching of thick gold films. Fig. 5.27 shows etched fingers separated by 12 microns in a gold film 3 microns thick.

Reactive ion etching (RIE) and plasma etching. These processes are not too dissimilar to ion beam milling except that chemical compounds are introduced into the equipment to improve the etch rate. It uses relatively high energy directive ions giving only a small amount of undercutting of the masking medium. It therefore has advantages over wet etch processes and is used for instance for the etching of via holes in GaAs section where a minimum amount of undercutting is highly desirable. For this purpose chlorine based compounds such as carbon tetrachloride are introduced into the RIE equipment. RIE is also being utilized more and more on multi-level

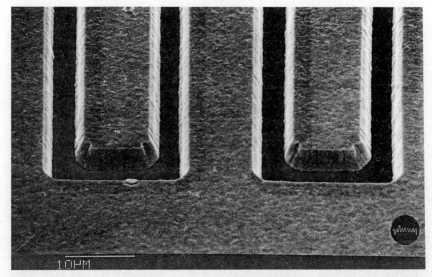

Fig. 5.27 Ion beam milled patterns in a 3 micron thick gold film.

resist processes because of its high anisotropic etching behaviour. In this process fluorine based chemicals are used.

The plasma etching process is very similar to reactive ion etching and they are distinguished from each other only by the operational pressures and voltages at which the equipments are operated. RIE operates at lower pressures ($\sim 10^{-2}$ torr) than plasma etching and is therefore somewhat more directional.

Fig. 5.28 summarizes these techniques with regard to the directionality of the etching process.

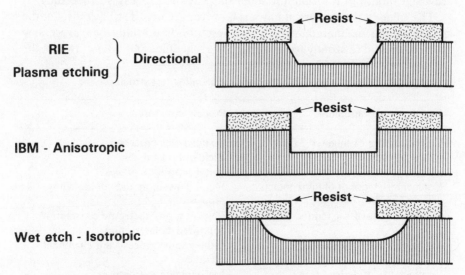

Fig. 5.28 Etch profiles for different etching processes.

It is important to note that although wet processes are rapidly being superseded by dry ones they are not without their problems. A major concern of the GaAs technologist is the damage caused by the dry process techniques which generally are high energy processes. It appears to be relatively easy to cause damage to the surface layers of crystalline GaAs which can cause serious degradation in electrical properties of the material and hence cause changes in the circuit performance.

5.5 PROCESS CONTROL

It is obvious from the preceding sections of this chapter that many technologies and process steps are necessary to fabricate a GaAs monolithic IC. If any one of these processes were not adequately controlled, then either the final circuits would not operate or the yield would be seriously impaired. For this reason many IC process lines maintain strict control by monitoring the process stages through the use of process control monitors. These are test patterns on the processed wafer, placed amongst the circuits being fabricated, which allow functional tests to be carried out as the processing proceeds. The functions tested are normally d.c. or low frequency tests on relatively simple structures backed up with visual measurements. From these results it is possible to monitor the degree of control of the fabrication process and to make judgements on whether to abort the process run if measured parameters drift outside previously specified limits.

This monitoring process is vitally necessary when setting up a process and can be used to advantage on mature processes to generate vital yield data.

Examples of the information gained from the process monitor test patterns is given in Table 5.10. Here we see that relatively simple tests can reveal a wealth of information about the way the process is proceeding.

The value of the process monitor is however greatest if strict control of the process steps is maintained. It is therefore important that processing personnel adhere strictly to the processing schedules and do *not* 'tweak' the

Table 5.10. Processes controlled by process monitor test structures

In-process measurement	Process monitored
Characteristics of ungated FET	Channel etch process
Characteristics of gated FET	Yield of FET process
Metallization resistances	Metal deposition process
Contact resistance of ohmic contacts	Metal deposition and contact alloy process
Measurements of capacitors	Thickness and dielectric constant of deposited dielectric film
Microscope measurement	Lithographic process and alignment accuracy
Conductivity of active layer	Ion implantation process

process when things appear to go wrong. In real-life situations the advent of computer-controlled equipment is making 'tweaking' more difficult. In these equipments the process steps are pre-programmed making interference with the planned schedule more difficult.

5.6 FUTURE TRENDS

It is never easy to predict the future in such a rapidly advancing area as GaAs microwave IC technology. However in order to meet the demands of the systems engineer in going to higher and higher frequencies and with improved performance at lower frequencies, it is certain that improvements in the technology are required.

As far as the materials technology is concerned a major area for research is that of the substrate. Relatively recent improvements in material quality have been made with the availability (though at a premium) of In-doped and ingot annealed GaAs boules. However the consistency in the quality of these and the 'standard' materials still leaves a lot to be desired. Surface quality of the polished wafer will also have to be improved if the fine line lithography required for the highest frequency circuit is not to be a yield limiting factor.

Lithography developments for sub-half micron feature sizes appear to be centred around the use of electron beam lithography. As described in the preceding text this is capable of very fine geometries but is at present limited in its use by poor throughput, equipment reliability and through extremely variable quality and non-standard electron sensitive resists. Increased throughput is one of the main topics for the electron beam machine manufacturer and many machines are currently being designed to turn them from laboratory research tools to high throughput manufacturing equipments. Resist developments are also carrying on apace; here adhesion to the GaAs surface, resistance to plasma processing and increased sensitivity are top items for improvement.

Other lithographic processes are also in development that could eventually aid the microwave circuit process technologist in reducing dimensional tolerances with minimal effect on yield. X-ray lithography is a theoretically attractive method for producing extremely fine line geometries. The short wavelength (2–10 Å) makes diffraction effects non-existent as is back scattering from the substrate. These properties mean that 'proximity' lithography in very thick resist is possible with very high definition.

There would also be no problems with dust on the mask as it would be transparent to X-rays. It would appear to be an ideal solution to a difficult problem. However it also has a number of distinct disadvantages. The first is in the manufacture of a suitable X-ray source. In the main these are big and very expensive though work is under way to develop lower-cost desk-top synchrotrons capable of producing the necessary flux of soft X-rays. The second is the difficulty in manufacturing the mask. At present very thin membranes are used coated with a metal film (gold) that is opaque to X-rays. This film is patterned to allow the X-rays to penetrate to the underlying resist-coated sample. The manufacture of the masking membranes is not a trivial matter and is generally carried out by first depositing the membrane on to a silicon slice (used as a dispensable substrate), coated with gold and then patterned. The silicon substrate is subsequently etched away to leave a very flimsy X-ray mask.

The optical stepper until the present day has been of somewhat limited use to the GaAs technologist mainly due to its resolution limits of around one micron. However recent dramatic improvements in lens design has pushed the resolution limits to a reproducible 0.5 micron and the prospects of reducing this further are very real. These recent developments have caused the GaAs technologist to reappraise the situation regarding the type of lithographic process he will use for 0.5 micron and smaller feature sizes and we shall find more and more use being made of the sub-micron capabilities of the optical stepper for the same reasons as those advanced by the silicon processing engineers at the one micron resolution level.

The move, in GaAs technology, away from wet etching processes has had tremendous gain with regard to fabrication yield, dimensional control etc., but not without a number of problems. Dry processing techniques based on energetic plasmas has introduced a challenge to the technologists that of damage to the semiconductor surface. This damage can cause severe degradation of circuit performance and lead to premature operational failure. New plasma techniques involving microwave frequency sources and large magnetic fields have been shown in laboratory experiments to cause far less surface damage than conventional dry process techniques. This technique based on electron cyclotron resonance (ECR) utilizes a powerful magnetic field into which a microwave (2.45 GHz) signal is propagated. At a critical field strength, a cyclotron resonance of the electrons occurs and subsequent collisions produce a plasma of very high ionisation density. The magnetic field extracts the reactive ions and at the low pressures used (10^{-5} torr) directional processing can be achieved without the need for the high accelerating voltages that are known to be responsible for the damage caused in the conventional processes. Early experiments have shown that etch rates can be achieved that are some five times higher than for normal plasma etching.

The use of lasers in the GaAs processing technology is opening up new areas for research. Already they are being used to condition the surface of the substrate prior to epitaxial growth by removing traces of unwanted oxide. They are also being used for the selective heat treatment of contacts. Here a finely focused steered laser beam is used to individually alloy contact areas without causing overall heating of the GaAs slice. In this way the stoichiometry of the non-heat-treated areas of the slice can be preserved leading to improved device and circuit performance. Laser drilling of via holes through the GaAs is also being used as an alternative to wet etching or reactive ion etching.

This chapter has dealt exclusively with a GaAs technology, but with developments such as those briefly outlined above it can be seen that existing technologies have a considerable amount of 'stretch' in them in terms of high frequency potential. However many new heterojunction materials structures with even higher frequency performance prospects are now being researched and indeed some have been developed to the stage where devices are already being commercially exploited. These materials will one day have a place in the microwave systems of the future, however, there is tremendous potential still remaining for GaAs MESFET-based integrated circuits and the temptation to jump from one improved material to another should be resisted if a mature compound semiconductor technology is to be achieved upon which circuit designers can design with the same confidence enjoyed by those working in silicon technology.

REFERENCES

1 Pengelly, R. S. and Turner, J. A. (1976) 'Monolithic broadband GaAs FET amplifiers' *Electron. Letters* **12** (10), 251−252.
2 Knight, J. R., Effer, D. and Evans, P. R. (1965) 'The preparation of high purity gallium arsenide by vapour phase epitaxial growth' *Solid State Electronics* **8** 178.
3 Pamplin, B. R. (1980) *Moleccular beam epitaxy* Pergamon Press, Oxford, England.
4 Gibbons, J. F., Johnson, W. S. and Mylroie, S., W., (1975) *Projected range statistics, semiconductors and related materials* (2nd Edition) Dowden, Hutchinson and Ross, Stroudsburg, PA.
5 Schwartz, B. (1969) 'Ohmic contacts to semiconductors' *Electrochem. Soc. Conf. Proc., Montreal 6−11 Oct. 1968.*
6 Cox, H. and Strack, H. (1967) 'Ohmic contacts for gallium arsenide devices' *Solid State Electronics* **10** 1213.
7 Fukui, H. (1979) 'Optimal noise figure of microwave GaAs MESFETs' *IEEE Trans. Electron Devices* **26** 1032.
8 Adachi, S. and Oe, K. (1983) 'Chemical etching characteristics of (001) GaAs' *J. Electrochem. Soc.* **130** 2427.
9 Pucel, R. A. (1981) 'Design considerations for monolithic microwave circuits' *IEEE Trans. Microwave Theory Tech.* **29** 513.
10 Pengelly, R. S. (1982) *Microwave field effect transistors − theory, design and applications* Wiley, New York.
11 Esfandiari, R., Maki, D. W. and Siracusa, M. (1983) 'Design of interdigitated capacitors and their application to GaAs monolithic filters' *IEEE Trans. Microwave Theory Tech.* **31** 57.
12 D'Asaro, L. A., DiLorenzo, J. V. and Fukui, H. (1978) 'Improved performance of GaAs microwave field effect transistors with low inductance via-connections through the substrate' *IEEE Trans. Electron Devices* **25** 1218.
13 Williams, R. E. (1984) *Gallium arsenide processing techniques* Artech House Inc., Dedham, M. A.
14 Hollahan, J. R. and Rosler, R. S. (1978) *Ion-beam techniques for device fabrication* Academic Press, New York.
15 Spencer, E. G. and Schmidt, P. H. (1971) *J. Vac. Sci. Technol.* **8** S52.
16 Coburn, J. W. (1982) *Plasma Chemistry and Plasma Proc.* **2** 1.

Chapter 6
GaAs Sampled Analogue Integrated Circuits

John A. PHILLIPS and
Stephen J. HARROLD

6.1. INTRODUCTION

The use of GaAs for implementing ICs in microwave and fast digital applications is well documented elsewhere in this book and is becoming increasingly more important. However there is one further area of application where GaAs can make a significant contribution which has not yet been so well described or demonstrated. This is the area of high performance sampled analogue ICs. This chapter describes the properties of sampled analogue signals and then gives details of how GaAs ICs may be designed to take advantage of these properties. The chapter concludes by describing two applications in the sampled-analogue domain.

One notable area where sampled analogue signals are processed is in the analogue-to-digital converter. This is an area where, in common with the sampled analogue ICs and techniques described later, relatively little work has yet been done using GaAs. The field is, however, so potentially large and important that the omission of coverage in this subject is deliberate, in order not to perpetrate an injustice on the subject. A number of references are included for those wishing to research the topic ([4], [6]–[9], [17], [18]).

In describing sampled analogue systems, the first question to be answered is 'Why sample?'. What advantages are gained by sampling a signal that are not also given by operating with a continuous signal? The answer lies in two main areas: the first concerned with the time multiplexing of signals, and the second concerning signal processing. These areas are described in more detail in the later parts of this introduction.

In the remaining section of this chapter, we develop (in Section 6.2) condensed theory of sampling and sampled signals, and cover signal flow graphs (SFGs) for describing filter circuits. In Section 6.3 we consider the practical aspects of sampling using GaAs devices, followed in Section 6.4 by examples of other analogue functions needed in sampled analogue circuits. In Section 6.5 we consider the design of two integrated circuits illustrating the use of techniques and methods described in previous sections.

6.1.1 Time division multiplexing

Time division multiplexing refers to the process of sending a collection of signals through a common transmission medium (such as a coaxial cable or a satellite link) by time sharing the use of the transmission medium between the signals. If a collection of continuous signals, having common frequency components, were to share a common transmission medium the resulting (received) signal would be made up of the interaction of all the signal and it would be impossible to recover any one of the transmitted signals from the received signal. This situation is obviously unacceptable, and one solution is to use a separate transmission medium (cable, satellite etc) for each signal. This solution suffers a severe cost penalty! The time-multiplexing alternative (Fig. 6.1) entails sampling the original signals in turn, and then transmitting these samples rather than the continuous signals. Provided the receiver has knowledge of the order and rate at which the signals were sampled, and the original signals occupy a limited bandwidth up to less than half the sampling frequency (see Section 6.2) then any of the original signals can be recovered from the received time-muliplexed signals. The condition that the original signals occupy a limited bandwidth is not generally a severe one.

6.1.2 Signal processing

Signal processing inevitably entails the passing of a signal through a filter of some kind in order to modify the signal and extract some information from

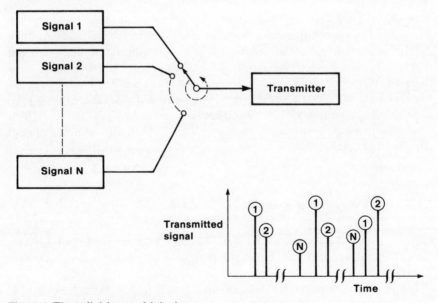

Fig. 6.1 Time division multiplexing.

it. This filter may comprise passive (RLC) components, or may be an active filter including some amplifying elements. In general, for continuous signals, the filter response depends quite critically upon the absolute values of the components used, and real component tolerances necessitate that some tuning of the component values is carried out after the filter is constructed. Furthermore, if a change in the filter response is required (i.e. some degree of programmability) then more tuning will be called for. This tuning must usually be carried out by hand by skilled operators as the frequency response depends on the interaction of all the component values and is thus a time-consuming and expensive task and is highly undesirable.

Since absolute component tolerances are generally much worse in integrated circuits than in discrete components, and the tuning of integrated components demands technically very difficult and expensive techniques such as laser trimming, it is also not generally possible to integrate continuous time filters in order to take advantage of the reduced fabrication costs and increased reliability usually associated with integrated circuits.

If the original (continuous) signals can be replaced with sampled signals (analogue or digital) then the whole situation changes. In the analogue domain it becomes possible to build filters whose response is determined by the sampling rate and by component ratios which can be defined very accurately in integrated circuits. Consequently the only tuning operation required becomes one of adjusting the sampling rate. Indeed, a significant degree of programmability is given simply by adjusting this one parameter. As with time-multiplexing, the bandwidth of the original continuous signal must be limited, but in exchange for this requirement it becomes possible to integrate the filter and thus to take advantage of all the benefits associated with integrated circuits.

In the digital (rather than analogue) domain the filter response is determined by the architecture, sampling rate, various control codes, and by the precision of the arithmetic. A high degree of flexibility can be achieved if it is possible to change the control codes through software commands, and this is the prime motive for the vast amount of effort presently being devoted to developing digital signal processing circuits. In order to take advantage of this flexibility an interface is required between practical signal sources which generally produce continuous analogue signals, and the sampled digital circuits; this interface again requires a sampled analogue circuit.

6.1.3 GaAs in sampled analogue ICs

Having answered the question of 'Why sample?', the next question is 'Why GaAs?'. The answer to this lies in the various conflicting demands made by integrated sampled analogue circuits.

The first requirement is for devices capable of acting as analogue switches. These devices must not allow any form of interaction between the control signal and the information signal which would alter the sampled value of the information signal and thus introduce distortion. Bipolar transistors are ruled out by virtue of the fact the base (control) electrode is not electrically isolated from the collector-emitter (information) path. In comparison, MOS field effect transistors make almost ideal analogue switches subject to the restriction that typical FET on-resistance and circuit capacitance values limit RC time constants to values above 1 ns, and thus sampling rates to well below 1 GHz. The second requirement, however, is for devices capable of building fast, high gain, amplifiers. In this respect MOSFETs suffer a severe disadvantage compared to bipolar transistors. Bipolar transistors can be fabricated with much smaller dimensions (in the direction of current flow) than can FETs, and consequently have a greatly superior frequency performance. Furthermore the ratio of transconductance (g_m) to output conductance (g_o) for bipolar transistors is also superior to that of FETs. Amplifier gain is critically dependent upon this parameter. Thus the automatic technology choice for high gain, fast, amplifiers is bipolar. The use of a FET technology, as made necessary by the requirement for analogue switches, results in a sampled analogue circuit whose performance is severely limited in comparison with an ideal technology capable of realising good bipolar devices and good FETs on the same chip.

The use of GaAs for integrating sampled analogue circuits is one way of achieving significantly higher performance without having to resort to the complexities of a merged silicon bipolar/FET technology, which in many cases compromises the properties of one or both types of device. Being a field effect device, the GaAs MESFET makes a good analogue switch (subject to the restriction that the gate-channel Schottky contact is not forced into forward conduction) and the physical properties of GaAs allow a significant speed advantage over comparably sized silicon devices, which enables fast amplifiers to be constructed. The improvement in performance arises from an electron mobility which is approximately six times that of silicon, and a peak electron velocity which under moderate electric fields is approximately twice as great, whilst the relative permittivities of GaAs and silicon remain approximately equal. A GaAs FET can thus be expected to exhibit a larger transconductance (and thus current drive) than a comparably-sized silicon FET, while maintaining a similar capacitance. Since the speed of a circuit is essentially determined by the current available to charge (and discharge) the capacitances present, a GaAs IC has an intrinsic speed advantage over an equivalent silicon IC. Differences in the processing technologies required (and also the relative immaturity of GaAs technology) introduce different parasitic elements (resistance and capacitance) in the GaAs and silicon FETs. These differences partially offset the intrinsic speed advantage of GaAs, but the essence of the above argument still holds true for sampled analogue ICs: fabricated devices do exhibit an advantage of between three and five times higher speed than equivalent silicon ICs.

6.2. SAMPLED SIGNALS AND FILTERS

The application of GaAs (and silicon) in the sampled analogue domain depends upon the properties of sampled signals, and the design of circuits requires a knowledge of the use of those well-known and much-feared mathematical tools, the Fourier transform and the Z-transform. For completeness, this section describes the properties of these transforms and of sampled signals, and then gives details of how these properties may be used to design signal processing elements such as the filters described in Section 6.5 of this chapter.

6.2.1 The Fourier Transform

Over a given interval, a function can be represented as a linear combination of an orthogonal set of different functions. Such a set is defined as comprising functions whose inner (or dot) product is zero. In other words, the integral of the product of any one function and the complex conjugate of any other function evaluates to zero over a given interval. The functions are all completely independent of each other over the interval. For example, the unit vectors $\bar{a}_x$, $\bar{a}_y$, $\bar{a}_z$ comprise a set of three orthogonal vectors in Euclidean three-dimensional space. The inner products are zero, and each unit vector is independent of the others. Similarly, the set of cosine and sine functions $\sin(2\pi f_o b)$, $\sin(4\pi f_o t)$, ..., $\cos(2\pi f_o t)$, $\cos(4\pi f_o t)$, ... also form an orthogonal set over the interval $t_o \leqslant t < t_o + 1/f_o$. Likewise, the set of complex harmonic components, made up of the function $\exp(j 2\pi n\, f_o t)$ where n is an integer, are orthogonal over the interval $t_o \leqslant t < t_o + 1/f_o$. This property of reducing an arbitrary function to a linear combination of a set of orthogonal functions, is the central pillar upon which the Fourier Transform rests. The Fourier Transform is restricted to those functions with a finite number of discontinuities, maxima and minima in the given period, and the integral of the magnitude of the function over the period must exist (i.e. not be infinite), but if these conditions are met then the function can be written as the Fourier series:

$$s(t) = \sum_{n=0}^{\infty} a_n \cos(2\pi f_o\, nt) + b_n \sin(2\pi f_o\, nt)$$

for $t_o < t \leqslant t_o + 1/f_o$

or, more compactly, as

$$s(t) = \sum_{n=-\infty}^{\infty} c_n \exp(j\, 2\pi f_o\, nt)$$

for $t_o < t \leqslant t_o + 1/f_o$

It should be noted that the expanded Fourier series is periodic with period $T = 1/f_o$ outside the interval $t_o \leq t < t_o + T$, even though only defined within this interval by the above equation. If, however, the given function $s(t)$ is also periodic outside of this interval then the Fourier series is unchanged but now applies for all time. In other words, the Fourier series for a periodic function can be found from the Fourier series of one period by simply removing the restriction that the series is only defined for the length of that one period.

Referring to the complex harmonic exponential representation of the Fourier series, the function $s(t)$ may also be interpreted as being a function of nf_o. The coefficient c_n represents the magnitude and phase of $s(t)$ at nf_o, and a plot of c_n versus nf_o would show the 'Complex Fourier Spectrum' of $s(t)$, noting that n is restricted to discrete, integer, values and the spectrum is thus a line plot.

Most real signals of interest are not restricted to a finite interval of time nor are they periodic, rather they extend for all time and are non-periodic. Under these conditions it becomes possible to replace the summation representation of the Fourier series with an integral representation, and the Complex Fourier Spectrum is transformed from a line spectrum to a continuous spectrum given by

$$S(f) = \int_{-\infty}^{\infty} s(t) \exp(-j2\pi f) \, dt$$

This last equation, defining the relationship between the Complex Fourier Spectrum $S(f)$ and the original function $s(t)$, is known as the Fourier Transform, and the Fourier Transform theorem states that given the Fourier Transform of a function, then the original function can always be uniquely recovered.

The most important property of this transform concerns the uniqueness of the transform. For any function $S(f)$, there is only one possible function $s(t)$ which transforms to $S(f)$ by the Fourier Transform. The Fourier Transform Theorem also gives the rules for recovering the original function:

$$s(t) = \int_{-\infty}^{\infty} S(f) \exp(j2\pi ft) \, df$$

This is known as the Inverse Fourier Transform and is also unique.

The widespread use of the Fourier Transform has arisen because of the need to carry out an operation known as convolution. It can be shown that the convolution of two functions can be achieved by multiplying the Fourier Transforms of the functions and then taking the Inverse Fourier Transform of the product. This sequence of operations is generally much easier to perform than the convolution operation, and essentially justifies the existence of the Fourier Transform. Exactly what convolution is and why it is carried out so frequently is described in the next section.

6.2.2 Convolution

Defining a system as a set of rules which associates an 'output' function with an 'input' function, and restricting our attention to systems which are time-invariant and obey superposition, then it becomes possible to define the response of the system to an arbitrary input in terms of the combined responses of a sequence of appropriately scaled impulses. For example, the arbitrary input function shown in Fig. 6.2 (a) can be considered as a sequence of scaled impulses as shown in Fig. 6.2 (b). If the response of the system to one impulse is as shown in Fig. 6.2 (c), then the output response to the input sequence is the sum of the sequence of impulse responses as shown in Fig. 6.2 (d).

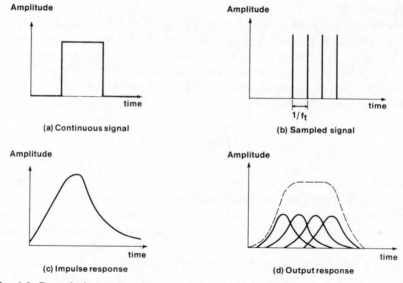

Fig. 6.2 Convolution.

Mathematically, the output response can be written as

$$s(t) \quad = \quad \int_{-\infty}^{\infty} r(\tau)\, h\,(t - \tau)\, d\tau$$
$$= \quad r(t)\, .\, h(t)$$

where $h(t)$ is the impulse response of the system and the symbol . denotes the convolution operation.

It should now be possible to see from the above that convolution is central to determining the response of any practical system (i.e. one that is time-invariant and obeys superpositon) to an arbitrary input, and that any process which makes evaluation of the convolution integral easier will be highly prized. As stated in the previous section, the Fourier Transform has this ability, replacing an integration operation with a simpler multiplication. Mathematically

$$s(t) = \int_{-\infty}^{\infty} r(\tau) \, h \, (t - \tau) \, d\tau$$

can be evaluated by taking the Inverse Fourier Transform of

$$S(f) = R(f) \, H(f)$$

where $S(f)$, $R(f)$ and $H(f)$, are the Fourier Transforms of $s(t)$, $r(t)$, and $h(t)$ respectively. It is this property which has led to the importance and widespread use of the Fourier Transform.

Note that $H(f)$ is the Fourier Transform of the system impulse response $h(t)$ and this is often referred to as the transfer function or frequency response of the system.

6.2.3 The sampling theorem

Up to this point in Section 6.2, we have dealt exclusively with continuous-time signals. Now we can introduce the concept of sampling and discuss the properties of sampled signals.

The sampling theorem (also known as Shannon's Theorem or Kotelnikov's theorem) states that if the Fourier Transform of a time function is zero for all frequencies above some frequency f_m, and the values of the time function are known for samples taken at a minimum rate of $2f_m$, then the original time function can be uniquely determined for all values of t.

The graphical proof of this theorem is shown in Fig. 6.3. Consider an arbitrary time function $s(t)$ (Fig. 6.3 (a)) whose Fourier Spectrum obeys the requirement of being zero above frequency f_m, as shown in Fig. 6.3 (b). If we now sample this time function at a rate f_s, as shown in Fig. 6.3 (c), we essentially introduce periodicity into the original, non-periodic, Fourier Transform $S(f)$. The period of the Fourier Transform of the sampled signal is the sampling rate f_s. It should be clear from Fig. 6.3 (d) that provided the frequency shifted images of the original Fourier Transform do not overlap, then any device which allows a given range of frequencies to be extracted from the sampled signal Fourier Transform can reconstruct the original Fourier Transform and thus the original continuous time signal. The condition that the frequency shifted Fourier Transforms do not overlap is met if

$$f_m < f_s/2$$

as already stated. The lower limit of f_s ($= 2f_m$) is known as the Nyquist frequency. Note that this frequency is a limit, and f_s must always be greater than (not equal to) this frequency in order to ensure non-overlapping shifted version of $S(f)$. A device which can extract a given range of frequencies from the Fourier spectrum is known as a filter.

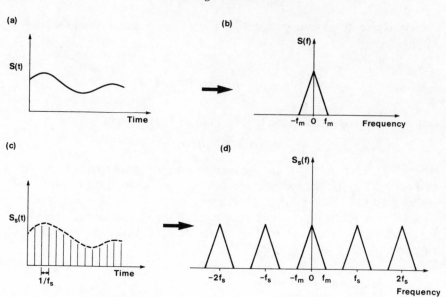

Fig. 6.3 The sampling theorem.

An alternative view of the sampled signal Fourier spectrum is to say that the sampling frequency is modulated by the original continuous-time signal. This method of shifting the frequency range of a signal is one way of producing amplitude-modulated (AM) signals for communication applications.

6.2.4 The Discrete Fourier Transform

Having dealt with the Fourier Transform of continuous signals, we can now look at the Fourier Transform of a sampled signal. Referring back to Section 6.2.1, it may be remembered that the Fourier Transform of a continuous, time-limited, signal with period $T = 1/f_o$ is a line spectrum having components at multiples of f_o. The spectrum is only defined in time for the duration of the time limited signal, and is shown in Figs 6.4 (a) and 6.4 (b) for a signal of duration $NT = 1/\Omega$. If the signal is periodic with period NT, then this line spectrum will be valid for all time.

If the continuous time limited signal is now sampled at a frequency $f_s = 1/T$, i.e. N samples are taken, the Fourier Transform becomes periodic with period f_s as described in Section 6.2.3 and shown in Figs 6.4 (c) and 6.4 (d). Note that the sampling theorem states that the original continuous signal can be recovered if the sampling rate is greater than twice the maximum frequency component of the original signal.

Mathematically, the Fourier Transform of the sampled signals can be written as

$$S(m\Omega) = \frac{1}{N} \sum_{n=0}^{N-1} s(nT) \, W^{-mn} \quad (0 \leqslant m \leqslant N - 1)$$

Here $S(m\Omega)$ is the nth component of the frequency spectrum, $s(nT)$ is the mth sample of the original signal, and W is the Nth complex root of unity, i.e. $W = \exp(j2\pi/N)$. The transform is also known as the Discrete Fourier Transform (DFT), and has properties similar to the Fourier Transform described in Section 6.2.1. Again its existence is justified by the need to perform convolution.

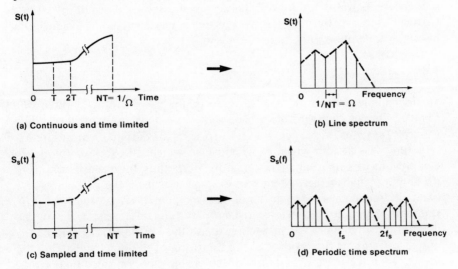

Fig. 6.4 Discrete Fourier transform.

The DFT, and hence convolution performed by the DFT, lends itself readily to computation by digital computer, although for signals involving many samples (N large), this may become very time-consuming and tedious. The Fast Fourier Transform (FFT) algorithm has been derived in an attempt to greatly reduce the computation time. This algorithm will not be described here as generally analysis of discrete signals and systems can be carried out more efficiently using the Z-transform described in the next section.

6.2.5 The Z-transform

The Z-transform is directly related to the DFT described above, and its existence is also justified by the need to perform convolution in discrete systems. Re-iterating the DFT, we have

$$S(m\Omega) = \frac{1}{N} \sum_{n=0}^{N-1} s(nT) \, W^{-mn}$$

If we replace W^m with the complex variable z, and the time limited sequence of samples $s(nT)$ with a non-limited sequence $f(n)$, then we can define the Z-transform as

$$F(z) = \sum_{n=0}^{\infty} f(n) z^{-n}$$

It can be shown that the discrete convolution of two sequences (e.g. a sampled signal and the sampled impulse response of a system) is equal to the inverse Z-transform of the product of the Z-transforms of each of the sequences. As was noted for continuous signals, this sequence of operations to determine the output is generally much easier than performing the discrete convolution operation.

6.2.6 Filters

Having finally arrived at the definition of the Z-transform in discussing sampled signals, we shall now show how it can be used to design real (discrete) systems, in particular we shall discuss the design of various filters.

A filter is a system which can remove or weight the various frequency components of a time function so that the output response is different from the intput signal to the system. An example might be the extraction of a band of frequencies from an amplitude modulated signal such as those shown in Fig. 6.3 (d), that is a band-pass filter. As was stated in the description of convolution in Section 6.2.2, the Fourier Transform of the output response is the product of the transform of the input signal and the transform of the impulse response of the system (the transfer function). In this example the input signal Fourier Transform is also shown in Fig. 6.5, and the key to the design of the filter is that the transfer function should ideally only be non-zero in the range of frequencies for which an output response is required.

In practice, real filters which operate in real time cannot have the ideal rectangular characteristics shown in Fig. 6.5 (b), as this would demand that the response to an input occurs before the input is presented to the system (the system is then referred to as being non-causal). Real (causal) filters will have finite bandwidth in the transition from pass-band to stop-band, and may also have finite ripple in the pass- and stop-bands, non-zero attenuation in the pass-band, and finite attenuation in the stop-band. In the continuous signal domain, these filters may be constructed from various combinations of resistors, capacitors and inductors, and much study has been devoted to the design of filters optimizing one or more parameters of the filter characteristics. Indeed, in general the system designer does not have to calculate the component values required for a particular filter characteristic, but may go to one of several standard references [22] giving tables of component values.

In the sampled-signal domain, one must somehow synthesize the sampled

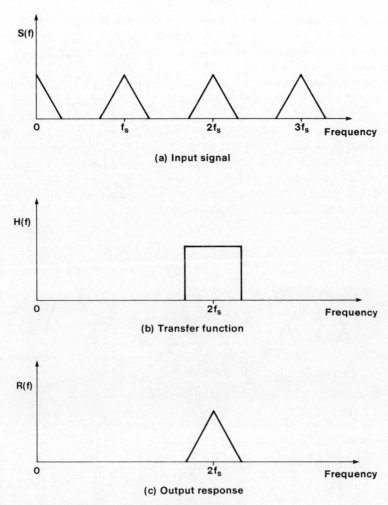

Fig. 6.5 Operation of an ideal band-pass filter.

version of the transfer function. This is commonly carried out by one of two processes, either by conversion from the transfer function of an equivalent continuous-time circuit using an appropriate transformation, or by an algorithmic technique using a computer to solve a set of linear or non-linear equations so as to minimize the difference between the designed discrete-time filter response and the required frequency response. The first method has the advantage that the vast amount of study applied to continuous-time filters can be taken advantage of. This technique is commonly used, for example, in the design of switched-capacitor filter circuits. The second method is more useful for cases where analogue equivalents do not exist to match arbitrary frequency response specifications. Several such algorithms exist ([23], [30]–[34]).

Due to the sampling action occuring in discrete-time systems, the filter transfer function will be repeated about multiples of the sampling frequency (Fig. 6.6). Input frequencies in the range of the higher frequency images will thus pass through the filter as well as the original required band of frequencies, indeed information present in these higher bands of frequency will appear to be shifted in frequency down to the original base-band. This frequency shifting effect is known as aliasing. In some systems the effect may be highly desirable, however in others precautions must be taken to ensure that no input signals are present in the range of the high frequency pass-bands. This generally entails the building of a continuous-time anti-aliasing filter to attenuate the undesirable frequencies.

The design and realisation of filters can be simplified using the concept of signal flow graphs. Before going on to describe the design processes in more

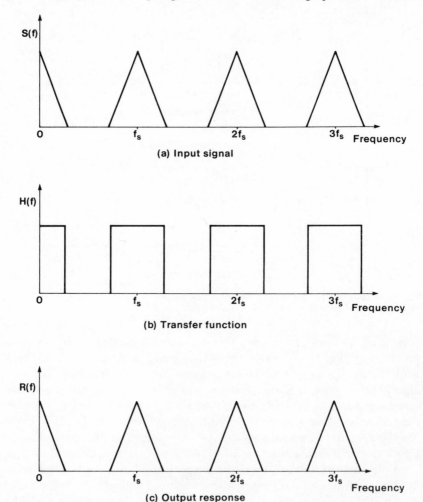

(a) Input signal

(b) Transfer function

(c) Output response

Fig. 6.6 Operation of a real band-pass filter.

detail, this is an appropriate point at which to discuss the properties and usefulness of these graphs.

6.2.7 Signal flow graphs

A circuit realisation of a system could be obtained from the mathematical manipulation of the system transfer function into a combination of much simpler terms which can then be realised from simple building blocks. Indeed, this is the basis on which many of the algorithmic methods of filter design work. However, for those system designers who, like the authors, are unable to factorize nth degree polynomials in their heads, it is generally much easier to manipulate the system transfer function by a graphical method. The graphs which are drawn are known as signal flow graphs. Even in those cases where a computer is used to factorize a complex transfer function, the use of signal flow graphs (SFGs) often eases the translation from the signal level to the circuit level.

The three basic operations required by a signal processing system are delay, multiplication and addition. These operations can be indicated by means of the symbols shown in Fig. 6.7 (a). The arrow indicates the direction of the signal flow, while the symbol alongside indicates the multiplication coefficient. Multiplication by z^{-N} corresponds to delay by N sampling periods. The absence of a coefficient implies multiplication by unity. Addition is signified by the convergence of two (or more) arrows at a node. Any system transfer function may be drawn as a signal flow graph and then manipulated to a different form by using the SFG equivalences shown in Fig. 6.7 (b). The aim of this manipulation is to rearrange the transfer function into a combination of building blocks which can be easily realised. Different arrangements will have different properties, for example in terms of the number of delay stages required, the maximum component value used, the component sensitivity, and the component costs. The final realisation of the system will depend upon the experience of the designer in choosing the form which optimizes the most critical parameters in the system.

6.2.8 Filter design based on continuous-time equivalents

Continuous-time filters are generally analysed using the complex frequency variable s, which is related to the real frequency variable $\omega(=2\pi f)$ by $s = \sigma + j\omega$. In order to synthesize a discrete-time version of a continuous-time filter, some transformation must be used which converts functions of s to functions of the discrete-time variable $z = \exp(j2\pi m/N)$.

Many such transformations exist, such as the LDI (lossless discrete integrator) transformation

$$s \rightarrow \frac{1}{T}\left(z^{1/2} - z^{-1/2}\right)$$

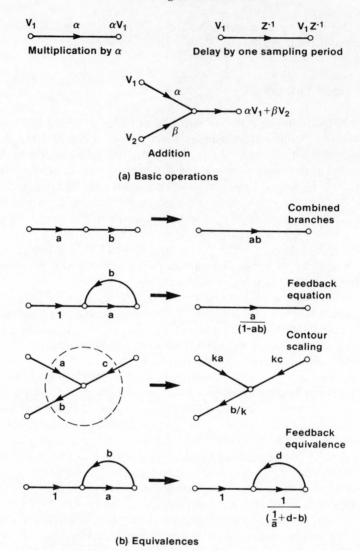

Fig. 6.7 Signal flow graphs.

and the bilinear transformation

$$s \rightarrow \frac{2}{T} \frac{(1 - z^{-1})}{(1 + z^{-1})}$$

where T is the sampling period. Other transforms exist having various desirable and undesirable characteristics. With the examples given, the LDI transformation is quick and easy to apply, but the discrete-time filter realisation is generally only an approximation of the original continuous time filter, particularly for frequencies approaching the Nyquist rate.

Conversely, the bilinear transformation produces more exact results, but is correspondingly more difficult to apply. Details of these and other transformations can be found in the literature ([24], [35]–[37]).

The general procedure to be followed in designing a discrete-time equivalent of a continuous-time filter is as follows. First, the transfer function of the original filter in the s-plane is derived, and from this the corresponding signal flow graph (SFG) is drawn. This continuous time SFG is then converted to a discrete time SFG using an appropriate transformation. The next stage in the process is to redraw the discrete time SFG so that it can be viewed as a collection of basic building blocks. For example, using the basic SFG rules described in Section 6.2.7 a finite impulse response (FIR) filter may be rearranged in combinations of the direct form or the cascade form of network as shown in Fig. 6.8.

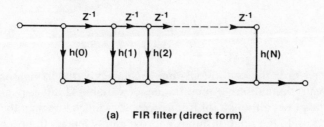

(a)   FIR filter (direct form)

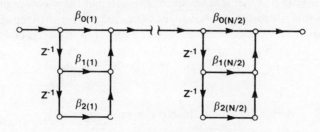

(b)   FIR filter (cascade form)

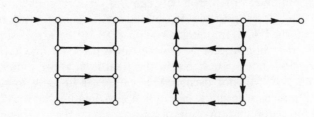

(c)   IIR filter (direct form)

Fig. 6.8 SFGs of various transfer functions.

The final stage in the design process involves the determination of the coefficients (i.e. the component values or ratios) in the discrete-time realisation. The coefficients of the SFGs will be specified in terms of the original filter components. In making the transformation from the continuous time domain to the discrete-time domain, frequency values are distorted so that the component values of the original filter cannot be used directly to determine the coefficient values. For example, considering the LDI transformation

$$s \rightarrow \frac{1}{T}(z^{1/2} - z^{-1/2})$$

if we replace s by $j\omega$ and z by $\exp(j\omega t)$ then the transformation becomes

$$j\omega \rightarrow \frac{1}{T}[\exp(\tfrac{1}{2}j\Omega T) - \exp(-\tfrac{1}{2}j\Omega T)]$$

$$\rightarrow \frac{j2}{T}\sin(\tfrac{1}{2}\Omega T)$$

$$\text{i.e. } \tfrac{1}{2}\omega T \rightarrow \sin(\tfrac{1}{2}\Omega T)$$

In other words the continuous-time frequency variable ω is not directly proportional to the discrete-time frequency variable Ω but dependent upon the sampling rate, although for frequencies very much less than the sampling rate ($\omega \ll 1/T$) the relationship is approximately linear. In order to design a discrete-time filter with a required cut-off frequency Ω_c and a sample rate of $1/T$ the corresponding continuous-time cut-off frequency ω_c must first be calculated by the above relationship. The original filter component values should fit this 'distorted' ω_c value, not the desired cut-off frequency value Ω_c. The coefficient values of the discrete-time SFG will then correspond to the 'distorted' original component values, and the discrete-circuit component values or ratios determined from these coefficients.

6.2.9 Filter design based on algorithmic methods

As an alternative to using a continuous-time filter as a starting point, discrete-time filters may be designed by algorithmic methods. The starting point this time is the ideal frequency response, and the aim is to realise a filter which minimizes the error between the ideal response and the final filter response. It is possible to immediately evaluate the Inverse Z-transform of the ideal frequency response so as to arrive at the filter coefficients, but this process is unnecessarily involved and difficult, and furthermore does not guarantee that the final filter representation will be either stable (i.e. will not oscillate) or causal (if filtering in real-time is required). Simpler techniques, based on approximating the ideal frequency response with a combination of simpler responses corresponding to different building blocks are often used. For example, the direct form of filter gives a transfer function of the form

$$H(z) = \sum_{n=0}^{N-1} h(n) z^{-n}$$

where $h(n)$ is the sampled impulse response of the block and the cascade form of filter a response of the form

$$H(z) = \prod_{k=1}^{N} (\beta_{0k} + \beta_{1k} z^{-1} + \beta_{2k} z^{-2})$$

Various algorithms exist, suitable for use in CAD [23], which are designed to approximate the ideal filter by terms corresponding to a combination of these and other building blocks. The different forms of building blocks produce systems with different properties, such as component sensitivity or phase shift, and as stated before the experience of the designer must be relied upon to use the form appropriate to the system requirements. Once the approximation has been optimized according to the various requirements of the filter (e.g. stop- and pass-band cut-off frequencies, allowable ripple, minimum stop-band attenuation, etc) then the discrete filter can be realised using the optimized coefficients to determine the filter component values or ratios in the building blocks.

6.3 GaAs SAMPLING CIRCUITS

The basis for all sampled analogue circuits is the sampling circuit itself. The theory relating to sampling and sampled signals is outlined in Section 6.2, and will not be iterated here. This section will develop, from a more practical point of view, methods for sampling analogue signals using GaAs MESFET-based circuits. The important problems peculiar to sampling in GaAs technologies will be covered, as well as some that will be familiar to those conversant with sampling using silicon devices.

The sampling of analogue signals using the silicon MOSFET is relatively simple. The gate of the three terminal device is well isolated from both the source and drain, and hence from the input and output ports of the switch. Interaction between the analogue signal and the switch control signal (commonly referred to as the 'clock') can occur due to coupling through the gate-channel capacitance, but such interaction (clock breakthrough) is independent of the analogue signal level. The effect of this interaction can either be removed by post-filtering, or cancelled by suitable circuit design (see Section 6.3.3). It should be noted that it is sometimes important to apply clock-cancelling techniques throughout a large circuit, rather than rely upon the use of a single clock cancelling circuit at the output; a build-up in the level of clock breakthrough can under some circumstances become great enough to significantly reduce dynamic range, and may prevent a circuit operating at any signal level.

The sampling of analogue signals using GaAs devices currently depends upon the GaAs MESFET; the absence (for the present at least) of a stable trap-free insulator in GaAs IC technology prevents the development of a GaAs MOSFET or MISFET to directly replace the popular silicon device. There is, however, more than a passing resemblance between the GaAs MESFET and the silicon MOSFET, and it is tempting to view the former as a higher performance replacement for the latter. Level-independent interaction between the clock and the analogue signal occurs in the same manner as with the MOSFET, and exactly the same clock-cancelling and filtering techniques familiar in MOS switch design can be applied to the GaAs MESFET to obtain a reduction in clock breakthrough levels. Independent of such clock signal interaction, the problems of clock driving are also the same in GaAs ICs as they are in silicon. Where problems arise, they concern the necessity to drive many switch circuits in parallel with minimum skew and with minimum power dissipation at the speed demanded.

In spite of these similarities between GaAs and silicon sampling circuits however, there are some significant differences. The GaAs MESFET is a junction FET having a Schottky barrier gate, and actually resembles the silicon JFET rather more closely than the MOSFET. Large gate-source or gate-drain currents may flow under conditions whether either of the junctions becomes forward-biased. Under these conditions interactions between the clock and the analogue signal may occur which are both non-linear and dependent on the analogue signal level. The non-linear products resulting from any gate current flow are more permanent than the level-independent products due to capacitive coupling, and cannot generally be removed by any amount of filtering. Their presence will generally not allow the original signals to be recovered intact. The onus of circuit design using GaAs MESFETs is therefore on the prevention, rather than the treatment, of these non-linear products. This consideration will emerge as a major theme in subsequent parts of this section. Methods of overcoming problems resulting from the differences between GaAs MESFETs and Si MOSFETs and of exploiting the potentially superior frequency response of the GaAs MESFET form a large part of the material in this section.

6.3.1 Forms of switching circuit

The standard form of series switching circuit using any normal semiconductor device has several possible faults, even if it is the simpler silicon MOSFET that is used as the switch. Amongst these are:

(1) Finite switch on-resistance.
(2) Incomplete isolation of input and output in the off state.
(3) Coupling of the clock waveform to the sampled or switched signal.

In sample and hold circuits, covered in Section 6.3.2, the first two of these, the finite on- and off-state resistances of the switching device, give rise to finite signal acquisition time and a finite signal hold time, respectively. The range of signal frequency that can be processed in such circuits is therefore limited. The third of the problems listed above is not usually too much of a difficulty if the sampling rate of the signal is well above the signal frequency. The use of a low-pass filter after the sampling circuit serves to remove any clock breakthrough. This is however only true if the sampling clock signal does not interact in a non-linear fashion with the sampled analogue waveform. Under these circumstances, no amount of filtering will recover the undistorted signal.

Take the GaAs MESFET switch J1 in Fig. 6.9 (a). Assuming it has a small a.c. signal of well-defined d.c. level at its source (V_i). The switching waveform at the gate of the MESFET is also well-controlled in level, and the gate-source and gate-drain junctions of the switching device can be guaranteed not to go into forward bias. Under these ideal conditions, this simple switching circuit can be used. However, if the amplitude of the signal V_i is large compared with the forward voltage of the FET gate junction, or varies in d.c. level by such a large amount, then the more elaborate circuit of Fig. 6.9 (b) must be used. The signal level at V_i is buffered by means of a

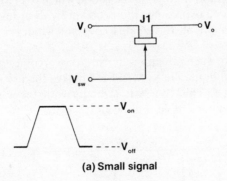

(a) Small signal

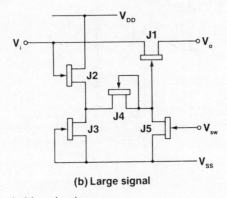

(b) Large signal

Fig. 6.9 MESFET switching circuits.

unity gain source follower, and used as the supply voltage for a switch driver circuit. Provided that the voltage at V_i never falls below the value $V_{SS} + |V_p|$, the control gate of the FET switch will be pulled down to approximately V_{SS}, turning the device off, or pulled up to the exact signal voltage, ensuring that the FET gate-source voltage is always almost exactly zero in the on-state. The switching waveform is now also independent of the switched waveform and its amplitude V_{sw} simply needs to swing from about V_p to zero or even to +0.7 V. It is now unimportant if J5 has a forward biased gate.

This circuit does however, still present a further problem which may call for more complex solution. The current in J1 depends upon the state of the switch. Hence so does the overall level shift introduced. This situation is covered in Section 6.4.2 in more detail, but ideally, the level shift must be zero. This condition must in any case prevail when the switch is in the on-state. To this end, the widths of J2 and J3 must be equal. In the off-state, however, the current through J4 passes through J2, causing the gate-source voltage to rise, possibly approaching a serious forward bias condition. If this is allowed to develop, then current will flow out of the J2 gate causing the input impedance of the circuit to fall, and probably corrupting the input V_i. Additional switching circuits can be found which overcome this problem. For example Fig. 6.10 overcomes the problem by ensuring that the current through the equivalent of J2 remains constant whatever the state of the switch.

The complexity of such circuits required to overcome the basic junction conduction problem in the GaAs MESFET makes the use of large numbers of them unattractive. This contrasts strongly with the case of the silicon MOSFET, where many simple and accurate switches may be present in a

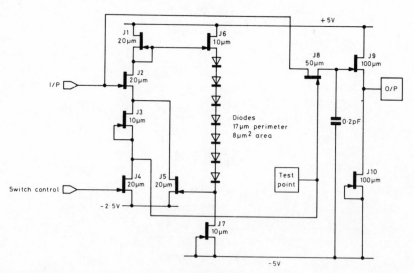

Fig. 6.10 Compensated sample-and-hold circuit.

circuit. Clearly the insulating gate of a GaAs MISFET would convey a great advantage. The GaAs MISFETs reported to date [25] do however remain in the off-state for a limited period only. This period is determined by the time constants associated with the charge trapping centres responsible for the depletion of the channel. In circuits where the minimum clocking rate can be above the limit set by this phenomenon, the MISFET can be of great advantage. Nevertheless, much more development is still needed before the MISFET can be seen as a sensible alternative to the MESFET.

6.3.2 Sample-and-hold circuits

Sample-and-hold circuits [14] in the simplest sense are just switches for sampling and a capacitor for holding the sampled signal until the next sample is taken. Sample-and-hold circuits based on the simple switch of Fig. 6.9 (a) can be rejected for all but the most specialized low signal level applications. The switch of Fig. 6.9 (b), however, is a practical proposition, and so is that of Fig. 6.10. An application based on the first of these two is described in Section 6.5.1.

The time required to acquire a signal to a given accuracy in a sample-and-hold circuit is determined by the product of the switch on-resistance, and the capacitance of the holding capacitor. This time constant T (the product of the capacitance of the hold capacitor C_{hold} and the switch on-resistance R_{on}) is the period in which about 63% of the difference between the signal being sampled and the signal presently being held can be transferred to the hold capacitor. Acquisition of 99% of the difference requires that a period of $5T$ be allowed to elapse.

Improvements in the acquisition times of sample-and-hold circuits are possible by the application of feedback and the incorporation of a wideband amplifier of low output impedance (Fig. 6.11). The speed limit of this circuit is determined by the slew rate of the amplifier when driving the hold capacitor. At the expense of some quite possibly considerable complexity, the performance of the sample-and-hold circuit is therefore made independent of the product of R_{on} and C_{hold}. In many cases, however, this sort of

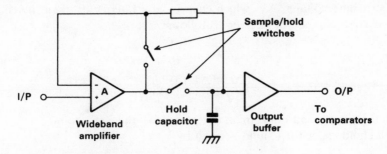

Fig. 6.11 Sample-and-hold circuit with negative feedback.

complexity is acceptable for only the simplest of circuits. In applications requiring many sample-and-hold circuits, the simpler circuits with their limited acquisition times must be accepted if complexity is to be kept within acceptable bounds.

The dynamic range of all of the above circuits is limited by a number of effects. At the highest signal levels, the power supply voltage forms the ultimate limit, although the circuit may not even allow signals to approach this maximum without significant distortion. With lower signal levels one of two effects may occur: noise from the switching circuits may exceed the noise inherent in the signal, and quantization errors may result from the storage of signals on small capacitors.

The first of these effects usually affects low frequencies. The GaAs MESFET suffers frequently from large levels of $1/f$ noise (see Section 6.4.5). This occurs since the $1/f$ noise corner of a typical GaAs MESFET may sometimes exceed some tens of megahertz. The second is a problem in high frequency circuits, where the minimum possible RC product is needed between the switch on-resistance and the hold capacitance in order to guarantee accurate acquisition of the signal within the sampling period. The quantization level is, of course, the voltage represented by one electron on the hold capacitor. A hold capacitor of 20 fF, for example, results in a voltage quantization of 8 μV. This is given by:

$$\triangle V \text{ (quantization)} = e/C_{\text{hold}}$$

$$e = 1.602 \times 10^{-19} \text{ Coulombs}$$

If the upper signal level is 0.1 V RMS, for example, the maximum dynamic range becomes approximately 82 dB.

6.3.3 Clock breakthrough and cancellation

The high speed signals that switch MESFETs on and off can get coupled rather easily to the signals that are being switched. The high rate of switching signal rise and fall, combined with the stray junction capacitance of the switch, is the primary cause. The amount of charge transferred via stray capacitance C_{ff} in one switching operation can be simply calculated. If i is the charging current that flows in the capacitor, Δt is the transition time of the switching voltage, ΔV, which changes at a rate dV/dt, then the charge transferred, Δq, can be derived as follows:

$$i = C_{\text{ff}} \frac{dV}{dt}$$

$$\Delta t = \Delta V/(dV/dt)$$
$$\Delta q = i \, \Delta t = C_{\text{ff}} \, \Delta V$$

In the sample-and-hold circuit of Fig. 6.9 (a), the clock voltage appearing on th ehold capacitor C-hold is given by

$$V_{\text{cb}} = \Delta q \, /C_{\text{hold}} = \Delta V \cdot C_{\text{ff}} \, / \, C_{\text{hold}}$$

In the interests of maintaining the minimum possible clock breakthrough, the value of C_{hold} should be maximized, and the value of C_{ff} minimized. The latter, usually achieved by making the switching device as small as possible, will generally reduce the speed of signal acquisition, since the on-resistance of the switch, R_{on}, will increase. If in addition the hold capacitance, C_{hold}, increases, the overall RC product of the sample-and-hold circuit will increase greatly, and the signal bandwidth handling capacity of the circuit will fall significantly.

The best way to reduce clock breakthrough remains to minimize the switching voltage required to turn the switch device on or off. In general, however, this will also increase the on-resistance of the switch. In addition, some means must be found to minimize the value of C_{ff} for a given on-resistance. This latter approach is a technological problem, and is beyond the scope of this section.

If the minimization of clock breakthrough cannot be carried far enough without reducing the overall bandwidth of the circuit too much, some form of block breakthrough cancellation must be found. Such techniques are common in silicon sampled analogue circuits. One such circuit is illustrated in Fig. 6.12. Here one of the simpler sample-and-hold circuits described in Section 6.3.2 has been augmented with an extra switch driver (J6 and J7) and a dummy switching FET (J8). The stray capacitance C_{ff} in the above analysis is the gate-source capacitance of J1 (C_{GS} in Fig. 6.12).

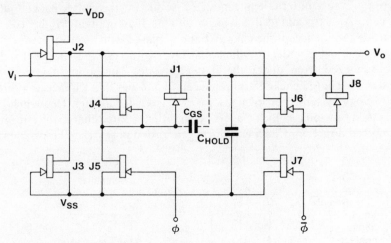

Fig. 6.12 Clock-cancelled sample-and-hold circuit.

Clearly, the requirements for clock cancellation make more onerous the problem of clock generation. Nevertheless, these problems can be solved, and the clock generator that was designed and used with this sample-and-hold circuit will be described in Section 6.3.5.

6.3.4 CCD sampling circuits

The explicit use of switches and capacitors as described in Sections 6.3.1 and 6.3.2 in a bucket-brigade-device (BBD) sample-and-hold configuration for storing a charge that represents a sample of a continuous analogue signal is only one way of achieving this result. Charge-coupled devices (CCDs) have been built in silicon over a number of years, and have now been seen in GaAs technologies ([12], [15], [16]). The difficulties in both making and using these devices at high clock rates are still great, but they have a number of unique and useful properties, and for the sake of completeness, a short description is given here.

In a CCD, charge packets are stored in cells in the semiconductor itself (Fig. 6.13), and transferred from cell to cell by special voltage waveforms applied to electrodes on the surface of the CCD which define the depth of the depletion later and thereby define the volume available in the cell in which to store the charge. The sampling process in a CCD involves the introduction of charge through an input diode (the input ohmic contact and the Schottky barrier input transfer gate G1 in Fig. 6.13) in such a way that the charge transferred into the first storage cell under $\phi 1$ is linearly dependent on the voltage on gate G2.

Transfer of charge in a symmetrical structure such as this is effected, in the direction desired, by introducing an asymmetry into the clock waveform (Fig. 6.14). The cell from which the charge is to be transferred has a negative voltage applied to its control electrode. The depletion layer beneath the electrode sweeps into the semiconductor, so forcing the charge packet out of the cell. An adjacent cell having a similar bias applied to the control electrode cannot accept the charge, but one biased with a zero volt potential has enough space under the depletion layer to accept the charge that has been squeezed from under the first electrode. The simplest system for transferring charge packets from one cell to another in a CCD uses a three-phase clock system, but in many cases it is easier at high frequencies to generate a four-phase clock (Fig. 6.14) in order to force charge to transfer in the desired direction. Clock waveform generation poses a similar problem in

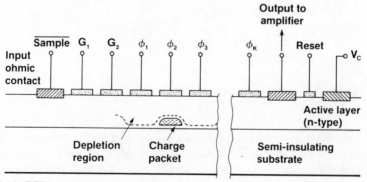

Fig. 6.13 CCD structure.

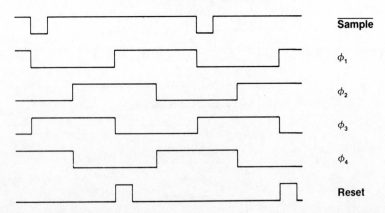

Fig. 6.14 Four-phase CCD clock.

CCDs as it does for BBDs. The problem is discussed further in Section 6.3.5.

The efficiency of charge transfer between sample-and-hold cells is typically much higher in a CCD than it is in a BBD. At a 1 GHz clock rate a charge transfer efficiency of well over 99% is obtained [1], with values of better than 99.9% possible. This compares to about 95% for the BBD ([15], [12]). Amplitude corrections to the delayed signal are therefore smaller in the CCD and are therefore less prone to inaccuracy.

The CCD in the role of a FIR filter involves the use of sampling electrodes placed above the charge packets in order to sense the amount of charge present. The signal flow graph (SFG) for this system (Fig. 6.15 (a)) is identical to that of Fig. 6.8 (a). The relative sizes of the sampling electrodes determine the weights (h(n)) applied to the sampled signal before summation takes place. An alternative, but equivalent SFG is shown in Fig. 6.15 (b) in which instead of being taken from within the CCD delay line, the samples for weighting are taken from the ends of a number of delay lines of differing length [1]. In this case, the width of the CCD delay line determines the value of h(n). This structure has been referred to as an 'Organ-Pipe' CCD. It has the advantage that sampling electrodes are not needed within the CCD delay line where they are difficult to co-site with the transfer control electrodes.

The CCD remains a potentially useful sampling and signal processing device. For its fabrication, however, it does need a process that differs from most standard MESFET analogue IC processes. For this reason, it remains something of a curiosity, and the more standard sampling circuits are likely to be much more popular.

6.3.5 Clock generation

The generation of clock waveforms of several hundreds of megahertz, with perhaps up to four phases in a well-defined timing relationship, and perhaps

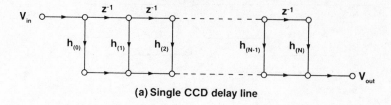

(a) Single CCD delay line

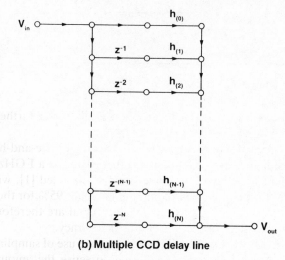

(b) Multiple CCD delay line

Fig. 6.15 CCD implementations of FIR filters.

also satisfying the requirement that pairs of clock signals do not overlap is a very difficult task, even with modern test equipment. The situation generally requires that such signals are generated using the same technology as the sampled analogue IC itself. At worst, a separate chip is tolerable if the pinout has been designed well enough to match the clock input pins of the main IC. The best approach, clearly, is to generate the clock waveforms on the IC itself.

The kinds of GaAs IC process suitable for making sampled analogue ICs also tend to be suitable for making relatively low power and low speed (up to 1 GHz) digital and mixed digital/analogue circuits for clock generation. Fig. 6.16 shows one such example. The IC for which this was developed will be described in Section 6.5.1.

The input clock waveform of Fig. 6.16 is biased at 0 V, and may be of small amplitude, say about 0.1 V p-p. This is amplified by a BFL inverter, and presented to the input of a differential amplifier. The second input of the differential amplifier is derived from the same amplified clock signal passed through a low-pass filter, whose RC product is set at a period rather lower than the minimum clock frequency desired. The d.c. level of the amplified input clock signal is therefore rendered unimportant, and the clock signal and its inverse emerge to be conditioned by a pair of 2-input cross-coupled BFL NOR gates, which provide non-overlapping clock signals. Each of these is applied to a separate differential amplifier in the same fashion at the clock signal in the first place, and not only do we have the non-overlapping clock waveforms that we desire for the switching of signals in a delay line, but we also have their inverses for driving the clock-cancelling switches described in Section 6.3.3.

This clock generator works up to about 1.3 GHz, using a standard 1V DFET process and provides the clocks for a transversal filter described in Section 6.5.1. Techniques for clock generation taken from silicon IC designs are also frequently useful, and may provide numerous alternatives to the circuit described here.

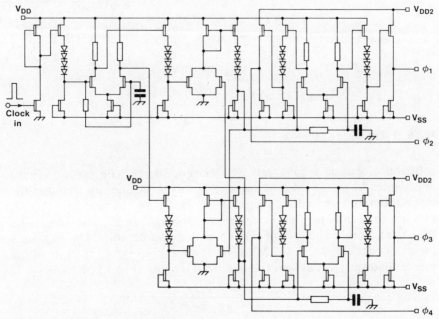

Fig. 6.16 Non-overlapping clock generator.

6.4 OTHER GaAs ANALOGUE CIRCUITS

Although the sampling circuit itself forms the heart of any sampled analogue IC, a number of other operations need to be performed as well in order to effect any useful signal processing function whether these are performed in a digital or an analogue fashion, the vital operations are:

(a) delay
(b) multiplication
(c) addition.

The implementation of these functions also requires a number of simple analogue circuit building blocks. In this section, the above functions are developed from the more elementary building blocks of analogue and sampled analogue circuits.

6.4.1   The source follower

High gain in an analogue amplifier is frequently accompanied by high impedance. Furthermore, such high gain circuits are often required to drive loads of high capacitance or low resistance and so buffering is required and typically performed by the use of a source follower circuit (Fig. 6.17).

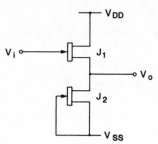

Fig. 6.17 The source follower.

In this circuit J2 ideally acts as a current source with infinite output impedance. In the absence of any static or dynamic currents at either the input or output terminals, and assuming that J1 is an identical device to J2, J1 is forced to carry the full current from J2 and so V_i and V_o are forced to be the same voltage. The gain of this circuit is therefore unity. Unfortunately the drain conductance of the MESFET is always finite, and the gain of the source follower (A_V) assuming equal sized devices J1 and J2 becomes

$$A_V \;=\; \frac{g_m/g_o}{(g_m/g_o + 2)}$$

where g_m is the mutual conductance of J1 and g_o is the drain conductance. For g_m to g_o ratios encountered in standard ion-implanted processes, typical values of A_V lie between 0.93 and 0.95.

It should be noted here that the voltage ($V_i - V_o$) appears across the gate-source junction of J1. If the gain of the source follower is much smaller than unity, or the input signal is large, this voltage can significantly forward bias the MESFET gate-source diode, so reducing the input impedance considerably and compromising the ability of this circuit to isolate a sensitive high impedance node from a low impedance load. This effect does not occur with the silicon MOSFET where under almost all conditions very little gate current can flow.

A similar effect can occur if the load driven by the source follower is highly capacitive and the input slew rate is high. Charging current for the load capacitance flows through the output terminal, forcing the source follower MESFET J1 to adopt a non-zero gate-source voltage. If the magnitude of the output current is sufficiently large compared to the mutual conductance of this device, then the input impedance will drop to a low value for positive going input transients, and J1 may completely turn off on negative-going edges. In both cases the effect will be to introduce dynamically varying distortions into the signal.

Both of the above situations are exaggerated by mismatches between the two MESFETs and by non-zero static values of load current not accounted for in the design. These effects introduce a permanent offset between the input and output voltages, so reducing the dynamic signal handling capacity of the circuit. In some applications the presence of the offset itself may introduce a problem.

6.4.2 Level shifting

Because the threshold voltage of the MESFET is typically non-zero, and usually negative in value, it is often necessary to adjust the d.c. levels of signals in an analogue circuit to ensure that the d.c. biasing of all parts of the circuit is correct. An extension to the source followers described in the previous section suffices to provide a restricted form of level shifting (Fig. 6.18).

For shifting the d.c. level of a signal in the negative direction it suffices simply to insert the appropriate number of forward biased Schottky diodes into the source of the source-follower MESFET (Fig. 6.18 (a)). Although it is possible to choose the current density in the diodes by specifying the diode area and the sizes of the MESFETs, the range of adjustment of the voltage across each is rather small, and we must accept the discrete nature of the

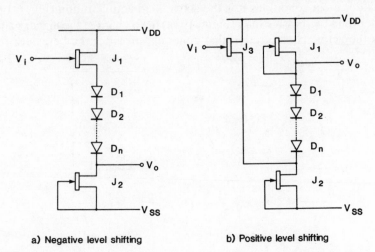

a) Negative level shifting   b) Positive level shifting

Fig. 6.18 Level shifting circuits.

value of the voltage offset obtained by this means. The diodes should in practice be chosen in such a manner as to ensure as low as possible an output impedance for the source follower, and so maintain a gain of close to unity and a high bandwidth.

To shift the signal level in a positive direction, the diode chain can be used 'in reverse' (Fig. 6.18 (b)). In this circuit the impedance at the foot of the diode chain is too low to connect to any normal circuit, so a source follower is used to raise the impedance. Under normal circumstances J2 acts as a current source balancing that of both J1 and J3. Its width should therefore be the sum of the widths of those MESFETS.

In both of these circuits, the precautions described in the previous sections with respect to the biasing of the source follower must be observed, and level shifting in approximately 0.65 V steps must be accepted.

6.4.3 Single-ended and differential amplifiers

Although more frequently used as the basis for a digital logic inverter, the single-ended inverting amplifier (Fig. 6.19) can provide a reasonable amount of inverting gain. For both of the configurations shown, the gain A_V can be expressed simply as

$$A_V = g_m/(g_o + g_1)$$

where g_m is the mutual conductance of the MESFET, g_o is the drain conductance of the amplifying MESFET and g_1 is the conductance of the load device. The expression for gain is, of course, bias-dependent, and so we must ensure that under quiescent bias conditions (typically with the output node at one half of the supply rail voltage difference) we get the gain that is desired.

Under most conditions it is the active load configuration (Fig. 6.19 (b)) rather than the passive load (Fig. 6.19 (a)) which gives the higher gain, and it is the active configuration that generally provides the best performance

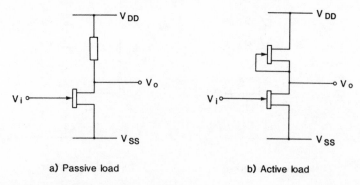

a) Passive load   b) Active load

Fig. 6.19 Single-ended inverting amplifiers.

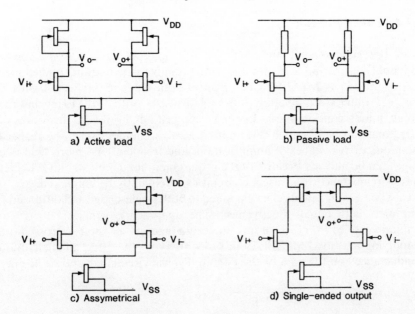

Fig. 6.20 Differential amplifiers.

when integrated, since the device to device matching between two MESFETs is usually rather better than the matching of a MESFET to a resistor. Such mismatches that do occur tend to force the d.c. bias at the input of the amplifier to shift either positive or negative of the desired voltage, both of which may reduce the designed gain, and if the bias shifts too far positive, the FET gate may conduct so reducing the input impedance and loading the driving stage.

The differential amplifier (Fig. 6.20) [10] is often used in preference to the single ended amplifier by virtue of its relatively large common-mode rejection ratio. The choice of active (Fig. 6.20 (a)) or passive (Fig. 6.20 (b)) loads is dictated by the gain requirement, and more importantly some amplifier saturation requirements. Because of the possibility of forward biasing the amplifier transistors, care must be taken in choosing the sizes of the MESFETs. If the current available from the common current source ever exceeds the total current available from the load devices, if for example one branch is allowed to saturate at the positive rail voltage, then the extra source current will flow through the gate of one of the amplifying transistors and may severely influence the stage driving the differential amplifier. This situation can of course occur dynamically, if there is a considerable capacitive load on the output of the differential amplifier, and the input slew rate is high. Some of these effects can be avoided by using an asymmetrical amplifier (Fig. 6.20 (c)) or the standard differential to single-ended amplifier arrangement (Fig. 6.20 (d)).

6.4.4 Cascode circuits

Both the single-ended and differential amplifiers described previously generally suffer from low gain due to the relatively high drain conductance (g_o) of the GaAs MESFET. The gain of both types of circuit is basically $k(g_m/g_o)$ where k is a function of the relative sizes of the MESFETs and the circuit under consideration. In many applications however, the value of g_m/g_o that can be obtained does not allow high enough gain to be realised using one or two stages of amplification, and frequently no more than two stages can be allowed because of the requirement for circuit stability. Under these circumstances the cascode circuit can frequently be employed.

Cascode arrangements can be applied to both single-ended and differential amplifiers with equal effectiveness. The approach is generally to try to maintain as constant a drain-source voltage as possible across both amplifying and load MESFETs in order to avoid the effect of their drain conductances on the gain of the circuit. Typical circuits are shown in Figs. 6.21 and 6.22.

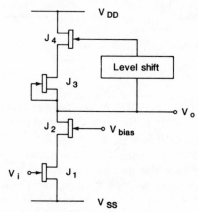

Fig. 6.21 Single-ended cascode amplifiers.

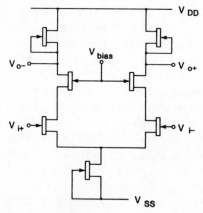

Fig. 6.22 Differential cascode amplifier.

The effect can best be illustrated with reference to Fig. 6.21. Here, J1 is the amplifying MESFET whose drain-source voltage is maintained at an approximately constant value by J2. The changes of current in J1 develop an output voltage across J3 and J4. The voltage across the load device J3, however, is maintained at a nearly constant level by J4, so ensuring a very high impedance at the output node and so a very high gain. Assuming that all MESFETs have equal sizes, the gain of this circuit is raised from $A_v = 1/2(g_m/g_o)$ to $A_V = 1/2(g_m/g_o) (g_m/g_o + 1)$. The use of a non-ideal level shifter will prevent the full benefit from being realised, but if the ultimate performance is required, cascode techniques can even be applied to this function as well.

6.4.5 Low-frequency effects in GaAs MESFETs

A now rather well-known set of effects in GaAs FETS, but with much less well-known causes, is the group of low-frequency anomalies observed in most devices. This group includes $1/f$ noise [19], low-frequency oscillation [13], backgating [26] and low-frequency variations in drain conductance [39]. All of these problems can have detrimental effects in sampled-analogue and other d.c.-coupled ICs. These effects, their causes and some possible cures are considered here. In all cases, the alleviation of their effects is based on a circumvention of the problem, rather than a cure of the fundamental cause. This cause would appear, at first sight, to be trapping of carriers in energy states close to the middle of the band-gap, in various parts of the MESFET but, as yet, little understanding at the real sources of such problems has emerged, and it seems likely that a considerable period will elapse before such understanding is obtained.

Noise with a $1/f$ component is a fundamental problem in GaAs devices. For many years, the noise corner in FETs has been known to be high, sometimes even extending to 100 MHz. The variation in noise corner can also be extreme, varying from device to device by orders of magnitude and little progress seems to have been made in tracing the sources of the problem. Some experiments [19] have shown that there may be two sources of $1/f$ noise in the GaAs MESFET. These are, broadly, the device surface and the device channel regions. The contribution of noise from these two regions is said to be roughly equal. The surface noise results from trapping of electrons in surface states, whilst the bulk noise is correlated with low-field mobility and deep level trap concentration. As yet, there seems to be no cure for broadband $1/f$ noise in GaAs ICs. In the case of the GaAs MMIC, this is of little consequence, but for the sampled-analogue IC, the dynamic range of signals that can be handled is restricted by this effect. In the near future it seems that the causes of this type of noise will not be found, and the effect must therefore be accepted.

Backgating and substrate conduction mechanisms appear to be inextricably linked with low-frequency oscillations in GaAs FETs and ICs. Observations of backgating effects seem to have shown up the fact that the low-level currents flowing in semi-insulating or implant-isolated GaAs substrates as a result of d.c. bias differences between adjacent devices have oscillatory components which may be in the hertz or kilohertz range. Whether these oscillations have their origins in the devices, or in the substrate is unclear. Nevertheless the observations seem to suggest that such oscillations are coupled throughout an IC, and if coupled to the input stage of a high-gain amplifier, for example, may then be amplified greatly and become the source of large-amplitude blocking waveforms or other non-linear behaviour. There are some solutions available to the circuit designer, however. Screening tracks placed between devices and appropriately biased are enough in many cases to prevent both backgating and oscillatory coupling currents. The disadvantage of this approach is in the expense of reduced device packing density. In addition, there does not yet seem to have been enough work conducted to allow the particular circumstances to be predicted under which device to device coupling can occur. Empirical methods are the only ones available presently.

The variation of FET drain conductance with frequency (see Fig. 6.23) is a major problem with design. In general, several frequencies are observed at which the drain conductance increases, and there is some correlation between the type of material on which the FET is made and the presence or absence of some of these conductance steps. This points to the growth mechanisms of crystals or the methods for preparing active layers as the possible sources of the problem.

By performing frequency-domain analysis of transient drain current waveforms in FETs at varying temperatures, both the activation energy and cross section of the trap responsible (assuming that a trap is responsible) can be obtained. Energies of 0.72 eV for the low frequency components, (f_1 in Fig. 6.23) and 0.28 eV for the higher (f_2) are typical of those obtained, and correspond to families of traps that have been identified in GaAs crystals elsewhere. Nevertheless, until the natures of the traps are discovered and means found to effect a cure, it is of academic interest to identify the trap

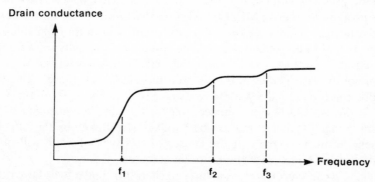

Fig. 6.23 MESFET drain conductance versus frequency.

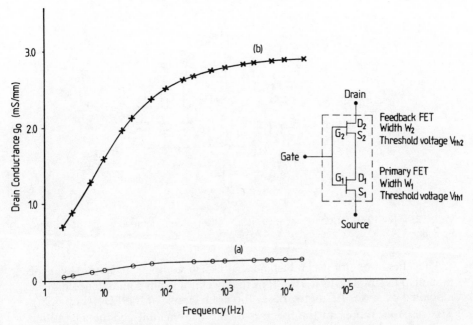

Fig. 6.24 Frequency dependence of small signal drain conductance (a) with and (b) without feedback FET.

energies involved. The key lies most likely in the material preparation conditions and procedures.

It is a major problem for the circuit designer that the drain conductance of a FET may increase with frequencies above a few Hz, and since the gain in FET-based amplifiers is usually inversely proportional to this parameter, it is difficult to design high gain circuits. Several techniques described earlier, including the use of cascode circuits, are available to counteract the problem. The general requirement is to make constant the drain-source voltage of the amplifying device. If this is achieved for all instantaneous input voltages, then the effective output impedance of the FET will be very high. Of course it is also necessary to achieve the same effect with the load impedance into which the amplifying device works, since otherwise the effort involved in raising the impedance of the amplifying device only, is largely wasted. A recent report indicated that the drain impedance could be controlled by employing the so-called self-bootstrapping technique to both the amplifying device and the active load [40]. Each amplifying device and load were made up of two FETs of different threshold voltages. The threshold voltage of the feedback FET was made more negative than the primary FET to insure that the latter was biased to operate in the current saturation region to realise the full benefit of this technique. Fig. 6.24 shows the basic circuit arrangement and the improvements in the effective drain conductance achieved.

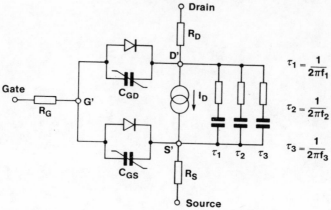

Fig. 6.25 MESFET model incorporating drain conductance variation with frequency.

The effects of drain conductance, although complex in nature, can be modelled in simple ways. The steps that occur in drain conductance may be modelled by series RC networks in the drain circuit of the FET (Fig. 6.25). The resistors represent the inverses of the incremental conductances, and each RC product is adjusted to cause the transition from low to high conductance at the correct frequency. The values of capacitance that derive from this type of model are however rather large. A fairly typical 1-mm wide FET may require a 133 ohm resistor and a 250 μF series capacitor to model the principal low-frequency drain conductance change. Clearly this value of capacitor is physically meaningless, except in that it could be said to represent the density of traps causing the effect that is modelled.

The techniques of Section 6.4.4 for increasing high frequency gain do however introduce their own problems. If adequate gain is obtained at high frequencies, then d.c. gain in an amplifier may well be much higher. Unless precautions are taken, for example rolling off the gain below the threshold frequency for drain conductance rise, then considerable instability may arise. This is particularly important if the coupling of low-frequency oscillations occurs. The combination of such oscillations and the very high gain that obtains at the same frequency is almost a guarantee of blocking behaviour in any circuit.

d.c-coupled circuits, amongst which many sampled-analogue circuits can be categorized, are particularly susceptible to the low-frequency effects outlined in this section. It is unfortunate that the source of the effect is poorly understood that a cure for the problem seems some way off and that circuit designers must, for the time being at least, find techniques for dealing with the problems that are introduced. The changes in drain conductance with frequency and the effects of backgating, both static and oscillatory, can largely be cured by the application of techniques mentioned here and described more fully in other sections of this chapter. The effects of $1/f$ noise, however, remain.

6.5. APPLICATIONS

Many possible analogue applications exist for GaAs ICs. However the greatest advantages lie with sampled-analogue applications making use of the ability of GaAs FETs to be used both as fast switches and as linear devices. This contrasts with silicon bipolar technology where the devices make poor switches and silicon MOS devices which have a limited frequency performance when compared to GaAs MESFETs. The wide range of sampled-analogue applications include analogue-to-digital converters (ADCs), digital-to-analogue converters (DACs), phase-locked loops (PLLs), switched-capacitor filters (SCFs), transversal filters, wideband amplifiers and many other areas where processing of sampled analogue signals is required. This section describes two GaAs ICs which have been developed for these applications.

6.5.1 Transversal filter

The wideband programmable transversal filter described here ([15], [21]) has been developed in response to requirements for complex filtering in modern radio systems. This chip represents a novel implementation of the transversal filter idea in a system containing a cascade of programmable analogue finite impulse response (FIR) filters allowing the conversion to baseband of a selected sub-band of the input frequency band. Full details of the implementation of the system are beyond the scope of this chapter and are not included. The outline given here, however, will give enough information to appreciate the operation and design of the chip described and other similar systems published recently ([1], [3], [12]).

The operation of the transversal filter is best explained in the time domain rather than the frequency domain. A continuous-time input is sampled and the sampled voltage is passed from one storage capacitor to another along a delay line at a rate determined by a clock. By Shannon's sampling theorem, the samples in a system sampled and clocked at a frequency f_s can adequately represent information with frequency components up to $f_s/2$. If an impulse of width $1/f_s$ is applied to the system, then at any one subsequent instant only one of the storage capacitors will be holding a signal whose magnitude will, ideally, be equal to the magnitude of the impulse applied. At each of the stages along the delay line, the signal amplitude is sensed and multiplied by a weighting factor by an analogue technique, such as capacitive voltage division, and then added together with the weighted signals from all other delayed samples at that instant. For the case of the impulse signal described above, the output of the chip in the time domain (i.e. the impulse response) will be a series of voltage steps determined exactly by the weighting factors used at the various stages along the line. In this way the appropriate weighting factors (for a given application) at each stage along the delay line are thus exactly equivalent to the magnitudes of the sampled impulse response of the desired filter.

If the sampled impulse response is represented by $h(n)$, then the filter transfer function is given by the Z-transform of $h(n)$, i.e.

$$H(z) = \sum_{n=0}^{N-1} h(n) \, z^{-n}$$

The upper limit in this summation is not infinite, but represents the length of the delay line in this case. It may be appreciated that this transform represents the direct form of filter described earlier in Section 6.2.8; the SFG was shown in Fig. 6.8 (a).

In the case of a non-impulse input, the output voltage is given by

$$V_{out}(z) = V_{in}(z) \cdot H(z)$$

The filter shown above requires $(N-1)$ delay stages and N weighting factors. Practical filters will require some means of sampling any applied continuous-time signal, and this sampling generally introduces a further delay. The resultant filter configuration then has N delay stages and N weighting factors.

For the case of an ideal continuous-time low pass filter with a cut-off frequency f_c, the impulse response (obtained from an Inverse Fourier Transform performed on the ideal frequency response) is proportional to

$$\frac{\sin (2\pi f_c \, t)}{(2\pi f_c \, t)}$$

One means of transforming a continuous-time filter design to a sampled filter design corresponds to defining the sampled impulse response as equally spaced samples of the continuous time impulse response. The sampled impulse response weighting factors $h(n)$ are then proportional to

$$\frac{\sin (2\pi \, n \, f_c/f_s)}{(2\pi \, n \, f_c/f_s)}$$

since $t = n/f_s$ where f_s is the sampling frequency. Fig. 6.26 (a) shows the continuous-time and sampled impulse responses for the case where $f_s = 4f_c$.

Remember that to avoid aliasing effects, the maximum bandwidth (f_{max}) must be restricted to half the sampling frequency, i.e. $f_s = 2f_{max}$. The impulse response shown in Fig. 6.26 (a) then corresponds to a filter with a cut-off frequency $f_c = \frac{1}{2} f_{max}$, i.e. this is a half-band low-pass filter.

An impulse response of the form $\sin(x)/x$ contains significant response values up to large values of t. Large values of n would be required for the sampled filter to reproduce accurately the ideal continuous-time filter shape, so requiring a very long delay line in the transversal filter implementation. Furthermore, the ideal filter is non-causal in that an output response is expected (both in the continuous and sampled domain) before the input signal is applied. Referring to Fig. 6.26 (a) again, the impulse input is applied at $t = 0$, yet an impulse response is required at negative values of t for the continuous-time filter, and at negative values of n for the sampled filter.

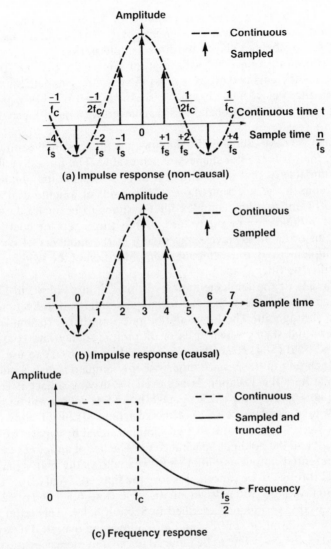

Fig. 6.26 Low-pass filter.

The non-causality of the ideal filter makes real-time implementations impossible. However, if the output response is delayed, the impulse response is effectively shifted in time and the use of negative values of n can be avoided. For small numbers of samples, the effect of delaying the output response relative to the input is not usually significant. Fig. 6.26 (b) shows the sampled impulse response of an ideal filter whose output is delayed by three sample periods relative to the input.

The use of very long delay lines can be avoided by truncating the sampled impulse response sequence, at the expense of reducing the stop-band loss and some discrepancy between the ideal filter response and the sampled response. Fig. 6.26 (c) shows the response of a low-pass transversal filter implemented with a truncated sequence of seven weighting factors, the output response being delayed by three sample periods to give a sampled impulse response coincident with that of Fig. 6.26 (b). Of the seven weighting coefficients two are zero, and the rest are symmetrical about the centre of the delay line. In the case of a half-band low-pass filter (i.e. $f_s = 4f_c$), a 30 dB stop-band loss is achieved with these seven weighting coefficients.

Although we have used a half-band low-pass filter as an illustration of the principle, many other filter shapes are achievable. The filter described here is programmable in that three different filter shapes can be obtained from the same chip by the selection of one of three sets of weighting coefficients. This is sufficient to allow complex filter systems to be realised. A certain degree of redundancy in the values of the three sets of coefficients is exploited in order to achieve some saving in the numbers of coefficients actually implemented, thus effecting a simplification of the switching system for selecting the filter shape.

Brief details of the implementation of this chip are shown in Fig. 6.27, from which it will be seen that the techniques described in Section 6.4 are employed throughout. The input and output amplifiers are simple passive load differential stages which drive a differential delay line consisting of sample-and-hold (S/H) cells (Fig. 6.27 (b)). It is necessary to use two S/H cells for each unit of delay since otherwise the sampled signals would smear into each other; the isolation is achieved by driving adjacent cells with antiphase non-overlapping clock signals. In the S/H cells it will be seen that the gate-source capacitance (C_{gs}) of the switching MESFET allows a clock frequency signal to distort the wanted sampled signal by capacitive division between C_{gs} and the holding capacitance C_s. The use of antiphase delay lines and a differential output amplifier, however, allows the major part of this clock breakthrough to be cancelled before the filtered signal leaves the chip.

In the S/H cell, the switch driver circuit of Sections 6.3.1 and 6.3.2 is used, as is the cascode technique described in Section 6.4.4. This latter method improves the gain of the buffers and also allows the overall d.c. level shift introduced by the delay line to be reduced to zero, so increasing the dynamic range of the circuit. The signal tapping circuit is a simple source follower (Fig. 6.27 (c)) which drives a capacitive weighting and summing circuit, the magnitude of the individual impulse response values being determined by the relative values of capacitors in the network. Negative weights are provided by attaching the tap driver to the opposite phase delay line.

Using a simple ion implanted MESFET process with a single threshold voltage of -1.0 V, and without resorting to self-aligned technologies, clocking rates of over 1000 MHz have been achieved, allowing an input signal bandwidth of d.c. to 500 MHz to be used. The accuracy of the passband filter shapes and of the stop-band attenuation values is generally maintained until very close to the maximum clocking rate of a chip. A photomicrograph of the chip and a comparison of the theoretical and achieved filter responses are shown in Fig. 6.28.

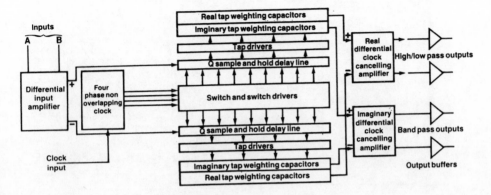

(a) Block Diagram

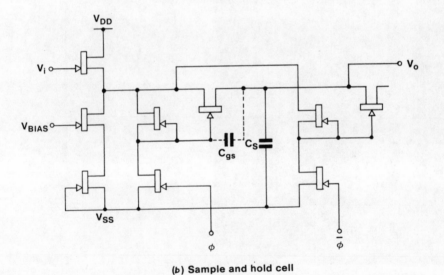

(b) Sample and hold cell

(c) Sample weighting circuit

Fig. 6.27 Transversal filter circuit and its key building blocks.

Amplitude (dB)

Frequency/Clock

(c) Band-Pass Output, Clock at 1000 MHz.

Amplitude (dB)

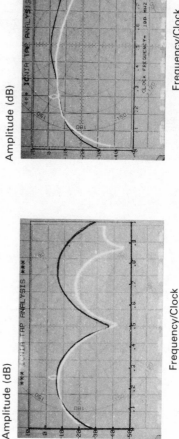

Frequency/Clock

(f) Band-Pass Output, Clock at 100 MHz.

(a)

Amplitude (dB)

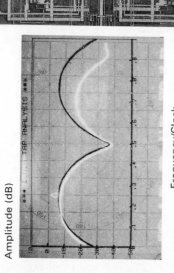

Frequency/Clock

(e) Band-Pass Output, Clock at 100 MHz.

Amplitude (dB)

Frequency/Clock

(b) Band-Pass Output, Clock at 10 MHz.

Amplitude (dB)

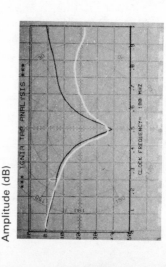

Frequency/Clock

(d) Low-Pass Output, Clock at 100 MHz.

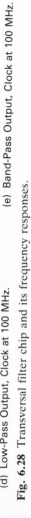

Fig. 6.28 Transversal filter chip and its frequency responses.

6.5.2 Switched-capacitor filter (SCF)

The use of switched-capacitor techniques for filtering applications is now well established. However, to date SCF applications have been limited to relatively low frequencies (sub-megahertz) by the speed limitations of the implementing technology (generally CMOS or NMOS). A complete review of the design techniques of switched-capacitor filter is outside the scope of this book; further details can be obtained from the literature ([27] – [29], [37]). The essence of the techniques involves designing active filters using amplifier, resistor and capacitor components and then to replace all resistors with a capacitor which is alternatively switched between ground and the circuit by different phases of the switching clock (Fig. 6.29).

Charge transferred into the capacitor on one phase of the clock is discharged into the opposite port during the second phase. Since there is a packet of charge transferred every clock period, the discrete equivalent of a continuous current flows from one port to the other. The amount of charge transferred in a given period is proportional to the voltage across the capacitor, to the value of the capacitance and to the switching frequency. The structure therefore behaves just like a resistance, where the resistance is inversely proportional to both the switched capacitance and the clock rate.

Where a switched-capacitor equivalent of an inductor (or even a capacitor) is required, an integrator is used in which the integrating resistor is, naturally, replaced by a switched-capacitor. In either of the reactive components, the voltage across the device and the current flowing through it are in phase quadrature. If the voltage across an inductor is represented by the voltage at the input of the integrator, then the voltage at the output of the integrator, lagging 90 degrees behind the input voltage can be used to represent the current in the inductor. A similar argument applies in the case of the capacitor where the voltage and current are represented by the integrator output and input voltages respectively.

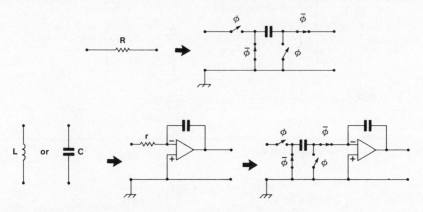

Fig. 6.29 Monolithic implementation of RLC components.

In replacing inductors and resistors by their switched-capacitor equivalents, a filter transfer function can be defined in terms of capacitor ratios alone rather than absolute component values. This is ideal in integrated circuits, where this technique permits untuned operation with high accuracy.

This section describes the design and performance of a two-pole bandpass filter IC implemented in GaAs ([11], [20]). Operation with centre frequencies up to 10 MHz has been reported. The extension of switched-capacitor techniques into the MHz range will allow the use of these techniques in applications such as clock recovery in data communications systems, agile filtering for radar and communications receivers, and video processing in television receivers. A higher level of integration in these systems may thus be achieved than has been possible in the past.

This filter was designed using the approximate LDI transform (see Section 6.2.8) on a simple RLC ladder prototype. The approximate LDI transform has the advantage of simplicity, and using an f_s/f_o ratio also eases the anti-aliasing and output smoothing requirements associated with sampling filters. Use of an RLC ladder prototype produces a component-economical filter whose response is relatively insensitive to component values.

The RLC prototype of the second-order band-pass filter is shown in Fig. 6.30. This filter has a transfer function given by

$$\frac{V_o}{V_i} = \frac{1}{1 + sRC + \dfrac{R}{sL}}$$

which can be re-arranged to give terms involving $1/s$ and constants only, i.e.

$$V_o = \frac{1}{sC}\left[\frac{V_i}{R} - \frac{V_o}{R} - \frac{V_o}{sL}\right]$$

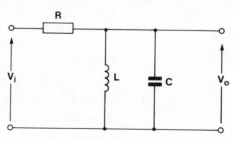

Fig. 6.30 Basic switched capacitor filter prototype.

The signal flow graph (SFG) for this expression can then be drawn and is shown in Fig. 6.31 (a).

In order to realise the switched-capacitor equivalent of this SFG, it is necessary to transform from the continuous-time complex frequency variable s to the discrete-time complex frequency variable z. Using the

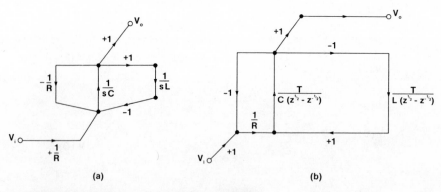

Fig. 6.31 Decomposition of SCF SFG using LDI.

approximate lossless discrete integrator (LDI) transformation for the reasons stated above.

$$s \rightarrow \frac{1}{T}(z^{1/2} - z^{-1/2})$$

where T is the switching period. The SFG of Fig. 6.31 (a) can now be drawn in terms of the variable z to give Fig. 6.1 (b).

This SFG can be further redrawn in terms of building blocks known as multi-input integrators. The circuit shown in Fig. 6.32 integrates during the even (E) phase of the switching clock and is thus called a multi-input E-integrator. The SFG of this circuit is shown in Fig. 6.33. Similarly, Figs 6.34 and 6.35 show the circuit and SFG of a multi-input O-integrator which integrates in the odd (O) phase of the switching clock. Note that the switched capacitors C_a and C_c in Figs 6.32 and 6.34 use the so-called 'parasitic insensitive' structure. The response of these integrators is unaffected by stray capacitance to ground which might otherwise modify the filter transfer function.

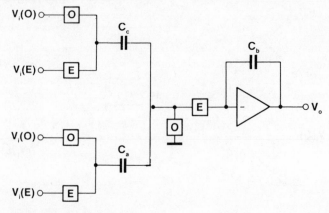

Fig. 6.32 Multi-input E-integrator.

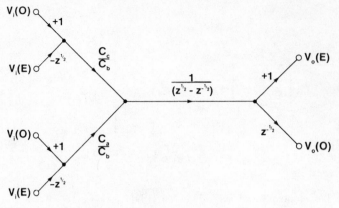

Fig. 6.33 SFG of multi-input E-integrator.

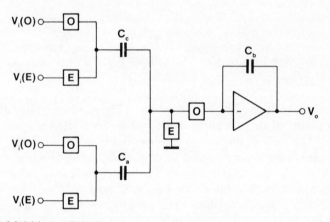

Fig. 6.34 Multi-input O-integrator.

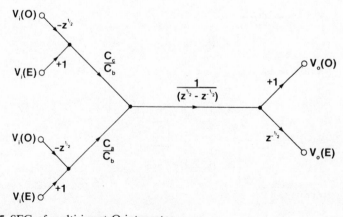

Fig. 6.35 SFG of multi-input O-integrator.

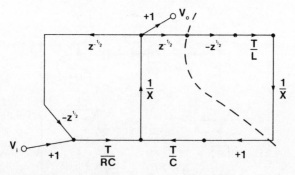

Fig. 6.36 Factorized SFG of SCF.

Both of these integrators require input branches with $-z^{1/2}$ terms, and output branches with $z^{-1/2}$ terms. The SFG of the bandpass filter (Fig. 6.31) (b)) can be redrawn to fit this format by factorizing -1 to $(-z^{1/2} . z^{-1/2})$. The result is Fig. 6.36. Comparing this with Figs 6.33 and 6.35, it is possible to see that this SFG can be approximated by one E-integrator and one O-integrator. Taking the equivalent circuits from Figs 6.32 and 6.34 we arrive at the circuit for the second order SCF (Fig. 6.37 (a)).

The $z^{-1/2}$ feedback branch at the output of the left-hand integrator cannot be realised as it must be sampled in the E-phase, as can be seen from the actual SFG of this SCF shown in Fig. 6.37 (b). This is the nature of the approximation, but if a large (f_s/f_o) ratio is maintained, in this case 25, the errors due to this approximation will be small (< 0.1 dB).

Comparing the SFG's for the SCF design (Fig. 6.37 (a)) and the LDI transform of the RLC prototype (Fig. 6.36) allows the SCF capacitor ratios to be defined for any chosen centre-to-switching frequency ratio, bearing in mind the values of R, L and C used to define the centre frequency (f_c) in the prototype filter must be modified to allow for distortion by the LDI transform. We saw in Section 6.2.9 that this transform equates

$$\omega_o \frac{T}{2} \quad \text{to} \quad \sin \left(\Omega_o \frac{T}{2} \right)$$

where Ω_o is the SCF centre frequency and $1/T$ is the switching frequency. Thus ω_o must be calculated to fit the required SCF centre frequency, and the RLC values must fit this 'distorted' value of f_o. For the second-order bandpass filter with a Q of 16 and a switch-to-centre frequency ratio of 25 (i.e. $\Omega_o T/2\pi = 1/25$), this yields (in relative units of capacitance):

$$C_{11} = 1$$
$$C_{21} = C_{12} = 15.9372$$
$$C_1 = C_2 = 63.3298$$

C_1 and C_2 have been scaled to give equal time constants for the two integrators.

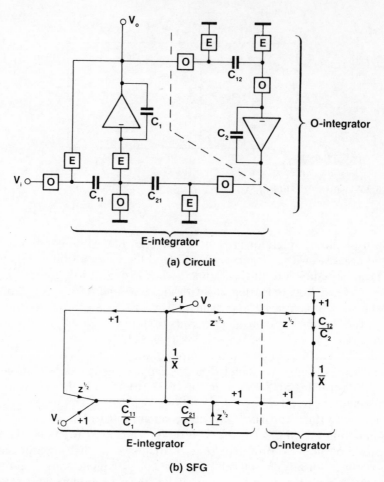

Fig. 6.37 Final SCF circuit and SFG.

In order to reduce the large spread in capacitor values for this filter (and a higher Q-factor would require a still larger spread in capacitance values), it is possible to transform the smallest capacitor (C_{11}) into a T-network made up of larger values (Fig. 6.38). Rescaling the capacitor values to give a new minimum value of unity and then gives:

$$C_{a1} = C_{c1} = 1$$
$$C_{21} = C_{12} = 1.76676$$
$$C_{b1} = C_1 = 7.02058$$

The capacitance spread and the total capacitance has thus been greatly reduced, with a consequent reduction in the amplifier settling time, although at the expense of a sensitivity to parasitic capacitance to ground (at the central T-node) and an increased sensitivity to capacitor mismatch. The final circuit diagram for the SCF is shown in Fig. 6.39.

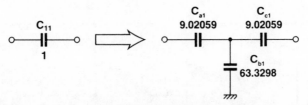

Fig. 6.38 T-network equivalent of the smallest SCF capacitor.

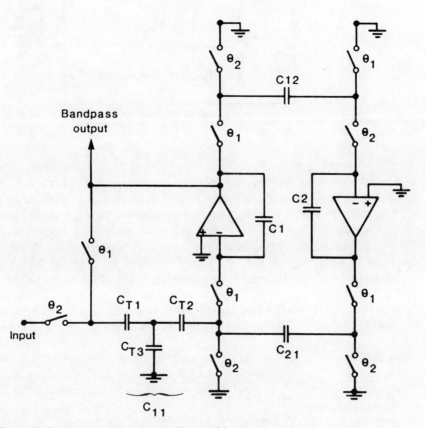

Fig. 6.39 Second-order band-pass filter architecture.

For good agreement between the RLC prototype response and the SCF response, the critical requirements are for high gain, fast settling amplifiers, fast switches (low RC time constant) and high capacitor ratio accuracy. The filter IC was designed with these requirements in mind using the design techniques described in Section 6.4, and by careful layout techniques. Further details are given below.

The amplifier circuit is shown in Fig. 6.40. This is a two-stage differential-input amplifier design. The first stage is an asymmetrical differential amplifier which drives a second-stage inverting amplifier via a level-shifting

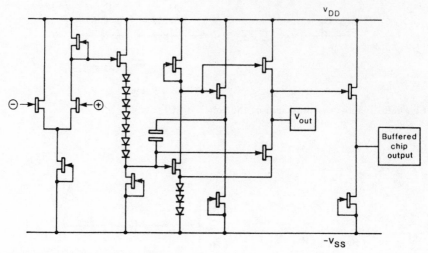

Fig. 6.40 Amplifier circuit diagram.

source follower. The negative rail for this second stage is arranged to be approximately two volts above V_{SS} by biasing through a chain of three large area diodes. This then ensures that the current-source FET in the level shifting network is kept in saturation so as to maintain a high transfer efficiency. The number of diodes in the level-shift network is calculated on the basis of requiring the output of the first stage to sit in the high-gain region of the transfer characteristics when the output of the second stage is mid-rail, and to minimize the input offset voltage defined in this condition.

In any two-stage amplifier design it is possible for the phase shift between the inverting input and the output to reach zero degrees at high frequencies while the gain is greater than unity. Any feedback from the output to the input could then cause the circuit to oscillate unless appropriate remedial action is taken. Such action is provided in this circuit by connecting a 'compensation capacitor' around the second stage; this capacitor introduces local negative feedback into the second stage at high frequencies, and ensures that the overall amplifier gain is rolled-off to unity well before the phase shift approaches zero. Undesirable feedforward effects (where the signal bypasses the second stage completely via the compensation capacitor) are reduced by driving the capacitor via a source-follower stage; this source follower allows feedback signals to pass from the second stage output to the input, but effectively shorts all feedforward signals from the second stage input to the V_{SS} rail.

Outputs from the amplifier are taken from a 'totem pole' push-pull stage for requirements within the filter IC, and from a subsquent source-follower buffer for off-chip requirements.

The switches used in the filter were standard MESFETs, the control voltage applied to the gate being derived from a source follower to ensure that the gate could never be forward-biased (see Fig. 6.9 (b)). Level-dependent interaction between the clock and the switched analogue signal is thus avoided. Clock breakthrough via the gate-channel capacitance is level-independent and simply adds a d.c. offset to the filter output without affecting the bandpass frequency response. The control voltage was obtained from a non-overlapping two-phase clock supplied by an external generator. The width of the switch FETs was chosen so that the switched-capacitors could charge and discharge through these switches with a time constant much shorter than the anticipated minimum switch closure time (approximately half the clock period).

Using a simple ion-implanted MESFET process with a single threshold voltage of approximately –1V and without resorting to self-aligned technologies, this bandpass filter has shown an insertion loss of approximately 5 dB, and a centre frequency accuracy error within ± 5% (varying with the switching frequency) up to 200 MHz. Operation at a switching frequency of 250 MHz has been seen at an increased accuracy error of − 13%, from which it can be inferred that the filter performance was limited at this frequency by the amplifier settling time of approximately 2 ns. The chip and its transfer characteristic at this frequency are shown in Figs. 6.41 and 6.42.

The relatively high insertion loss and the centre frequency accuracy errors of this IC are predominantly due to a relatively low amplifier gain of approximately 42 dB. As described in Section 6.4, GaAs MESFETs exhibit a relatively low value of g_m/g_o which limits the maximum gain which can be achieved by simple amplifier designs. New filter designs are now emerging,

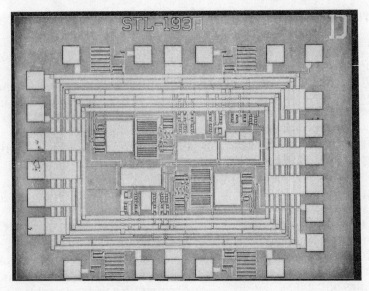

Fig. 6.41 Photomicrograph of the band-pass filter IC.

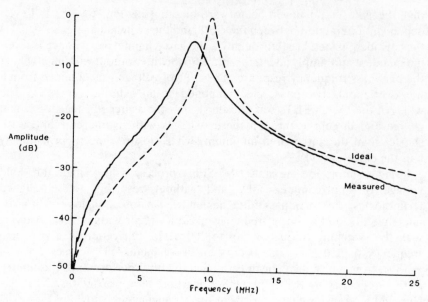

Fig. 6.42 Measured and ideal transfer characteristics of GaAs band-pass filter at 250 MHz switching frequency.

however, that are relatively insensitive to the gains of the amplifiers used, so allowing higher accuracy to be achieved with presently available simple amplifier designs.

6.6. CONCLUSION

The use of GaAs in sampled analogue circuits represents a relatively new and potentially powerful technique. Of course, such applications using silicon technology are themselves in their infancy. It is possible, then, that we may find silicon versions of the chips that have been described in this chapter appearing in the near future, and challenging the performance of the GaAs IC in the same way that recent advanced silicon ECL has done in the case of GaAs DCFL circuits. The situation with sampled analogue ICs is, however, different to that of digital circuits. If the use of low-resistance switches is the key to the advantage of the GaAs sampled analogue IC, then the mobility advantage of the GaAs MESFET will remain, and competition from silicon will never become severe.

LF noise and other effects such as those described in Section 6.4.5 do present a problem that can be difficult to endure. The attentions of crystal growers seems to be required here in order to effect a cure for this problem and in the meantime, some circuit techniques are available to alleviate the symptoms of the bulk and surface carrier traps that are responsible for such effects.

The performance of ADCs, and the demand that they place on circuits and processes, are only just beginning to be understood in the GaAs IC field. The general problems of data conversion for the growing army of digital signal processing applications will mean that this is a field in which much growth can be expected. It is not without its problems. The low-frequency effects that are important in so many sampled-analogue applications are no less crucial here and matching, reproducibility and other manufacturing problems play an important role.

The key to exploiting GaAs d.c.-coupled and sampled analogue circuits is to recognize that the architectures for silicon circuits that have been rejected as too low in speed for the requirements of today are likely to show a considerable performance improvement if implemented in GaAs. In addition, other architectures that make use of the advantages in speed, bandwidth and switch on-resistance will undoubtedly be discovered yet.

REFERENCES

1. Sovero, E. A., Hill, W. A., Sahai, R., Higgins, J. A., Pittman, S., Martin, E. H. and Pierson, R. L. Jr. (1983) 'Transversal filter application of a high speed gallium arsenide CCD' *IEEE GaAs IC Symp.* Phoenix, Arizona, USA.
2. Hoskins, M. J. and Hunsinger, B. J. (1983) 'Buried channel accoustic charge transport devices in GaAs' *IEEE GaAs IC Symp.* Phoenix, Arizona, USA.
3. Sovero, E. A., Hill, W. A., Higgins, J. A. and Sahai, R. (1984) 'Microwave frequency GaAs charge-coupled devices' *IEEE GaAs IC Symp.* Boston, Massachusetts, USA.
4. Greiling, P., Lee, R., Winston, H., Hunter, A., Jensen, J., Beaubien, R. and Bryan, R. (1984) 'GaAs Technology for high speed A/D converters' *IEEE GaAs IC Symp.* Boston, Massachusetts, USA.
5. Barker, G., Phillips, J. and Mun, J. (1986) 'A high-speed programmable transversal filter' *IEEE GaAs IC Symp.* Grenelefe, Florida, USA.
6. Weiss, F. (1986) 'A 1 Gs/s 8-bit GaAs DAC with on-chip current sources' *IEEE GaAs IC Symp.* Grenelefe, Florida, USA.
7. Fawcett, K., Hafizi, M., Lin, S., Myers, D., Esfandiari, R., Kim, M., Williams, T. and Englekirk, M. (1980) 'High-speed, high-accuracy, self-calibrating GaAs MESFET voltage comparators for A/D converters' *IEEE GaAs IC Symp.* Grenelefe, Florida, USA.
8. Ducourant, T., Binet, M., Baelde, J.-C., Rocher, C. and Gibereau, J.-C. (1986) '3 GHz 150mW, 4 bit GaAs analogue to digital conversion' *IEEE GaAs IC Symp.* Grenelefe, Florida, USA.
9. de Graaf, K. and Fawcett, K. (1986) 'GaAs technology for analogue-to-digital conversion' *IEEE GaAs IC Symp.* Grenelefe, Florida, USA.
10. Larson, L., Jenson, J., Levy, H. and Greiling, P. (1985) 'GaAs differential amplifiers' *IEEE GaAs IC Symp.* Monterey, California, USA.
11. Harrold, S., Vance, I., Mun, J. and Haigh, D. (1985) 'A GaAs switched-capacitor bandpass filter IC' *IEEE GaAs IC Symp.* Monterey, California, USA.

12. Hill, W., Sovero, E., Higgins, J., Martin, E. and Pittman, S. (1985) '1 GHz Sample-Rate GaAs CCD Transversal Filter' *IEEE GaAs IC Symp.* Monterey, California, USA.

13. Miller, D., Bujatti, M. and Estreich, D. (1985) 'Low Frequency Oscillations in GaAs ICs' *IEEE GaAs IC Symp.* Monterey, California, USA.

14. Barta, G. and Rode, A.G. (1982) 'GaAs Sample-and-Hold IC using a 3-Gate MESFET Switch' *IEEE GaAs IC Symp.* New Orleans, Louisiana, USA.

15. Burke, B., Nochols, K., Kelliher, J. and Murphy, R. (1982) 'Fabrication of GaAs CCDs for High-Speed Spatial Light Modulators' *IEEE GaAs IC Symp.* New Orleans, Louisiana, USA.

16. Higgins, J., Milano, R., Sovero, E.A. and Sahai, R. (1982) 'Resistive Gate GaAs Charge Coupled Devices' *IEEE GaAs IC Symp.* New Orleans, Louisiana, USA.

17. Meignant, D. and Binet, M. (1982) 'A High Performance 1.8 GHz Strobed Comparator for A/D Converters' *IEEE GaAs IC Symp.* Phoenix, Arizona, USA.

18. La Rue, G. (1983) 'A GHz GaAs Digital to Analogue Converter' *IEEE GaAs IC Symp.* Phoenix, Arizona, USA.

19. Folkes, P. A. (1986) 'Characteristics and Mechanism of 1/f Noise in GaAs Schottky Barrier Field-Effect Transistors' *Appl. Phys. Lett.* **48**(5) (Feb. 3).

20. Harrold, S. J., Vance, I. A. W. and Haigh, D. G. (1985) 'Second Order Switched Capacitor Bandpass Filter Implemented in GaAs' *Electronics Letters* **21**(11) (May 23) 494–6.

21. McKnight, A. J., Mun, J. and Vance, I. A. W. (1984) 'High Frequency GaAs Transversal Filter' *Electronics Letters* **20**(2) (Jan 19) 84–5.

22. Zverev, A. I. (1967) *Handbook of Filter Synthesis* Wiley, New York.

23. Steiglitz, K. (1970) 'Computer-Aided Design of Recursive Digital Filters' *IEEE Trans. Audio Electroaccoust.* **AU–18** (June).

24. Golden, R. M. and Kaiser, J. F. (1964) 'Design of Wideband Sampled-Data Filters' *Bell System Tech. J.* **43**(4) part 2 (July) 1533–45.

25. Lee, W. S. and Swanson, J. G. (1983) 'Pulsed Field-Effect Channel Current Transients in Plasma-Anodised Alumina/GaAs Metal/Insulator/Semiconductor Field-Effect Transistors' *Thin Solid Films, Special Issue on III-V MIS* **103**(1/2) (May 13) 177–91.

26. Lee, W. S. (1987) 'Sensitivity of GaAs Analogue Circuit Building Blocks to the Effect of Back-Gating' *IEE Electronics Letters* **23**(11) (May 21) 587–9.

27. Haigh, D. G. and Singh, B. (1982) 'A Brief Overview of Switched-Capacitor Filter Design' *Workshop on Design and Fabrication of Integrated Circuit Filters* Imperial College, London, UK.

28. Choi, T. D. and Broderson, R. W. (1980) 'Considerations for High Frequency Switched-Capacitor Ladder Filters' *IEEE Trans. on Circuits and Systems*, **CAS–27**(6) (June) 545–52.

29. Petersen, B. (1981) 'Switched-Capacitor Filters – Analog Signal Processing in a Digital Age', *Teleteknik* **I–2**.

30. Deczky, A. G. (1972) 'Synthesis of Rescursive Digital Filters using the Minimum p-error Criterion' *IEEE Trans. Audio Electroacoust.* **AU–20**. (Oct.) 257–63.

31. Shanks, J. L. (1967) 'Recursion Filters for Digital Processing' *Geophys.* **32**(1) (Feb.) 33–51.

32. Gold, B. and Jordan, K. L. (1969) 'A Direct Search Procedure for Designing

Finite Duration Impulse Response Filters' *IEEE Trans. Audio Electroacoust.* **AU–17**(1) (March) 33–6.

33. Rabiner, L. R. and Schafer, R. W. (1971) 'Recursive and Non-Resursive Realisations of Digital Filters Designed by Frequency Sampling Techniques' *IEEE Trans. Audio Electroacoust.* **AU–19**(3) (Sept.) 200–7.

34. Herrman, O. and Schuessler, H. W. (1970) 'Design of Nonrecursive Digital Filters with Minimum Phase' *Electronics Letters* **6**(11) 329–330.

35. Papoulis, A. (1957) 'On the Approximation Problem in Filter Design *IRE Conv. Record* Part 2 175–85.

36. Rader, C. M. and Gold, B. (1967) 'Digital Filter Design Techniques in the Frequency Domain' *Proc. IEEE.* **55** (Feb.) 149–71.

37. Allen, P. E. (1984) *Switched Capacitor Circuits* Van Nostrand Reinhold Electrical/Computer Science and Engineering Series.

38. Makram-Ebeid, S. and Minodo, P. (1983) 'Side-Gating in GaAs Integrated Circuits: Surface and Bulk-Related Phenomena' *IEEE GaAs IC Symp.* Phoneix, Arizona, USA.

39. Camcho-Penalosa, C. and Aitchison, C. S. (1985) 'Modelling Frequency Dependence of Output Impedance of Microwave MESFETs at Low Frequencies' *IEE Electronics Letters* **21** 528–9.

40. Lee, W. S. and Mun, J. (1987) 'An Improved Negative Feedback Technique to Reduce the Drain Conductance of GaAs MESFET for Precision Analogue ICs' *IEE Electronic Letters* **23** 705–7.

Chapter 7

Heterojunction Integrated Circuits

William A. HUGHES, Ali A. REZAZADEH and
Colin E. C. WOOD

7.1 INTRODUCTION

Since the first GaAs metal semiconductor field effect transistor (MESFET) was reported in 1967 there has been a growing interest in GaAs-based transistors as discrete devices and in analogue and digital integrated circuits. This has principally been due to the high electron mobility and low interconnect capacitance associated with the high resistivity of semi-insulating GaAs substrates compared with silicon. The GaAs MESFET has been the workhorse of GaAs discrete transistors and ICs and has continued to outperform worldwide the silicon-based counterparts. More recently MESFET technology has matured sufficiently to enable its use as the principle active device in commercial hybrid and monolithic integrated circuits and as a standard 'foundry' device.

The demand for more control and sophistication in simple GaAs MESFETs has stimulated much advanced materials growth activity. Techniques such as molecular beam epitaxy (MBE) and metal-organic chemical vapour deposition (MOCVD) have proven invaluable in improving devices and circuits. The evolution of MBE and MOCVD to today's level of sophistication has, for the first time, made possible a generation of devices which utilize heterojunctions between the alloy III-V, aluminium gallium arsenide (AlGaAs) and the binary III-V compound, GaAs. The band gap increases with aluminium concentration which leads to a difference (discontinuity) in the band-gap energies at GaAs/AlGaAs heterojunctions. The ability to so engineer the band structure of transistors has opened up new and exciting possibilities in transistor materials structure design, which has subsequently led to the inception, development and rapid demonstration and exploitation of new heterostructure devices.

The most notable of these new devices are two generically different operational types: the high electron mobility transistor (HEMT or MODFET) and the heterojunction bipolar transistor (HBT). In concept the HEMT is a sophisticated MESFET (analogous to the Si MOSFET) depending upon lateral unipolar transport. On the other hand the HBT is a III-V analogue of the silicon bipolar transistor.

Of these two main heterostructure devices, the HEMT has received most attention due to its comparative ease of material preparation and its potential as a low-noise analogue or low-power digital device. Several acronyms are used for the HEMT transistor family, in addition to HEMT, the most common being: MODFET (modulation doped field effect transistor), TEGFET or 2 DEGFET (two-dimensional electron gas field effect transistor) and SDHT (selectively doped heterojunction transistor). A humourous alternative explanation of the latter acronym is 'single Dingle heterojunction transistor' after Ray Dingle who was the co-discoverer with Horst Stormer of the modulation doping effect. Each of these names refers to a different aspect of the device's construction or operation which will be dealt with in the following section.

The performance advantage of HEMTs compared with MESFETs derives from the separation of the carriers (electrons) from their parent donor atoms across the AlGaAs/GaAs heterojunction which virtually eliminates ionized impurities as a carrier scattering mechanism. This carrier-ionized donor separation is only possible by the use of heterojunction, which creates a significant conduction band discontinuity at the material interface. Thus accumulation of electrons as a sheet or a quasi two dimensional ($\sim$80 Å wide) electron gas (2DEG) occurs.

Following the discovery of the modulation doping phenomena in 1978 the first HEMT device was fabricated in 1980. The practical and technological developments of this device and its variants has been so rapid that the HEMT family now holds performance records in most fields of low-noise analogue and high-speed digital devices and circuits. HEMT logic gates switch faster than any other semiconductor technology and sub-micron HEMT devices hold records for low-noise performance at microwave and millimetre wave frequencies. These devices have now extended the domain of transistor amplification up to 94 GHz for the first time. Control of material quality and complexity together with processing technology for these devices continues rapidly to improve, and new material structures building upon the HEMT principle are pushing HEMT performance still further.

The HBT concept, long recognized for its potential in high-speed applications, has recently stimulated an increased level of research activity particularly in the GaAs/AlGaAs material system. This interest stems largely from predictions of excellent high-frequency performance, both as discrete devices and in integrated circuit applications. One important application of HBTs is likely to be in analogue-to-digital converter circuits where very high sampling rates are desirable and the circuits require the high driving power (large transconductance) of an HBT. Digital ICs based on HBT devices is expected to provide a variety of advantages in comparison with GaAs MESFET approaches. These advantages include: better threshold voltage uniformity in bipolar which is determined by the band gap of the material rather than by process parameters; high output current drive capability (both within a chip and off chip) leading to low sensitivity of propagation delay time to capacitive loading and fan-out; increased speed due to the high cut-off frequency, f_T, possible with HBTs.

Applications for ultra high performance devices such as the HEMT or the HBT are manifold. Large-scale digital integrated circuits based on these devices will undoubtedly be in the active core of future generations of supercomputers. Levels of integration of HEMT based circuits are currently sufficiently mature that 4 K static random access memories (SRAMs) have been fabricated. Applications also exist for smaller scale integrated circuits (SSI) as very high speed prescalers or multiplex/demultiplex combinations. These circuits will perform the high-speed front-end functions such as multiplexing and down conversion for systems with slower but more complex silicon circuitry performing the lower frequency systolic processing.

Analogue applications for HEMT and HBT devices are arguably more immediate. In particular, HEMT devices are ideally suited for the critical front-end amplification stage of a low noise microwave receiver system. Such amplifiers could be hybrid circuits based on discrete HEMT devices or more advanced monolithic microwave integrated circuits (MMICs). Already, HEMT devices have achieved lower noise figures than any other transistor types for comparable frequencies.

It is increasingly apparent that HEMT and HBT devices offer large improvements in performance compared to alternative conventional semi-conductor technologies, either silicon or gallium arsenide based.

In this chapter we will review the evolution and technological development together with the basic theory of heterojunction devices. A comparison is then made of the operational frequency limits, gain, noise figure etc. performances of these devices with their leading alternative device types.

7.2　THE HIGH ELECTRON MOBILITY TRANSISTOR (HEMT)

The high electron mobility transistor (HEMT) has been receiving increasing attention worldwide during the last few years due to its outstanding performance as a discrete device and in integrated circuits. Digital ICs based on the HEMT hold performance records for propagation delay and RAM access time, and sub-micron HEMT devices have recorded the lowest noise figures of any technology at microwave and millimeter wave frequencies.

The existence of an electron accumulation layer at the interface of certain heterojunctions was first proposed by Anderson [1] in July 1960. He considered a heterojunction of germanium on gallium arsenide in which the narrow band gap material (in this case germanium) has a higher electron affinity than the wide band gap material (gallium arsenide). Electrons are attracted towards the narrow band gap material and accumulate at the interface between the two materials.

In 1970, during their studies on superlattices, Esaki and Tsu [2] suggested that ionized donor impurities and free electrons could be spatially separated so reducing the coulombic interaction between them. In October 1978

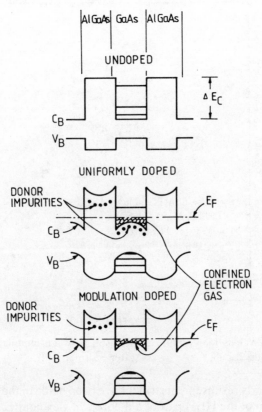

Fig. 7.1 Energy-band diagrams for undoped, uniformly doped and modulation doped GaAs/AlGaAs superlattices (after Dingle [3]).

Dingle and co-workers [3] grew the first 'modulation-doped' GaAs/AlGaAs superlattice in which the donor impurities were present only in the wide band gap AlGaAs layer (Fig. 7.1). The resulting reduction in ionized impurity scattering in the GaAs layers led to electron mobilities higher than previously observed in doped GaAs and were particularly enhanced at low temperatures (Fig. 7.2). This was in agreement with the idea of reduced ionized impurity scattering as this would be particularly pronounced when phonon scattering was reduced at low temperatures.

Mimura and co-workers [4] were the first to fabricate a device which they called HEMT (high electron mobility transistor) using the new structure in May 1980. They observed a mobility enhancement of 30% in the HEMT compared with the MESFET at 300 K, and a mobility 5 to 6 times higher than that of the MESFET, at 77 K. This was followed in August 1980 by Delagebeaudeuf *et al.* [5] who called the device TEGFET (two-dimensional electron gas FET) and by Wira *et al.* in January 1981 [6] who called it MODFET (modulation doped FET). The first enhancement mode device was reported by Mimura *et al.* [7] who reported a transconductance, g_m, of 409 mS/mm at 77 K, the highest g_m of any field effect device at that time.

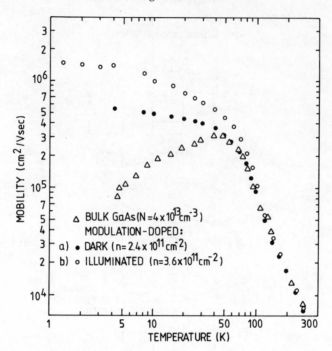

Fig. 7.2 Mobility versus temperature comparison between modulation doped and unintentionally doped bulk GaAs layers (after Stormer [20]).

Since these early results, great strides have been made in the understanding of the operation of the HEMT from a theoretical viewpoint, as well as in the fabrication of discrete devices and ICs. Low-noise HEMT devices are now sold commercially and HEMT digital ICs have reached such maturity that 4 K SRAMs have been successfully fabricated [8].

It is becoming apparent that the HEMT provides a significant performance advantage over other IC technologies (silicon bipolar and gallium arsenide MESFET) in both digital and analogue circuit functions. The aim of this section is to explain the operation of the device and present theoretical and practical results which demonstrate its potential. In addition HEMT fabrication technology will be described and future developments discussed.

7.2.1 Modulation doping

In conventional GaAs MESFETs the high electron concentrations necessary for device operation are obtained by doping the channel layer n-type with, for example, silicon donor atoms. The resulting 'free' electrons then share the same physical space with the ionized donors and so interact with them as they move from source to drain under the influence of an applied field. This Coulombic interaction, known as ionized impurity scattering, is the dominant scattering mechanism in MESFETs and limits the electron mobility to relatively low values.

This scattering can be reduced dramatically by the technique of modulation doping. A heterojunction composed of AlGaAs and GaAs can be grown such that the n-type donors are introduced only into the wide band-gap AlGaAs layer. The electrons diffuse from the AlGaAs conduction band to the GaAs conduction band as the latter has a lower energy. The resulting depleted AlGaAs layer forms a positively charged region which attracts the electrons back to the interface. The resulting electric field perpendicular to the interface causes severe band bending as shown in Fig. 7.3. The electrons are confined to the narrow (approximately 100 Å)

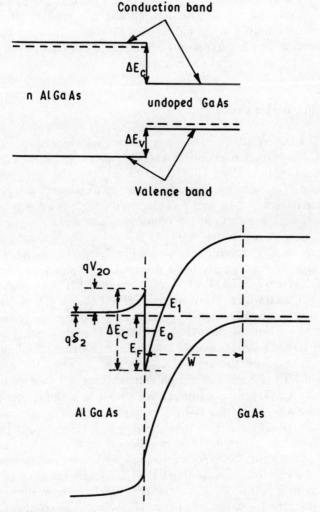

Fig. 7.3 Band diagram of the n-AlGaAs/GaAs heterojunction before (upper) and after (lower) equilibrium is established (after Delagebeaudeuf [41]). The electrons diffuse from the n-doped AlGaAs to the undoped GaAs and accumulate in the potential well.

potential well where they move parallel to the interface as a two-dimensional electron gas (2DEG).

Motion perpendicular to the interface is restricted by the band bending which leads to quantization of the electron energies as discussed later. As a result discrete energy sub-bands are formed in the quantum well. A constant Fermi level is maintained across the heterojunction, the position of which determines the occupancy of each energy sub-band.

Perhaps the most important result of modulation doping is that the electrons are physically separated from their parent donor ions, and move in undoped material. Indeed they may be further separated by introducing an undoped AlGaAs 'spacer layer' between the GaAs and doped AlGaAs. This tends to increase the electron-ionized donor separation and therefore reduces ionized impurity scattering further, thereby enhancing the mobility of the structure. This has important consequences as regards the transport properties of the structure as discussed later.

7.2.2 HEMT device structures

Section 7.2.1 described the basic constituents of modulation doping: an undoped GaAs layer, an undoped AlGaAs spacer and a doped AlGaAs layer. To realise practically a HEMT device, a doped GaAs capping layer is also required. This serves to provide a low resistance access region to the gate, which is recessed to make contact with the doped AlGaAs layer (Fig. 7.4 (a)). It is also used for ohmic contacts and to prevent surface depletion and oxidation of the AlGaAs layer.

An alternative layer scheme is shown in Fig. 7.4 (b) in which the structure of Fig. 7.4 (a) has essentially been reversed. This structure is called the inverted HEMT or IHEMT. To date, the IHEMT has received little attention compared with the conventional HEMT shown in Fig. 7.4 (a). This has been due principally to the technological problem of growing good quality GaAs on top of the AlGaAs n-doped layer, leading to a poor mobility in IHEMT structures [9]. The highest reported mobilities for this structure, 92 000 cm^2/Vs at 77 K [10] are significantly lower than conventional HEMT values, although enhanced mobility values have been recently reported [11] using a superlattice structure in the spacer layer.

Despite these problems the IHEMT may provide certain advantages over conventional HEMTs. In particular the doped AlGaAs layer provides a potential barrier to carrier injection into the substrate from hot electrons in the 2DEG. This would be expected to improve the output conductance of the device. In addition, because the 2DEG resides on the GaAs side of the heterojunction, it will be closer to the edge of the gate in the case of the IHEMT structure by a distance $2\Delta d$ [12] where Δd is the separation between the centre of the 2DEG and the heterojunction. This would be expected to increase the transconductance, g_m, of the device which is proportional to the inverse of the gate–2DEG separation. Indeed the highest

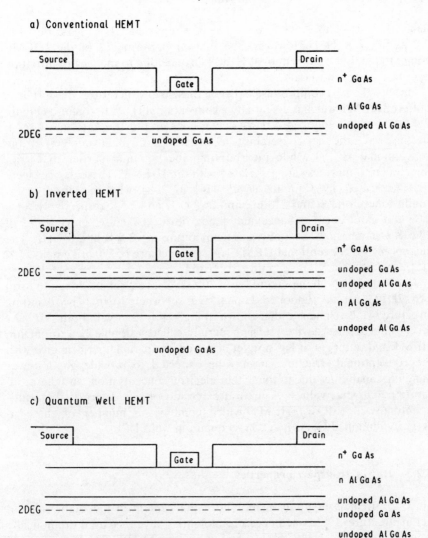

Fig. 7:4 Cross-sectional view of various HEMT devices (a) conventional HEMT (b) IHEMT (c) quantum well HEMT.

transconductance of any field effect device was obtained from an IHEMT by Cirillo *et al.* [12] who obtained transconductances of 1180 and 1810 mS/mm at room temperature and 77 K respectively. This high g_m was obtained by reducing the gate–2DEG separation. However, this also leads to an increased gate capacitance. As a result of this high capacitance, the estimated value of the f_T of these devices was only 7.5 GHz for an effective gate length of 1.7 μm. The best reported IHEMT ring oscillator was reported by Kinoshita *et al.* [10] who measured 26.3 ps gate propagation

delay at 77 K. This is well above the best-reported HEMT gate delay of
5.8 ps at 77 K [13]. However, the inferior performance of the IHEMT
compared with the conventional HEMT may be due to the lack of attention
paid to this device to date.

Perhaps the best compromise between these two structures is the AlGaAs
buffered HEMT which is essentially a conventional HEMT structure (Figure
7.4 (a)) with an undoped AlGaAs layer beneath the 2DEG (Fig. 7.4 (c)).
In this way, the device retains the material growth advantages of the
conventional HEMT while incorporating the potential barrier to carrier
injection into the substrate associated with the IHEMT. These devices were
first fabricated by Camnitz *et al.* [14] in 1984 who measured transconductances of 125 and 210 mS/mm at 300 K and 77 K respectively for a
0.35 μm gate length enhancement mode device. A noise temperature of
10.5 K was measured at a physical temperature of 12.5 K at 8.3 GHz. The
current record conventional HEMT transconductance of 570 mS/mm at 77 K
[15] is held by an AlGaAs buffered HEMT.

Alternatively, both upper and lower AlGaAs layers can be doped to form
two 2DEGs in the undoped GaAs layer between them. The resulting
structure is then similar to that shown in Fig. 7.4 (c) but with both AlGaAs
layers doped. This device [16] is generally called a double heterojunction
HEMT and, although it has not yet achieved comparable performance with
the conventional structure, it may be expected to provide good power
handling capabilities due to the higher electron concentration, and therefore
saturated current, values. A natural extension of this device is the multi-
quantum well (MQW) HEMT which combines a number of these n-
AlGaAs/undoped GaAs/n-AlGaAs quantum wells [17].

7.2.3 HEMT transport properties

Since the first observation of mobility enhancement in modulation-doped
heterostructures [3] low temperature mobilities have increased dramatically
from 20 000 cm^2/Vs in 1978 [3] to 1.95×10^6 cm^2/Vs [18]. The best reported
2DEG mobility is now approximately one order of magnitude greater than
that achieved in highest purity bulk GaAs [19]. This is principally due to
ionized impurity scattering [20] which limits the mobility of bulk GaAs,
especially at low temperatures (Fig. 7.5). In the HEMT, the effects of
ionized impurity scattering are reduced by carrier-donor separation and by
carrier screening in the 2DEG. Both mobility and velocity enhancement of
modulation-doped structures over conventional GaAs MESFET structures
have been reported. These enhanced transport properties are critical in the
operation of the HEMT and affect material design for optimized performance.

(a) **Low field mobility.** Despite the spacial separation of electrons and
donors via modulation-doping, many scattering mechanisms are present in
the HEMT which limit mobility enhancement of the device to some extent.

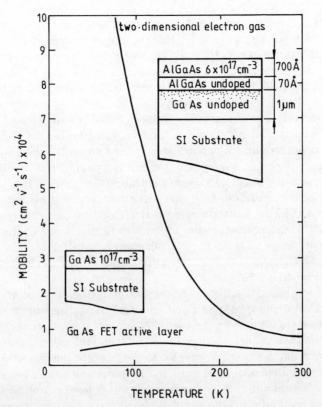

Fig. 7.5 Mobility versus temperature comparison between modulation doped layers and GaAs MESFET layers (after Delagebeaudeuf [41]).

The most important of these is bulk phonon interactions, both optical and acoustical, which cannot be prevented at finite temperatures [19]. At temperatures above 80 K, mobility data on modulation-doped structures is of the same order as that for highly pure GaAs (8500–9000 cm^2/Vs) which sets an upper limit to the mobility obtainable in modulation-doped structures at 300 K [21]. At low temperatures, however, phonon scattering in both modulation-doped structures and bulk GaAs is reduced. This leads to a dramatic increase in mobility in the modulation-doped structure. However, the mobility in a GaAs MESFET active layer does not attain such high values due to ionized impurity scattering (Fig. 7.5).

Apart from phonon scattering, the principal remaining scattering mechanisms in the HEMT are (a) ionized impurity scattering – by remote impurities in the doped AlGaAs and by impurities in the GaAs, (b) interface roughness scattering and (c) inter-sub-band scattering which are discussed below.

At low temperatures, the scattering rate depends on the separation of electrons from the ionized scattering centres which are located principally in the doped AlGaAs. It may be expected, therefore that increasing the spacer layer thickness (di) would lead to an increase in the low temperature mobility. Some workers have found that the mobility increases monotonically with spacer layer thickness ([18], [22]), whilst some have found that a peak in the mobility is obtained at spacer thicknesses of 50–100 Å ([23], [24]). The position of this peak has been found to depend on aluminium mole fraction ratio [23] and on temperature [24].

The differences in the reported results of mobility dependence on spacer thickness is due to the effect on the scattering rate of other parameters, such as 2DEG concentration, which are also dependent on the spacer thickness. As the spacer thickness is increased, the transfer of electrons into the 2DEG is reduced and the sheet carrier concentration decreases ([18], [24]). The sheet carrier concentration (n_s) affects the mobility in a number of ways. The ionized impurity scattering rate is decreased by carrier screening [25] which leads to an increase in the mobility with increasing n_s. In addition, increasing n_s leads to a higher Fermi energy reducing ionized impurity scattering which becomes less effective for higher energy electrons according to theoretical studies [26].

For Coulombic scattering only, therefore, including impurities in the doped AlGaAs, undoped spacer and undoped GaAs, the mobility would be expected to increase with increasing sheet carrier concentration. Indeed, by using illumination to increase n_s, Hiyamizu et al. [18] measured an increase in mobility from 1.95×10^6 cm^2/Vs to 2.12×10^6 cm^2/Vs at 5 K. This complicates the dependence of mobility on spacer thickness as increasing di leads to higher mobility via increasing electron donor separation and lower mobility via reduced n_s. If the latter mechanism is dominant at large di then a peak is observed in the mobility dependence on di.

Due to the quantization of the 2DEG potential well, a number of sub-bands exist which represent discrete electron energy states. As the sheet carrier concentration is increased, states in the lowest available sub-band become filled. Eventually states within the first excited sub-band become available for electron scattering leading to an increase in scattering rates and a decrease in mobility. This process is known as inter-sub-band scattering. The onset of inter-sub-band scattering occurs at a sheet carrier concentration of approximately 7×10^{11} cm^{-2} according to theoretical predictions [19], above which the mobility will begin to decrease with increasing sheet carrier concentration before ultimately increasing again [26]. This is in good agreement with experimental evidence [18].

The effect of interface roughness on mobility has been studied theoretically [26] but is very difficult to measure experimentally. However, Stormer [20] has suggested that indirect evidence of this effect comes from the dependence of mobility on crystal growth temperature. Optimum mobilities are obtained at growth temperatures between 600 and 700°C [27]. This temperature range coincides with the range of temperature over which the interface roughness in GaAs–AlGaAs superlattices has been found to be lowest [28].

The effect of the various scattering mechanisms on overall mobility has been studied theoretically [19] and the results are summarized in Fig. 7.6. The total mobility resulting from the contributing of all scattering mechanisms is shown as a solid line and compared with experimental results (solid circles).

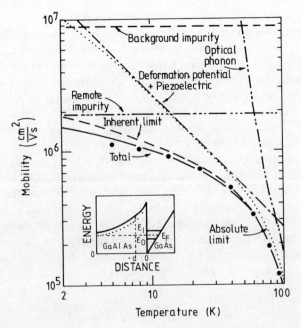

Fig. 7.6 Theoretical calculation of mobility versus temperature showing the effect of the various scattering mechanisms on mobility (after Walakiewicz [19]). The resulting theoretical curve agrees well with the best experimental data of Hiyamizu [18].

(b) **High field properties.** The enhancement of mobility in modulation-doped structures over conventional MESFET material is now well established, especially at low temperatures. This improved mobility leads to lower parasitic source and drain resistances. However, in short-channel FETs device performance is limited by the effective electron saturation velocity in the high field regions of the device. We must therefore consider the high field transport properties of the device, in addition to low field mobility, when determining the performance of practical HEMT devices.

At very high electric fields (greater than 3 kV/cm) velocity–electric field (v–E) characteristics are more appropriate than mobility in determining device performance. Drummond *et al.* [29] measured the HEMT v–E characteristics using pulsed Hall measurements up to an electric field of 2 kV/cm. They measured carrier velocities of 1.7×10^7 cm/s and 2.24×10^7 cm/s at 300 K and 77 K respectively compared with typical doped GaAs saturated velocities of 1×10^7 cm/s at room temperature. These v–E characteristics were in good agreement with Monte Carlo calculations based on undoped GaAs [30]. Lower velocity values were obtained by Masselink *et al.* using a geometric magnetoresistance method [31]. Velocities of 1×10^7 cm/s and 1.4×10^7 cm/s were obtained at 300 K and 77 K under an electric field of 3 kV/cm for channel lengths of greater than 6 μm. The highest reported velocity of 2D electrons of 3.8×10^7 cm/s at 77 K was obtained at an electric field of 2.4 kV/cm by a pulsed Hall technique [32].

HEMT effective satured velocities have also been inferred from measured device characteristics (usually from the intrinsic transconductance) using device models. For a 1 μm gate length device at room temperature, Lee *et al.* [33] reported a velocity of 1.5×10^7 cm/s, Hirano *et al.* [34] obtained 1.4×10^7 cm/s and Drumond *et al.* [35] obtained values of 2×10^7 cm/s at 300 K and a higher value of 3×10^7 cm/s at 77 K. These velocity values all exceed the doped GaAs velocity implying a velocity enhancement of HEMTs over GaAs MESFETs.

The velocity-field characteristics of HEMTs has been studied theoretically by Cappy *et al.* [36] using a one-dimension numerical model. They found that the average velocity under the gate can be higher than the static peak velocity giving rise to velocity overshoot. This is enhanced under the high electric fields in short gate length devices giving rise to an increase in g_m as the gate length is reduced [36]. Velocities of up to 2.5×10^7 cm/s at 300 K were obtained theoretically [36] for a gate length of 0.8 μm (see Fig. 7.7) supporting the observed velocity enhancement.

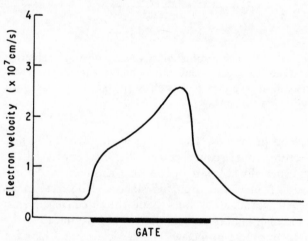

Fig. 7.7 Electron velocity as a function of position for a HEMT from calculations based on a numerical model (after Cappy [36]). Peak velocities of over 2×10^7 cm/s were obtained.

7.2.4 Low temperature degradation and persistent photoconductivity

The device and circuit performance of HEMTs can improve dramatically at low temperatures (less than 100 K) due to the improvement in device transport properties at these temperatures as discussed above. However, problems of drain current collapse, persistent photoconductivity and threshold voltage shifts can limit the usefulness of these devices at cryogenic temperatures. Drummond *et al.* [37] first reported such problems in December 1983. They measured threshold voltage shifts of about 0.2 V at 77 K with respect to room temperature. This threshold voltage shift was found to reduce dramatically on illumination with white light, and persisted when the illumination was removed. These effects were believed to be due to a charge injection/trapping process in which hot carriers injected into the AlGaAs layer are trapped by DX centres (deep electron trap levels) in the AlGaAs long enough to give rise to a reduced 2DEG concentration. This leads to threshold voltage shifts and also to a dramatic reduction in the drain current characteristics (I–V collapse) of the device [38].

The DX centres are effectively empty at room temperature due to thermal energy but fill up at low temperature. Illuminating the devices empties the traps and tends to reduce low-temperature degradation. Electrons optically excited from the traps are not readily recaptured due to a large potential barrier to capture leading to a persistent photoconductivity (PPC) effect [39]. This may last many minutes. At low temperatures drain current collapse generally occurs when a small drain potential is applied (less than 0.5 V), while voltages greater than about 1V tend to restore the characteristics [38]. This is consistent with large potential barriers to capture and emission.

The early problems in cryogenic operation of HEMTs described above have been reduced today by careful control of material growth and processing conditions ([37], [38]). However, by replacing bulk doped AlGaAs with a superlattice replacement of n-GaAs/AlGaAs or n-GaAs/AlAs (see Fig. 7.8), it is possible to eliminate such problems as DX centre formation which is a characteristic of doped AlGaAs. By using such a superlattice Fischer *et al.* [40] observed a reduced degradation of characteristics at 77 K and no noticeable light sensitivity. This replacement of bulk doped AlGaAs offers the possibility of exploiting the enhanced transport properties of the HEMT at low temperature without the deleterious effects associated with the DX centre of bulk AlGaAs.

7.2.5 Device theory and modelling

In recent years major advances in the device and circuit performance of HEMTs have increased interest in using these devices and stimulated an interest in the modelling of such structures. The first models were analytical and one-dimensional ([41], [42]). Two dimensional models have been developed in order to describe more accurately a number of phenomena that could not be analysed with more simple models ([36], [43]). More recently, Monte Carlo models have been developed [44]. In this section, some of the more significant analytic expressions will be presented and discussed in

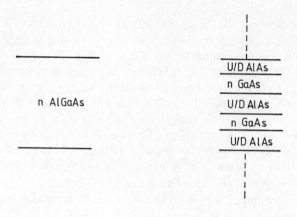

Fig. 7.8 Doped AlGaAs and the undoped AlAs/doped GaAs superlattice replacement.

order to aid understanding of device operation and to provide useful design and diagnostic tools.

Fig. 7.3 shows the band structure at the GaAs/n–A1GaAs heterojunction both before and after equilibrium is established by setting the Fermi levels on both sides of the heterojunction equal. The band bending shown arises when the Fermi levels are made equal and is due to the charge separation induced. Electrons leave the doped A1GaAs and accumulate in the quantum well at the GaAs side of the heterojunction, leaving behind positively charged donors near the heterojunction in the A1GaAs.

The quantum mechanical confinement of the electrons in the potential well on the GaAs side of the heterojunction leads to the formation of discrete energy sub-bands, each sub-band corresponding to a discrete electron energy. A useful approximation to the potential well is to consider it to be perfectly triangular ([41], [45]). The solutions for the wavefunction are Airy functions with eigenvalues [45]

$$\text{En} \approx \left(\frac{h^2}{2M_1}\right)^{1/3} \left(\frac{3\pi q E_{iG}}{2}\right)^{2/3} \left(n + \frac{3}{4}\right)^{2/3} \quad (1)$$
$$n = 0, 1., 2 \ldots$$

where M_1 is the longitudinal effective mass in the 2DEG and q is the charge of an electron. E_{iG} is a quasi constant electric field at the heterojunction interface in the GaAs. This electric field may be related to the sheet carrier concentration n_s in the 2DEG by

$$\varepsilon_{GaAs} E_{iG} = q n_s \quad (2)$$

and hence the positions of the energy levels in the 2DEG can be calculated from the sheet carrier concentration. By integrating the number of occupied states in the 2DEG, Delagebeaudeuf and Linh [41] showed that the sheet carrier concentration is given by

$$n_s = \frac{D_s kT}{2} \log\left[(1+\exp\{q/kT(E_F-E_0)\})\ (1+\exp\{q/kT(E_F-E_1)\})\right] \quad (3)$$

where D_s is the 2-dimensional density states in the 2DEG, k is Boltzmann's constant, T is the temperature, E_F is the Fermi level position and E_0 and E_1 are given by equation (1). Note that Fermi-Dirac rather than Boltzmann statistics are used in the derivation of this equation [41].

The sheet carrier concentration may also be obtained by solving Poisson's equation in the doped A1GaAs layer [41].

$$qn_s = \sqrt{(2q\ \varepsilon_A\ N_d\ v_{20} + q^2\ N_d^2\ di^2)} - q^N{}_d\ di \quad (4)$$

where ε_A is the permittivity in A1GaAs, N_d is the A1GaAs doping density and di is the spacer layer thickness. From Fig. 7.3, the band bending in the A1GaAs layer is given by

$$v_{20} = \Delta E_C - \delta_2 - E_F \quad (5)$$

where δ_2 is the difference between the A1GaAs conduction band edge and the Fermi level position in the A1GaAs.

Equations (3) and (4) may be solved to give the 2DEG concentration n_S as a function of layer parameters. Fig. 7.9 shows n_s plotted against the A1GaAs doping density with the undoped spacer layer as a parameter. This figure demonstrates quantitatively the decrease of n_s with spacer layer thickness discussed in Section 7.2.3.

Fig. 7.10 shows the band diagram of the HEMT under the influence of a Schottky gate with barrier height V_{bi}, placed directly on to the A1GaAs. A potential V_G is applied to the gate which is usually negative for a normally-on device. If V_G is sufficiently negative, or the A1GaAs layer is sufficiently thin, the gate depletion and junction depletion regions will overlap. The A1GaAs layer is then fully depleted and the device is in the so-called 'charge control regime'. By integrating Poisson's equation in the doped A1GaAs layer between the gate and the heterojunction, it can be shown that the sheet carrier concentration under the influence of a Schottky gate is given by ([41], [42])

$$n_s = \frac{\varepsilon_A}{q\ (d_2 + \Delta d)}\ (V_G - V_p) \quad (6)$$

where V_p is the pinch-off voltage given by

$$V_p = V_{bi} + E_F - \Delta E_C - V_{p2} \quad (7)$$

and

$$V_{p2} = \frac{qN_d}{2\varepsilon_A}(d_2 - di)^2 \quad (8)$$

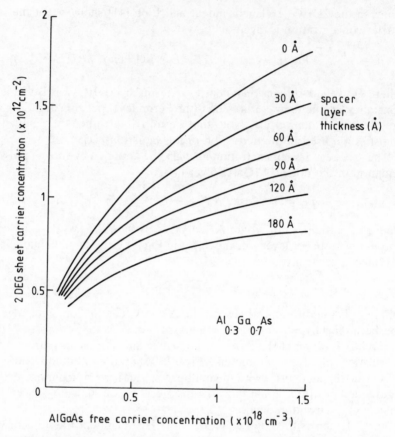

Fig. 7.9 Theoretical calculations for the sheet carrier concentration versus the AlGaAs free carrier concentration with the spacer layer thickness as a parameter (after Delagebeaudeuf [41]).

Here d_2 is the thickness of the AlGaAs layer, di is the AlGaAs spacer layer thickness and d is the distance between the centre of the 2DEG and the heterojunction, generally taken as 80 Å [42] (see Fig. 7.10).

From these equations we may determine the doped AlGaAs thickness required to achieve a given pinch-off voltage. This is shown in Fig. 7.11 which constitute useful design curves. This set of curves enables the AlGaAs thickness d_2 to be determined from the measured pinch-off voltage. This AlGaAs thickness is important in determining device transconductance and capacitance as discussed later.

If we now consider the device operating as a FET, then if $V_C(x)$ is the channel potential at a distance x from the source end of the gate then the equation for charge control becomes [41]

$$n_s = \frac{\varepsilon_A}{q(d_2 + \Delta d)}(V_G - V_C(x) - V_p) \tag{9}$$

The channel current at x is then given by

$$I(x) = q \, n_s \, W_g \, v(x) \tag{10}$$

If we assume that velocity saturation occurs exactly at the drain edge of the gate, then for a short gate length device we obtain [41]

$$I_{DS} = \frac{W_g \, \varepsilon_A \, v_s}{(d_2 + \Delta d)} \, (V_G - V_p - R_s \, I_{DS} - \Delta E_C L_g) \tag{11}$$

from which I_{DS} depends linearly upon the gate potential. We may now obtain the transconductance, g_m, by differentiating equation (11) to give

$$g_m = \frac{\partial I_{DS}}{\partial V_G} = \frac{g_{mo}}{1 + R_s \, g_{mo}} \tag{12}$$

where

$$g_{mo} = \frac{W_g \varepsilon_A v_s}{(d_2 + \Delta d)} \tag{13}$$

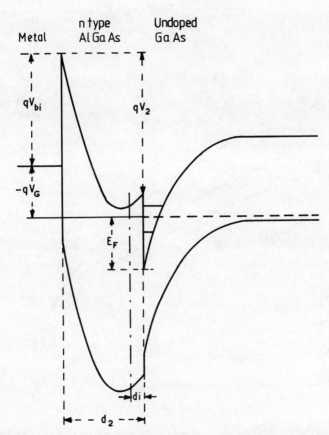

Fig. 7.10 Band diagram of the n-AlGaAs/GaAs heterojunction under the influence of a Schottky gate on the AlGaAs (after Delagebeaudeuf [41]).

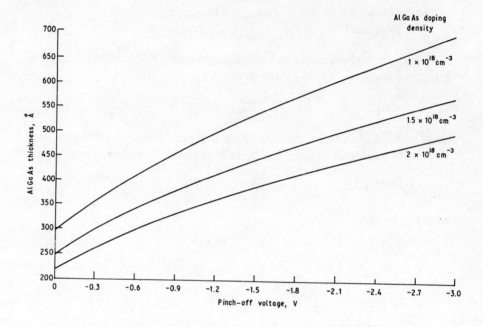

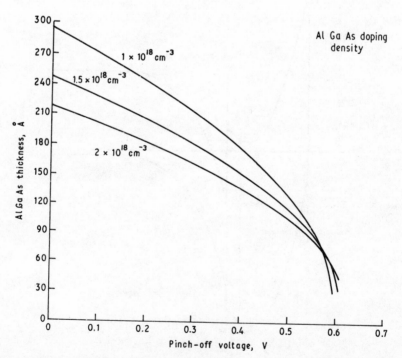

Fig. 7.11 Calculated dependence of pinch-off voltage on AlGaAs thickness for (a) normally-on and (b) normally-off HEMTs with AlGaAs doping density as a parameter.

The gate-source capacitance of the device may be approximated from equation 3.

$$C_{gs} = Aq \frac{\delta n_s}{\delta V_G} = \frac{A\varepsilon_A}{(d_2 + \Delta d)} \qquad (14)$$

where A is the effective area of the gate. These equations are similar in form to those for MESFET devices if we replace the constant quantity $(d_2 + \Delta d)$ by the Schottky depletion depth which, for a MESFET, depends on doping and applied bias. Equations (11)–(14) represent useful, if rough, approximations for determining trends in I_{DSS}, g_m and C_{gs} which, to a large extent, determine device performance. The value of d_2 can be calculated from the pinch-off voltage of the device from equation (4).

The g_m of the device can be increased by reducing the AlGaAs thickness (for example by increasing the AlGaAs doping or reducing the pinch-off voltage). However, this also increases the capacitance. In practice it is normally the transition frequency (f_T) which we require to be maximized which is given approximately by

$$f_T = \frac{g_m}{2\pi C_{gs}} \qquad (15)$$

Increasing the f_T of the device increases the maximum available gain according to the relation [46]

$$\text{MAG} = \frac{1}{4}\left(\frac{f_T}{f}\right)^2 \frac{1}{(g_d + 2\pi f_T C_{gd})(R_s + R_g + R_i + 2\pi f_T L_s)} \qquad (16)$$

where g_d is the drain conductance, C_{gd} is the gate drain capacitance, L_s is the source inductance and R_s, R_g and R_i are the source gate and input resistances.

To increase f_T we must reduce the gate length of the device and increase v_s, the average saturated transit velocity. Reported values of v_s deduced from material and device measurements exceed the GaAs MESFET value of 1×10^7 cm/s by up to a factor of two at 300 K (see Section 7.2.1). This tends to increase the g_m/C_{gs} ratio of HEMTs and hence improves the f_T. In addition, the maximum operating frequency of digital logic circuits is determined by the rate at which the input FET of the following gate can be charged by the preceeding output FET. This drive capability is also determined by the g_m/C_{gs} ratio.

The noise figure NF of the HEMT can be estimated from the well-known Fukui equation for MESFETs [47],

$$NF = 1 + K_F \cdot 2\pi f C_{gs} \left(\frac{R_s + R_g}{g_m}\right)^{1/2} \qquad (17)$$

where K_F is an empirically determined constant. For MESFETs, K_F is typically 2.5 whereas for HEMT devices, K_F values of around 1.5 have been obtained [48] implying lower noise figures. It should be noted that the noise figure depends on the ratio $C_{gs}/\sqrt{g_m}$, or $(C_{gs}/g_m)\sqrt{g_m}$ and therefore low

g_m–low C_{gs} devices would be expected to give lower noise figures. This is in contrast with logic circuits where, although it is the ratio g_m/C_{gs} which is important, high g_m–high C_{gs} combinations would be expected to be less sensitive to parasitic interconnect capacitance. From these arguments we may conclude that for low noise figure, moderate doping and large AlGaAs thickness are preferable whereas for logic applications we require high doping and small AlGaAs thickness. In practice, additional parameters such as parasitic pad capacitances, breakdown voltage, device uniformity requirements and substrate current complicate the analysis and more elaborate modelling techniques coupled with extensive experimental data are required before the device can be designed with confidence for a given application.

One extra geometrical consideration when designing a HEMT device is the gate width. Since, as for the MESFET, g_m and C_{gs} both scale with gate width, then for logic applications we may expect no dependence of switching speed on gate width. However as the gate width is reduced, parasitic interconnect capacitance becomes more significant and the switching time is increased. Larger gate widths overcome this but increase power dissipation due to the increase in I_{DSS}. The choice of gate width is therefore a trade-off between power dissipation and speed. For low noise application, we may re-write equation (17)

$$NF = 1 + K_F 2\pi f \; \frac{C_{gs}}{g_m} \; \sqrt{[g_m (R_s + R_g)]} \qquad (18)$$

The ratio C_{gs}/g_m is independent of gate width, g_m and R_g increases with gate width and R_s decreases. The noise figure therefore increases with gate width. However, at very small gate widths parasitic pad capacitance becomes comparable with C_{gs} and R_g becomes insignificant compared with increasing R_s. The optimum gate width will depend on the exact structure of the device including process-dependent parameters such as normalized source resistance, gate resistance and parasitic capacitances. Many low-noise HEMT devices employ multi-gate figures to reduce R_g without increasing R_s.

7.2.6 Fabrication techniques

(a) **Material growth.** HEMT material is usually grown by molecular beam epitaxy (MBE) which is a method of growing semiconductor materials by effusion in an ultra high vacuum ([49], [50]). The constituent atomic species, such as Ga, As, Al and Si and Be for n- and p-type doping, are heated independently in Knudsen cells and effuse onto the heated GaAs substrate. They then combined to form the semiconductor layer. Constituent elements can be incorporated into the layer being grown by opening the appropriate shutter in front of the cell. The amount of any given element in the material can be adjusted via the cell temperature which changes the flux from that cell. In this way, stoichiometry, A1 mole fraction and doping level can be varied [49].

Growth rates in MBE are typically around 1 μm per hour which corresponds to a rate of less than 1 monolayer per second. As this is well below the speed of operation of the shutters, layer dimensions in MBE can be controlled to within monolayer accuracy [50]. Interfaces grown by MBE are consequently inherently abrupt, making MBE ideal as the growth technique for HEMT materials, where the abruptness of the undoped GaAs/A1GaAs heterojunction is critical to the operation of the device. To date, the best device and circuit performance results have been achieved using MBE ([13], [50], [54]) and the highest reported mobilities for HEMTs [18] have been for MBE grown layers.

Although MBE has been the most popular growth technique for HEMTs to date, metal organic chemical vapour deposition (MOCVD) has recently been receiving increasing attention. In this growth method, the group three elements in the form of metal alkyls, and the group five elements as hydrides, are passed over a heated substrate when they combine to form the III–V compound. Although control of the layer thicknesses and interface abruptness is not inherently as good as in MBE [49], there may be advantages in using MOCVD such as superior wafer throughput and surface quality [50]. Recent noise figure results obtained by MOCVD [60] rival the best reported results for MBE and demonstrate that good performance can be obtained using MOCVD.

The design of the constituent layers of a HEMT structure is important in determining the transistor performance, ease of fabrication and even device uniformity. Fig. 7.12 is a schematic cross-section through a typical device showing the various layers. The design trade-offs relating to gain, noise and

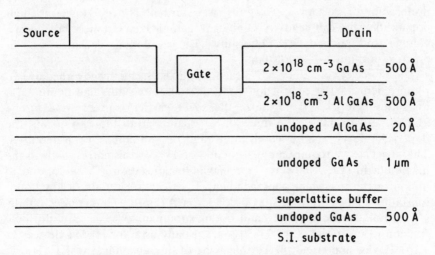

Fig. 7.12 Schematic cross-section through a typical HEMT showing the various material layers.

digital logic performance were discussed in the previous section. More practical aspects relating to material growth and device fabrication will be discussed below.

The growth sequence from the substrate begins with a few hundred ångstroms of undoped GaAs in order to provide an atomically smooth surface. This may be followed by a superlattice buffer (SLB) which typically comprises layers of alternating undoped GaAs/undoped AlGaAs (each approximately 50 Å thick). It is believed that defects migrating from the substrate during layer growth congregate at the GaAs/AlGaAs interfaces resulting in improved material quality of the overlying GaAs layer.

To ensure good carrier confinement of the 2DEG in the quantum well, it is useful to make the buffer layer slightly p-type as the natural depletion associated with the 2DEG/p-type buffer interface will give rise to a steep potential barrier. Nominally undoped GaAs grown by MBE is inherently p-type with a background density of around 10^{14} cm^{-3} [49] which is ideal for this purpose.

The spacer layer consists of undopedd AlGaAs and is typically 20 Å thick. Thicker layers would lead to higher electron mobility but lower sheet carrier concentration. In digital logic circuits, the time to charge and discharge the input capacitance of the following device is critical and hence g_m and I_{DSS} must be maximized. For this reason, it is preferable to maximize the sheet carrier concentration.

Once the dopant density and AlGaAs mole fraction ratio for the doped AlGaAs layer have been chosen, the thickness of the layer to achieve a required pinch-off voltage is calculated from equation (7). The device pinch-off voltage is chosen depending upon the device application. High dopant densities lead to smaller AlGaAs thicknesses for a given pinch-off voltage and hence lead to high g_m – high C_{gs} combinations which are preferable for certain applications as described above. However, high dopant density can lead to significant gate leakage, lower breakdown voltage and poor device uniformity. Typical doping densities are, in practice, around $1-2 \times 10^{18}$ cm^{-3}.

The choice of aluminium mole fraction ratio is essentially governed by two requirements: large conduction band discontinuity and small donor level ionization energy. To maximize the sheet carrier concentration in the 2DEG, and to inhibit the injection of carriers from the 2DEG into the doped AlGaAs, a large conduction band discontinuity ΔE_c is required. This suggests we require a mole fraction of 43% as this corresponds to the maximum in ΔE_c. However, for maximum ionization of donors, and to minimize the persistent photoconductivity effect, we require a low donor ionization energy E_d. For silicon donors in AlGaAs, E_d increases rapidly above mole fractions of 23% and is a maximum at 43%. As a result, mole fraction ratios in the range 23–30% are usually used for HEMT devices.

(b) **Device and circuit processing.** One of the advantages of HEMTs, as replacements for GaAs MESFETs, is their similarity in structure and processing requirements. Both devices are FETs with source and drain ohmic contacts and a gate Schottky contact. Once the material has been grown, an existing MESFET fabrication line can accommodate a HEMT process with a few minor adjustments. The ohmic contact requirement for the HEMT can be different from the MESFET if the GaAs cap layer is too thin such that the A1GaAs layer below could play a significant part in the contact. However, by using a thick (> 500Å) heavily doped cap layer, the contact is made only to the GaAs cap and in which case standard MESFET contact technology may suffice. Transfer of electrons to the 2DEG can then take place via the doped A1GaAs layer. The use of a thick cap layer also serves to reduce parasitic source and drain access resistances. The devices may also be isolated in the same way as MESFETs by mesa etching or by implantation isolation.

The etching of the gate recess is similar to MESFET etching in that the I_{DSS} through the device may be monitored to determine the end point of the etch. However, to aid uniformity a selective etch should be chosen such that the etch rate in A1GaAs is less than that in GaAs. This method can be extended to produce etch stops at a pre-determined position in the A1GaAs by incorporating a stop layer, such as A1As, in which the etch used has a negligible etch rate. In this way the pinch-off voltage of the device can be determined at the growth stage.

Ion implantation of the n^+ source-drain regions of the device may be employed in a similar way to MESFETs. However, since optimal operation of the HEMT relies on the abruptness of the GaAs/A1GaAs interface, care must be taken during the anneal of the implant to avoid excessive interdiffusion of the n–A1GaAs and undoped GaAs layers. Any diffusion of the silicon or aluminium in the n–A1GaAs layer into the region of the 2DEG would degrade the performance of the device. To avoid this problem, rapid thermal anneal techniques are generally employed to reduce the anneal time and hence minimise any inter-diffusion. Ion implantation using self-aligned techniques may be carried out in analogy with the GaAs MESFET but again care must be taken with the subsequent anneal.

7.2.7 Device performance

The main interest in discrete HEMT device performance has been in microwave and millimeter wave noise and gain performance. Thomson-CSF was first to report the noise figure of the devices in July 1981, measuring 2.3 dB noise figure with 10.3 dB associated gain at 10 GHz [51]. In August 1981, Fujitsu fabricated the first HEMT integrated circuit; a 1.7 μm gate HEMT ring oscillator with 17.1 ps switching delay at 77 K [52]. Fujitsu also demonstrated the first HEMT analogue amplifier in February 1983 [53].

Mobilities (at 5 K) are now as high as 2.12 $\times$ 10^6 cm^2 V^{-1} s^{-1} [18]. Transconductances have climbed to 480 mS mm^{-1} at 300 K [54] and 580 mS mm^{-1} at 77 K [55] the highest values reported for non-inverted field effect devices. In the microwave low-noise FET area, using a 0.55 μm gate technology, researchers at Thomson-CSF obtained noise figures of 1.26, 1.7 and 2.25 dB at 10, 12 and 17.5 GHz with associated gains of 12, 10.3 and 6.6 dB, respectively [56]. At low temperatures, the noise performance is enhanced substantially. For example, Fujitsu have fabricated a 0.5 μm gate device with 0.35 dB noise figure at 100 K with 12 dB associated gain [52]. More recently Mishra and co-workers at GE have fabricated a 0.25 μm gate length device with a noise figure of 1.3 dB at 18 GHz with 12 dB associated gain [58]. The unity current gain cut off frequency of this device was 70 GHz.

At millimeter wave frequencies, the best noise figure performance has been reported by Duh *et al.* [54] who obtained noise figures of 1.5, 1.8 and 2.7 dB at 30, 40 and 62 GHz with associated gains of 10, 7.5 and 3.8 dB. This device had a maximum frequency of oscillation f_{max} of 165 GHz, the highest reported for any HEMT device. Very recently transistor amplification at 94 GHz has been reported for the first time [59]. Using a single stage HEMT amplifier a gain of 3.6 dB was obtained at 94 GHz and an output power of 3.4 mW with 2 dB power gain achieved.

Fig. 7.13 shows the best reported noise figures for 0.25 μm and 0.5 μm gate length HEMTs and MESFETs ([54], [58], [60]–[68]). The lowest noise figures are for 0.25 μm HEMTs. It is important to note that the results obtained for 0.5 μm HEMTs are comparable with 0.25 μm MESFETs and significantly better than 0.5 μm MESFETs. This demonstrates the significant noise advantage of HEMTs compared with MESFETs.

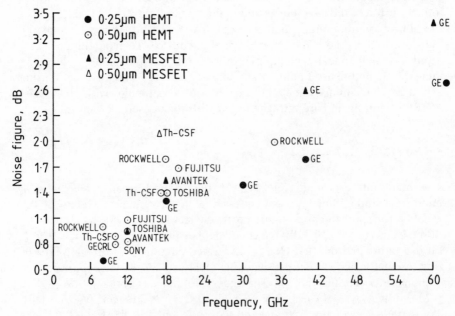

Fig. 7.13 Noise figure results from published work comparing 0.25 μm and 0.5 μm HEMTs and MESFETs.

7.2.8   Circuit performance

The first HEMT integrated circuit was a 27-stage ring oscillator fabricated by Fujitsu in August 1981 [52] which gave 17.1 ps propagation delay per gate at 77 K, a record at that time. In 1982 Fujitsu reduced the delay to 12.8 ps at 77 K [69]. Current best ring oscillator results, 5.8 ps gate delay at 77 K for a 0.35 μm gate device [13], is the fastest switching speed reported in any semiconductor technology.

In 1983 HEMT digital circuits progressed from ring oscillator test circuitry to frequency dividers. Bell Laboratories reported a D type flip-flop divide-by-two circuit based on 1 μm gate length devices operating at 3.7 GHz with 2.4 mW/gate power dissipation and 38 ps propagation delay at 300 K, and 5.9 GHz with 5.1 mW/gate power dissipation and 18 ps propagation delay at 77 K [70]. Nishiuchi *et al.* obtained divide-by-two frequencies of 5.5 and 8.9 GHz from a master/slave flip-flop at 30 K and 77 K respectively using 0.5 μm gate length devices [71]. The fastest HEMT divide-by-two circuit, reported in October 1984, is an AND-NOR gate master/slave flip-flop with a maximum frequency of 10.1 GHz at 77 K [72], a record for static frequency dividers at that time.

In April 1984 the first SRAM results were reported by Rockwell who obtained a 1.1 ns access time 4-bit SRAM [73]. This was followed by Fujitsu with a 1 K SRAM with 0.9 ns access time at 77 K [74], and more recently, with a 4K SRAM giving 2.0 ns access time at 77 K [75]. This latter circuit is the largest reported HEMT IC to date (mid 1987).

The improvements in the digital IC performance of HEMTS has stimulated an increased level of activity in high speed MESFET and silicon bipolar technologies. The trend has been to smaller device geometries and self-aligned technology such as the SAINT (self-aligned implantation for N^+-layer technology) MESFET and the SST (super self-aligned process technology) for the Si bipolar. Current record speeds for the various technologies are most easily compared by means of ring oscillator gate delays. MESFETS and HEMTS are comparable at room temperature: 9.9 ps for 0.4 μm SAINT MESFET [76] and 10.2 ps for 0.35 μm HEMT [13], but the HEMT gate delays are reduced dramatically to 5.8 ps at 77 K [13]. Silicon bipolars have significantly higher gate delays, the best reported being 25.8 ps for a 0.35 μm SST process [77]. GaAs HBTS also have demonstrated lower gate delays than their silicon counterparts with 16.5 ps using a modest 2 μm self-aligned emitter [78]. The gate delay times of HEMT, MESFET, HBT and Si bipolar technologies are summarized in Fig. 7.14 [79].

A comparison between GaAs and Si technologies has also been made for more typical logic circuits [79] Fig. 7.15 shows the multiplication times reported for various bit width multiplier circuits. A comparison is more readily made for 8-bit multipliers where the shortest reported multiplication time was obtained from a HEMT-based circuit.

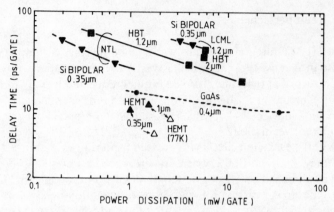

Fig. 7.14 Logic gate delay and power dissipation comparison between HEMT, GaAs MESFET, silicon bipolar and HBT technologies (after Sugeta [79]).

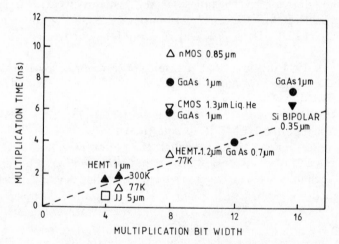

Fig. 7.15 Comparison of multiplication times for various technologies (after Sugeta [79]).

7.3 THE HETEROJUNCTION BIPOLAR TRANSISTOR (HBT)

The heterojunction bipolar transistor (HBT), long recognized for its potential in high speed applications [80], has recently stimulated an increased level of research activity, particularly in the GaAs/GaAlAs hetereostructure system. This interest stems largely from predictions of excellent high frequency performance, both for discrete devices and in integrated circuit applications. Bipolar transistors with wide band gap emitters are known to have a number of potential advantages over homojunction devices for high speed applications. These advantages include lower base resistance, lower emitter-base capacitance, and the ability to function well at high current densities. HBTs have additional advantages in comparison with Si homojunction transistors due to the higher electron mobility, the lower parasitic interconnect capacitance associated with the high resistivity of GaAs substrate and the expected superior radiation hardness and high temperature performance.

In comparison with GaAs ICs implemented with MESFETs, a GaAs heterojunction bipolar technology is expected to benefit from the relative ease of making small structures intrinsic to vertical devices, from the lower sensitivity to capacitive loading and lower voltage swings which are typical of bipolar logic, and from the increased availability of computer aided design tools which already exist for Si bipolar technology. In addition, digital circuits based on HBTs have better threshold voltage control than the (doping dependent) pinch-off voltage of MESFETs because the threshold voltage of an HBT is primarily dependent upon the bandgaps of the emitter and base materials. HBTs based on GaAs/GaAlAs technology can also benefit from the availability of advanced expitaxial techniques for this material system and from the possibility of utilizing transient velocity overshoot effects to achieve increased speed as discussed in Section 7.2.3.

These advantages will allow the HBT to play a dominant or possibly leading role in future digital applications. Microwave characteristics have only recently been measured for HBTs and device structures have not yet been optimized. Consequently, the small amount of available data is strictly preliminary information. Nevertheless, this preliminary data is quite encouraging and results achieved with the HBT have been impressive. A current-gain cut-off frequency of 45 GHz for an emitter width of 2.0 μm has recently been reported [81]. Non threshold logic (NTL) ring oscillators have been operated at 16.5 ps/gate and current mode logic (CML) ring oscillators at 27.6 ps/gate [81]. Frequency dividers have also been fabricated operating with an input frequency of 11 GHz [81].

These encouraging early results coupled with the great potential of the HBT as a logic device suggest that the HBT could provide certain performance advantages over other IC technologies based on Si bipolar or GaAs MESFETs. In the following sections a simple theoretical background for the HBT is given together with the experimental results which have already been demonstrated for HBT technology.

7.3.1 Principle of operation

In homojunction transistors, the gain required for device operation is derived from a doping imbalance between base and emitter which ensures a large injection of electrons from emitter to base. The resulting low base doping, however, leads to a large base resistance which degrades the transistor performance. The concept of using a heterojunction in a bipolar transistor to improve its performance is almost as old as the bipolar transistor itself [8]. Kroemer [82] recognized that by using a heterojunction

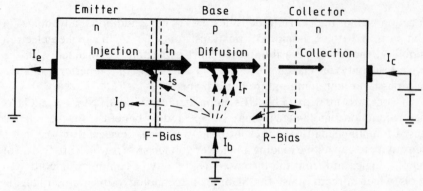

Fig. 7.16 Schematic diagram of an n-p-n bipolar transistor connected in common-emitter configuration showing the various current components.

structure, the transistor performance could be improved because a much higher base doping could be maintained. This is possible because the band discontinuity between the wide band gap emitter and low bandgap base enhances electron injection from emitter to base and reduces hole injection from base to emitter. As a result, useful current gain can be obtained from HBTs even when the base doping is much higher than the emitter doping.

In a bipolar transistor (BT) the operating principle is the injection of electrons from the emitter region into the base, and their subsequent collection by the collector (Fig. 7.16). The base of the BT can be maintained electrically neutral since the charge due to the injected electrons in the base will be compensated for by an increase in hole concentration. This quasi-neutral condition means that large currents can be transported across the base for small voltage drops. The direct control of the emitter-base potential barrier height by the emitter-base voltage V_{be} leads to the well-known equation for the collector current [83]:

$$I_c = I_s \left(\exp(qV_{be}/kT) - 1 \right) \tag{19}$$

where I_c is the collector current and I_s is the saturation current given by

$$I_s = \frac{qD_n A n_i^2}{P_b W_b} \tag{20}$$

where D_n is the diffusion coefficient for electrons in the base, n_i is the intrinsic carrier concentration, P_b is the base doping, W_b is the base width and A is the emitter/base area.

The operation of the BT is in a vertical direction and the electrons and holes in the base of the device are spatially co-incident. This gemoetry means that greater charge, and therefore current per unit width of emitter is available. The pertinent quantity is the charge integrated perpendicular to the direction of current flow. For example, a bipolar with a 1 μm emitter stripe width and 1×10^{18} cm^{-3} doping concentration in the base can contain up to 1×10^{14} cm^{-2} electrons. This charge can be developed in the device for a small change in V_{be} resulting in a large transconductance. Further discussion referring to HBT operation will be given in Section 7.3.3.

7.3.2 HBT device structures

Attempts to realise HBT structures were not fruitful until the successful development of the GaAs/AlGaAs material system ([84], [85]). LPE growth was initially used to demonstrate the high d.c. performance potential of HBTs. With the development of more advanced epitaxial methods, particularly MBE and MOCVD in the mid 1970s, growth technology improved in precision and ability to handle ternary and quarternary III–V semiconductors [86] a new era began for semiconductor heterojunction devices.

Table 7.1 describes the vertical structure of a typical HBT and Fig. 7.17 shows a schematic cross-section through the device. The low doping of the emitter minimizes emitter capacitance, yet provides the necessary current densities to minimize voltage drops or current saturation. The aluminium mole fraction is chosen so that the hole injection is minimal ($x \sim 0.25$ to 0.3). Further details will be given in the fabrication techniques section.

Table 7.1. Typical epitaxial structure of HBT

Layers	Thickness (Å)	Type	Doping (cm^{-3})	Al/As fraction
Cap	500	n^+	2×10^{18}	0
Grading	200	n^+	2×10^{18}	0 to 0.30
Sub-emitter	610	n^+	2×10^{18}	0.30
Emitter	3000	n	2×10^{17}	0.30
Grading	300	n	2×10^{17}	0.30 to 0
Spacer layer	100		Undoped	0
Base	1000	p^+	1×10^{19}	0
Collector	5000	n	5×10^{16}	0
Collector contact	5000	n^+	2×10^{18}	0
Substrate		SI	undoped	

(a) **Emitter/base grading.** Fig. 7.18 shows the energy band diagram of an HBT with graded (Fig. 7.18 (a)) and abrupt (Fig. 7.18 (b)) heterojunctions. In an abrupt heterojunction, there is a potential spike at the emitter/base interface which reduces injection into the base and hence reduces current gain. However high velocity electrons are injected into the base as a result of the spike which tends to reduce transit time. To increase the current gain, the conduction band spike may be eliminated by grading the aluminium concentration over a distance of 200–300 Å or more. Hayes *et al.* [87] have shown a simple procedure to determine the optimum emitter/base grading.

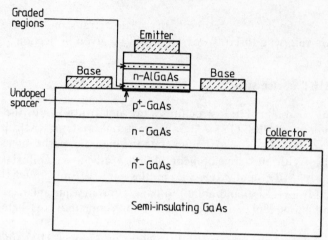

Fig. 7.17 Schematic cross-section of an n-p-n AlGaAs/GaAs heterojunction bipolar transistor.

In the literature there is some confusion as to the importance of grading the emitter/base band gap. The spike may or may not limit current gain, depending on the rate of emission of electrons over the spike, their rate of thermalization in the base, and the rate at which the thermalized electrons are carried away by diffusion.

(b) **Graded base structure.** A recent innovation in the design of HBT structures is to grade the base region in order to improve the base transit time [88]. A quasi electric field can be generated in the base region by creating an Al gradient in the base of a GaAs/AlGaAs HBT, decreasing in

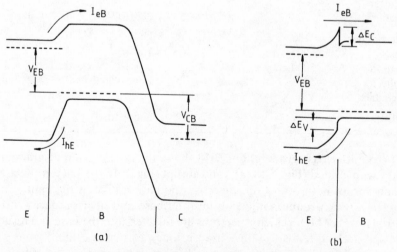

Fig. 7.18 Energy band diagram of a heterojunction bipolar transistor (a) with a graded emitter/base junction and (b) with an abrupt emitter/base junction.

Al content from emitter to collector. Even grading from 15% Al to zero over 1000 Å of base width will provide a quasi electric field of 7.5 KV cm^{-1}. Recent experimental results obtained on graded band gap base HBTs demonstrate that when the Al mole fraction in the base varies by only 0.1 over a base thickness of 1500 Å the base transit time is reduced by a factor of three [88]. Pico-second laser pulse/probe experiments [89] have verified the effect of a grading induced field in improving electron velocities in graded band gap base HBTs.

The use of graded base HBTs provides the device/circuit designer with another degree of freedom in reducing the base resistance by using a thicker graded base resulting in a reduced base sheet resistance instead of a conventional thin homogeneous base, with both having equivalent base transit times.

(c) **Collector-up (inverted) HBT.** The achievement of very-high-frequency performance in an HBT depends critically on minimizing device parasitic time delays. In a normal HBT configuration the collector area is larger than the emitter area, as shown in Fig. 7.17, so the collector capacitance-base resistance time constant in an HBT may be a factor limiting the frequency response, unless this is addressed in the design. Kroemer [90] has shown how this may be reduced by use of an inverted (collector-up) design in which the current flow is restricted to the small collector region of the structure by the barriers of the wide-gap n-p+ junction near the base.

(d) **Double structure HBT (DHBT).** HBT devices with a wide band-gap material in both the collector and the emitter are termed double structure HBT (DHBT). This structure offers the promise of very high speed integrated circuit performance both for emitter coupled logic (ECL) and for saturated logic families such as integrated injection logic (I^2L). The basic advantage of DHBT can be summarized as follows: (1) suppression of hole injection into the collector in saturation logic, (2) emitter collector interchangeability leading to simplification of the layout of circuits such as ECL gates, (3) a better control of the offset voltage Δv_{ce} and saturation voltage. In pursuit of these advantages, there is an increasing interest in these devices ([91], [92]) and d.c. current gains of up to 1650 have been reported [92].

7.3.3 Device theory and modelling

In order to predict the potential advantage of HBT over Si bipolar and GaAs MESFET an increased interest has been stimulated in recent years in the modelling of HBT devices. A suitable theory of a wide band-gap emitter for transistors is given by Kroemer [82]. Using this theory a simple example of a more general central design principle of HBTs will be given to demonstrate the potential of a wide band-gap emitter for improving the performance of HBTs.

(a) **Current gain capability.** Fig. 7.16 shows carrier movement in an n-p-n transistor, where electrons are injected from emitter to base. Some of these electrons recombine at the emitter-base interface, some in the base region and some reach the collector. A common figure of merit for the current handling capability of a bipolar transistor is the common-emitter gain h_{FE} which is given by [90]:

$$h_{FE} = \frac{I_c}{I_b} = \frac{I_n - I_r}{I_p + I_r + I_s} \tag{21}$$

where I_n is the electron current injected from the emitter to the base, I_p is the hole current injected from the base to the emitter, I_s is the current due to electron-hole recombination in the emitter-base space-charge region and I_r is the electron injection current which is lost to recombination within the base region. Provided the base-width W_b is less than about one tenth of the diffusion length of electrons in the base, the effects of I_r on current gain can be ignored.

Assuming that I_s is also small relative to I_p then equation (21) can be simplified to

$$(h_{FE})\text{max} = \frac{I_n}{I_p} \tag{22}$$

I_n and I_p can be expressed in terms of emitter-base voltage V_{be} [83]:

$$I_n = \frac{qAD_{nb}N_{ob}}{L_{nb}} \ (\exp \ (qV_{be}/kT) - 1) \tag{23}$$

$$I_p = \frac{qAD_{pe}P_{oe}}{L_{pe}} \ (\exp \ (qV_{be}/kT) - 1) \tag{24}$$

where D_{nb} and D_{pe} and L_{nb} and L_{pe} are the minority carrier diffusion coefficients and minority carrier diffusion lengths for electrons and holes in the base and emitter regions. Substituting equations (23) and (24) in (22) the current gain is given by:

$$(h_{FE})\text{max} = \frac{D_{nb}L_{pe}}{D_{pe}L_{nb}} \times \frac{N_{ob}}{P_{oe}} \tag{25}$$

where N_{ob} and P_{oe} are the equilibrium carrier concentrations in the base and emitter respectively. These two values can be expressed in terms of n_i, the intrinsic carrier concentration, P_b and N_e, the doping level of base and emitter respectively and the net result is given in equation (26):

$$(h_{FE})\text{max} = \frac{D_{nb} L_{pe}}{D_{pe}L_{nb}} \times \frac{N_e}{P_b} \times \left(\frac{m_{nb}^* \ m_{pb}^*}{m_{ne}^* \ m_{pe}^*}\right)^{3/2} \exp \ (\Delta E_g/kT) \tag{26}$$

where m_{nb}^*, m_{pb}^*, m_{ne}^* and m_{pe}^* are the effective masses for electrons and holes in the base and emitter respectively. From equation (26) it can be seen that for higher current gain either the emitter doping should be larger than the base doping or ΔE_g should be larger than kT.

For a homojunction bipolar transistor, equation (26) can be simplified to:

$$(h_{FE})max = \frac{D_{nb} L_{pe}}{D_{pe} L_{nb}} \times \frac{N_e}{P_b} \tag{27}$$

Considering a homojunction and a heterojunction bipolar transistor with similar doping profile one can combine equations (26) and (27) to give

$$(h_{FE})_{max}\text{hetro}/(h_{FE})_{max}\text{homo} = \left(\frac{m_{nb}{}^* \, m_{pb}{}^*}{m_{ne}{}^* \, m_{pe}{}^*}\right)^{3/2} \exp\left(\Delta E_g/kT\right) \tag{28}$$

The advantage of using a heterojunction is clearly demonstrated in equation (28). In HBTs where large values of $\Delta E_g/kT$ are readily obtainable, a very large value of $(h_{FE})_{max}$ can be obtained. It is clear from equation (26) that in the HBT, even when the base doping is higher than the emitter doping, very large d.c. current gains can be obtained because of the heterojunction $\exp\left(\Delta E_g/kT\right)$ term.

(b) **Threshold voltage control.** In digital circuits the control of the logic level and the turn-on characteristic is of prime importance, since a change in device threshold voltage can adversely affect the noise margin, speed and power of digital logic circuits. In the bipolar transistor, the threshold voltage V_t can be defined as the voltage necessary to sustain a given collector current I_c and can be obtained from rearranging equation (19)

$$V_t = V_{BE} = (kT/q) \ln (I_c/I_s), \tag{29}$$

Substituting for I_s from equation (20) and using the fact that $n_i^2 \sim \exp(-E_g/kT)$, where E_g is the energy band-gap in electron volts, equation (29) can be written as

$$V_t = \frac{E_g}{q} + \frac{kT}{q} \ln \left(\frac{P_b W_b k I_C}{q A D_n}\right) \tag{30}$$

The energy band-gap is dependent only on the semiconductor material used. The V_t is only logarithmically dependent on the device geometry and the doping concentration. Therefore, the sensitivity of the threshold voltage to process parameter variations is rather low for bipolar transistors.

(c) **High frequency capability.** As a basis for comparison of the transistors the cut-off frequency, f_T, is widely used as the figure of merit. The f_T expresses the frequency dependence of current gain (the frequency at which the current gain is unity) and is given by

$$f_T = \frac{1}{2\pi\tau_{ec}} \tag{31}$$

where τ_{ec} is the total emitter-to-collector delay time, which can be expressed by

$$\tau_{ec} = (\tau_e + \tau_b + \tau_d + \tau_c) \tag{32}$$

Here τ_e is the emitter-base junction capacitance charging time which is given by

$$\tau_e = (r_e + r_{ee})(C_{je} + C_{jc} + C_p) \tag{33}$$

where $r_e = kT/qI_c$ is the intrinsic emitter resistance, r_{ee} is the emitter series resistance, C_{je} is emitter capacitance, C_{je} is the collector capacitance and C_p is the parasitic capacitance associated with the emitter.

τ_b is the base transit time which is given by

$$\tau_b = \frac{W_b^2}{\eta D_n} \tag{34}$$

where $\eta = 2$ for a uniformly doped base layer.

τ_d is the base collector depletion layer transit time and is given by

$$\tau_d = \frac{x_d}{2V_{sat}} \tag{35}$$

where x_d is the width of the collector depletion region and V_{sat} is the saturation velocity of carriers in the collector.

τ_c is the collector resistance-capacitance charging time which is given by

$$\tau_c = r_c C_{jc} \tag{36}$$

where r_c is the collector series resistance.

Substituting equations (32) to (36) in equation (31) gives the transition frequency

$$f_T = \frac{1}{2\pi} \left[(kT/qI_c + r_{ee})(C_{je} + C_{jc} + C_p) + \frac{W_b^2}{2D_n} + \frac{x_d}{2V_{sat}} + r_c C_{jc} \right]^{-1} \tag{37}$$

In order to obtain a high f_T each time constant contributing to the τ_{ec} in equation (32) must be minimized. The transistor should therefore have a narrow base width, a narrow collector region and should be operated at a high current level. We shall discuss below how the heterojunction bipolar transistor yields a higher f_T than a homojunction transistor. Comparing a homojunction bipolar and an HBT with identical dimensions, the HBT can permit a very low emitter-base charging time, τ_e, at low current density due to the lower emitter doping compared with the base. Thus a higher f_T can be obtained for the HBT at low current densities. The base transit time, τ_b, is another factor in the total transit time, and it varies with the square of the base thickness. Thus, this time can be reduced by decreasing the base thickness. At an equivalent doping level, GaAs has a diffusion constant D_n four times greater than silicon. As a result, with an identical base thickness, τ_b can be reduced by a factor of about 2, while achieving a base resistivity about 5 times smaller than that in silicon transistors.

Finally, the last two terms in τ_{ec} are, respectively, functions of the collector thickness and doping level. Depending on the applications, a compromise must be sought between the base-collector breakdown voltage and the collector doping level which together determine the depletion region thickness and the base-collector transition capacitance. Two points may be noted here: first the use of very thin high mobility epitaxial collector layers tends to result in very low series collector resistances. Second, the velocity overshoot in GaAs is higher than in Si. Thus the GaAs/AlGaAs heterojunction transistor allows a considerable overall gain to be achieved.

Another figure of merit in bipolar transistors is f_{max}, the maximum frequency of oscillation which is defined as the frequency at which the power gain is unity. This is particularly dependent on the lateral geometry of the transistor. The f_{max} is given by

$$f_{max} = (\alpha_0 f_T / 2\pi R_b C_c)^{1/2} \qquad (38)$$

where R_b is the effective base resistance, C_c is the effective collector capacitance and α_0 is the d.c. common base current gain. The effective base resistance is an important parameter in determining the frequency performance of bipolar transistors, and design and manufacturing details affect its value to a greater extent than they do the other terms in equation (38).

From equation (38) it is clear that high values of f_{max} may be obtained in HBTs due to the lower values of R_b obtained as a result of the higher base doping.

(d) **Switching time.** In digital circuits the switching time (τ_s) is of particular importance since this parameter determines the circuit performance. In circuits based on HBTs, the switching time has been approximately estimated [84] as:

$$\tau_s = \frac{5}{2} R_b C_c + \frac{R_b}{R_L} \tau_b + (3C_c + C_L)R_L \qquad (39)$$

Here R_L and C_L are the load resistance and capacitance of the circuit. Equation (39) indicates the relative significance of the most important transistor parameters. It is clear that significant improvements on τ_s can be achieved by reduction of the base resistance which is associated with the base doping. Since in HBTs high base doping can be readily achieved, a substantial reduction in τ_s can be expected. In fact, two of the three terms in equation (39) depend linearly on R_b rather than with the square root as does the f_{max} (see equation 38)). This means that as long as those terms dominate τ_s, a reduction of R_b is even more effective in a digital switching transistor than in a microwave transistor. A much larger improvement in τ_s would result from a reduction of the transistor collector capacitance, obtained by using an inverted HBT structure.

7.3.4 Fabrication techniques

In order to exploit the real advantage of HBT devices and circuits it is of paramount importance to minimize the parasitic elements in the device. This requires careful device design and advanced fabrication techniques.

(a) **Material growth.** Most early work on HBTs employed liquid phase epitaxy (LPE) for the material growth. Impressive characteristics with high current gain have been reported [85]. However, LPE grown material suffers from poor layer-to-layer uniformity which virtually excludes integrated circuit applications. For use in integrated circuits it is essential to employ growth techniques capable of good uniformity, large growth area and good control of layer thickness and doping. Recent progress in MBE and MOCVD technology now allows the layer thickness and doping profiles to be controlled precisely with excellent uniformity across the wafer and good layer-to-layer reproducibility. Devices with very thin base regions and either abrupt or graded heterojunctions are readily prepared by these two techniques. A typical HBT epitaxial layer grown by MBE is given in Table 7.1. The n- and p-dopants were Si and Be respectively. The layers are grown on undoped semi-insulating substrate in order to reduce parasitic capacitances and facilitate the isolation of individual devices. The doping in the collector is relatively high in order to avoid base widening at high current density. This is of particular importance since in digital circuits, operation at high current density (10^4–10^5 A cm^{-2}) is desirable to reduce the emitter-base dynamic junction resistance. The base region is heavily doped to reduce the intrinsic base sheet resistance. The AlGaAs emitter layer has an aluminium content of about 30% and this has a band-gap approximately 330 meV greater than that of the base. Intentional grading of the heterojunctions has been used to minimize the effects of the conduction band spike which occurs in abrupt junctions. A GaAs cap layer is used on top of the AlGaAs emitter layer to facilitate ohmic contact fabrication.

(b) **Device and circuit processing.** Several processing technologies have been developed for fabrication of discrete devices and integrated circuits. The three commonly used processes are: (a) double mesa process, (b) ion-implanted process and (c) self-aligned process.

(i) *Double mesa process.* This is the simplest method of fabricating HBT devices. In this process contact is made to the base region after etching through the GaAs cap layer and the emitter AlGaAs layer by wet chemical etching. Care has to be taken to control the etching in order to stop at the top surface of the base region. A non-selective etchant may be used to remove the cap layer; this is followed by a selective etchant which removes the AlGaAs emitter region but stops at the GaAs base. Ohmic contact to the base region is usually made using Ti/Zn/Au or Zn/Au metallization systems. The collector region is also reached by wet chemical etching for ohmic contact formation. Ohmic contacts for the emitter and collector usually consist of AuGe/Ni/Au metallization system, similar to GaAs MESFETs. Isolation of the devices is by proton bombardment or by mesa etching. Fig. 7.19 shows a schematic transistor structure fabricated by the double mesa process.

(ii) *Ion-implanted process.* This method offers better control than the double mesa process and the finished devices have greater yield and lower parasitics. In this process, the base contact is achieved by selective

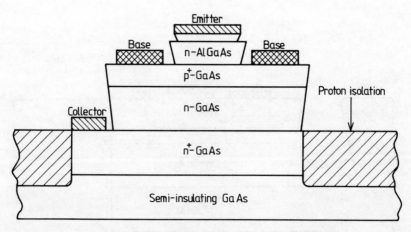

Fig. 7.19 Schematic HBT structure fabricated by double mesa process.

implantation of Be or Mg ions in the extrinsic base regions. The implant is subsequently activated by annealing after encapsulating the wafer to prevent out diffusion of As at the high annealing temperatures. High electrical activation of the implant tends to require high-temperature annealing. On the other hand, it is vital to minimize diffusion of the implanted species and of the epitaxially grown doping profiles. To solve this problem, rapid thermal annealing is normally used. Using this technique, the implants are activated in heating cycles of 20 or 30 seconds duration, which is too short to cause significant dopant diffusion. The lateral parasitic p-n diode formed between the extrinsic base contacts and the cap/emitter contacts is eliminated either by wet chemical etching or by boron implantation. Ohmic contact to the p-regions is made by Ti/Zn/Au evaporation. The collector and emitter contacts are formed in the same way as in the double mesa process. Device isolation is achieved by implantation isolation, typically using protons or boron. Fig. 7.20 represents a typical transistor fabricated by ion-implantation.

In the HBT structure the extrinsic base-collector capacitance, originating from the junction of the p^+ base implant and the collector epitaxial layer, can have a significant limiting factor in the speed of the device. Reduction of the extrinsic base-collector capacitance for Si bipolar transistors by the use of sidewall base contacts [94] or minimized extrinsic base areas [95] has led to improved performance for these devices. Oxygen ions (in sheet concentrations of the order of 10^{14} cm^{-2}) have been introduced to GaAs

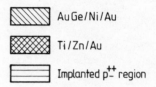

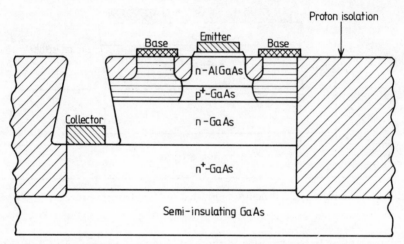

Fig. 7.20 Schematic HBT structure fabricated by ion implantation process.

HBT by implantation from the surface into the low doped collector region below the extrinsic base. Semi-insulating spacer layers are therefore formed at the base-collector junction which decrease the base-collector capacitance [96].

(ii) Self-aligned process. In order to develop circuits with higher speed and lower power consumption, it is necessary to scale down the active device dimensions to reduce device parasitics. To obtain such a scale reduction in HBT, a self-aligned process technology, in which the base is self-aligned to the emitter, is required. This process is analogous to the SAGFET process for MESFETs and requires a T-shaped emitter structure to act as a mask for the base implant (Fig. 7.21). Employing self-aligned techniques, Si bipolar transistors have made rapid progress during recent years [97]. With GaAs HBTs, the potential improvement in performance from self-aligned processes is even greater since the HBT intrinsic device speed can be significantly faster than that in Si bipolar transistors, and the need to reduce device parasitics is correspondingly stronger. In this process refractory alloys such as Germanides (GeMo or GeMoW) have been employed for the emitter n-GaAs contact [98]. This metallization will simultaneously withstand the high temperature required during p-type implant anneals and also provide a low ohmic contact to the n-GaAs emitter layer.

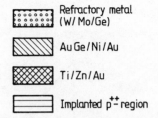

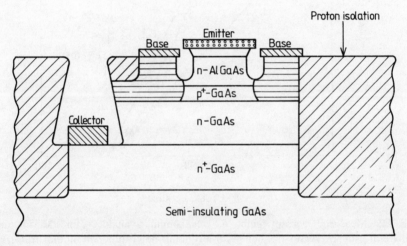

Fig. 7.21 Schematic HBT structure fabricated by self-aligned process.

7.3.5 Device performance

Encouraging results obtained from HBTs to date have demonstrated that these devices promise quite outstanding speed performance. It is likely that the HBT will be one of the dominant high speed logic devices in the future as it promises all the advantages of a Si bipolar – i.e. high transconductance, good threshold uniformity and a high off-chip drive capability – as well as the advantage of a heterojunction structure.

The highest d.c. common emitter current gain (h_{FE}) of 12 500 has been achieved on n-p-n HBTs fabricated on LPE material [99]. n-p-n DHBTs prepared by MOCVD have also demonstrated high h_{FE} values of 5000 [91]. HBT devices have also been fabricated on MBE grown material with h_{FE} values in the range of 3000 [100]. Although for digital circuits a d.c. current gain of about 100 would be adequate, the above records for h_{FE} demonstrate the capability of growth techniques in providing excellent material for HBT fabrication. Although much research work has concentrated on the n-p-n structure, p-n-p HBT devices have also been reported ([101], [102]).

High frequency performance of HBT devices has recently been reported

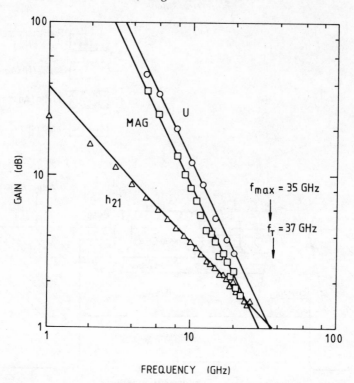

Fig. 7.22 Measured current gain, h_{21}, maximum available gain, MAG, and unilateral gain, U, versus frequency for a (digital) heterojunction bipolar transistor with 1.2 μm emitter width (after Asbeck [103]).

and the results achieved so far have been very impressive considering the moderate size of the transistor used. Rockwell has fabricated GaAs/ AlGaAs HBTs on MBE grown material [103]. These devicess have 1.2 μm emitter width and were fabricated using the ion-implantation process technology described in the last section. Fig. 7.22 shows the current gain (h_{21}), maximum available gain (MAG), and unilateral gain (u) of these devices computed from measured S-parameters. This data reveals a current-gain bandwidth, f_T, of 37 GHz and an f_{max} of 35 GHz. This f_T is two times higher than the f_T of the fastest Si bipolar transistors of similar geometries. HBTs fabricated using the self-aligned process have demonstrated greater speed performance. For example, HBTs based on a self-aligned base contact process and proton implantation, to reduce the extrinsic base-collector capacitance, have shown a current-gain bandwidth of 45 GHz at $I_c = 25$ mA for an emitter width of 2.0 μm [81]. This frequency corresponds to an emitter-collector transit time of only 3.5 psec. These results are particularly encouraging as the emitter widths used are relatively large.

Although the HBT structure has centred around GaAs/AlGaAs systems, these devices have also been fabricated in other compound semiconductors.

Useful transistor operation at temperatures up to 550°C has been demonstrated in an n-AlGaP/p-GaP/n-GaP material system. This high temperature operation may be of value for geothermal and other specialized applications [104]. Recent results [105] have shown that the use of p-InGaAs as a base region instead of conventional p-GaAs can improve significantly the transistor performance. For an emitter geometry of 2.5 μm × 4.5 μm $f_T = 40$ GHz and $f_{max} = 45$ GHz have been achieved from a p-InGaAs base device. This improved performance is due to the inherent superior properties of InGaAs. In particular the electron mobility and peak velocity are much higher in InGaAs than GaAs, and also the band-gap difference between the AlGaAs emitter and base is increased by the use of a InGaAs base. This permits greater base doping levels to be used, as well as a lower emitter aluminium concentration, which can lead to lower deep level concentrations [105].

7.3.6 Circuit performance

During the last five years, there has been a growing interest in the use of GaAs/AlGaAs HBTs for digital integrated circuits. Ring oscillators and frequency dividers are most commonly used to demonstrate the capabilities of HBT technology. The ring oscillators are of the non threshold-logic (NTL) type, as well as of the current-mode logic (CML) and classical emitter-coupled logic (ECL) types. The NTL logic gates have limited fan-out capability and are not suitable for use in complex digital systems, but their characteristics are most directly related to the transistor characteristics. In the NTL circuit approach the transistor saturation is avoided by the appropriate choice of the emitter and collector resistors as well as by the power supply voltage.

The first NTL ring oscillator implemented with GaAs/GaAsAl HBTs consisted of 17 inverters fabricated in 1984 [106] which gave 52 ps gate delay. In 1986 the propagation delay times decreased to 16.5 ps/gate with a power consumption of 11 mW/gate [81]. This result is remarkable considering the large emitter dimensions of HBTs used (2.5 × 4.5 μm^2). The 16.5 ps/gate is the current best ring oscillator result in HBT technology (mid 1987).

In the CML circuit approach, operation at different speed/power levels is possible with a constant logic swing by varying the collector load resistor value and the value of the current per stage. In 1984 a 19-stage CML ring oscillator with a propagation delay time of 40 ps was demonstrated [107] using HBT devices with 1.5 × 1.5 μm^2 emitters. Current best CML ring oscillator results are the 19-stage ring oscillator from Rockwell [81] which has a propagation delay per stage of 27.6 ps with power dissipation of 5.8 mW/gate. An ECL ring oscillator has also been fabricated using graded-bandgap base GaAs/AlGaAs HBTs. NTT of Japan [88] in 1986 reported ECL ring oscillator circuits with 11 stages giving a delay time of 65 ps/gate (obtained with HBTs having 3 × 9 μm^2 emitters).

In 1984 the first HBT frequency dividers were fabricated in an emitter-coupled logic circuit configuration. Frequency division (divide by 2 and double by 4) was obtained up to input frequencies of 4.5 GHz [108] using HBTs with 1.6×5 μm^2 emitter dimensions. Rockwell International obtained a frequency division of 8.5 GHz from a master/slave flip-flop with a total power consumption of 210 mW [107]. The fastest HBT frequency divider circuit, reported in October 1986, is a divide-by-four configured from two cascaded divide-by-two stages based on master-slave flip-flops with a maximum frequency of 11 GHz and a total power consumption of 315 mW [81].

The recent introduction of self-aligned fabrication technologies into silicon bipolar processes has led to a dramatic improvement in circuit performance. Gate delays of 25.8 ps for NTL and 46.3 ps for CML ring oscillators have been reported, with divider operation up to 10.4 GHz [109]. An analogous approach has also been used in the development of GaAs HBT technology. Here gate delays of 16.5 ps for NTL and 27.6 ps for CML have been reported, with divider operation up to 11 GHz [81]. With such similar gate delays being reported for these two technologies, it would clearly be of interest to quantify the difference in performance for circuits with identical transistor geometries, fabricated under equivalent conditions. Such comparison was reported [110] for silicon bipolar and GaAs/AlGaAs heterojunction technologies for high-speed ECL circuits. The gate delays for state of the art technologies were calculated using an analytical equation which expresses the gate delay in terms of all the time constants of the circuit with a fan-out of three to simulate realistic circuit operating conditions. For an emitter width of 2 μm, gate delays of 63 and 31 ps were predicted for silicon bipolar and GaAs HBT respectively. Similarly for an emitter width of 0.5 μm gate delays of 26 and 18 ps were predicted. This factor of 1.4 difference in performance was somewhat lower than the factor of 2.0 obtained at 2 μm geometries, and reflected the increased importance of load capacitance at sub-micron geometries. Sugeta *et al.* have compared the simulated gate delay time as a function of IC complexity for various logic circuits and technologies [79]. Again a realistic fan-in and fan-out of three was used but with an advanced lithography of 0.2 μm minimum feature size for all the technologies used in the comparison. The conclusions are summarised in Fig. 7.23 where GaAs HBT can achieve the shortest delay time per gate (under 10 ps) for 1K gate circuits. Maximum chip power dissipation was set at 3W for all the technologies.

7.4 CONCLUSIONS AND FUTURE TRENDS

In this chapter we have described the operation, device theory and experimental results of HEMT and HBT devices. In addition their suitability for analogue and digital integrated circuit applications has been discussed and circuit results presented. HEMT and HBT devices offer

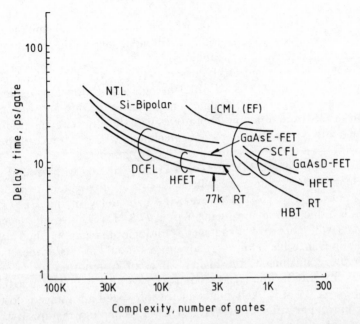

Fig. 7.23 Gate delay against complexity NTL = non-threshold logic; LCML (EF) = low-power current-mode logic (emitter follower), essentially ECL; SCFL = source-coupled FET logic; DCFL |=| direct-coupled FET logic; HFET |=| heterostructure FET; HBT = heterojunction bipolar transistor; RT = room temperature; gate length = 0.2 μm; emitter width = 0.2 μm; fan-in/fan-out = 3/3; 3W per chip (after Sugeta [110]).

potential advantages in microwave, millimeter wave and high speed-low power digital applications. These advantages stem mainly from the use of a heterojunction to improve the transport properties of the device. In the HEMT, modulation doping of the wide band gap layer facilitates separation of the channel electrons from the ionized donors, so reducing scattering and increasing the electron mobility. In HBTs, the use of a wide band gap emitter increases electron injection into the base whilst reducing hole injection into the emitter. The gain of the device is thereby increased.

Noise figure results from 0.5 μm HEMTs are comparable with the best results obtained from 0.25 μm MESFETs ([54], [58], [60]–[69]). HEMT devices have also given gain at 94 GHz from a 0.25 μm gate length device [59]. These impressive microwave results should lead to HEMT devices forming the basis of the next generation of low noise amplifiers, whether hybrid or monolithic, and also to their use as millimetric wave devices. HBT devices have been less successful to date as microwave components. However, this has been largely due to the relatively large dimensions of most reported HBT devices to date: emitter geometries of 2.5 μm $\times$ 4.5 μm giving an f_T of 40 GHz [105] whereas 0.25 μm HEMT devices have demonstrated f_T's of around 70 GHz [58]. Smaller dimension HBT devices may well be useful as power devices for microwave and millimeter wave applications, if low noise is not a requirement, due to the high breakdown voltage and large current capabilities.

Although HEMT and HBT devices may be expected to have potential advantages over the GaAs MESFET in logic applications, the experimental data has been inconclusive to date. This is due to the difficulty in comparing circuit results from different laboratories with differing circuit design, circuit loading, and device and interconnect technology. Current record speeds for the various technologies are most easily compared by means of ring oscillator gate delays. MESFET and HEMT circuits with around 0.4 μm gate length have comparable gate delays of approximately 10 ps at room temperature ([13], [76]). However, the gate delay of the HEMT based circuit is reduced dramatically to 5.8 ps at 77 K [13]. This result suggests that HEMTS will only prove advantageous for logic applications when cryogenic operation is possible. However, advantages should also be evident at room temperature judging by the higher values of f_T reported for HEMTS ~ 70GHz for a 0.25 μm HEMT compared with ~55 GHz for a 0.25 μm MESFET [111]. It is likely that some of this speed advantage is lost in the interconnect parasitics. However, as HEMT technology matures it would be expected that some speed advantages will manifest in logic circuits, even at room temperature. HBT devices have lagged somewhat behind HEMT and MESFET based ICs, principally because of the moderate emitter sizes obtained to date. The current record HBT gate delay of 16.5 ps [78] was obtained from a 2 μm emitter geometry. It is likely that reductions in emitter geometry to submicron dimensions will be possible in the next few years which should make the HBT a leading contender for high speed logic. Silicon bipolar devices tend to be slower than any of the three major GaAs technologies with gate delays of 25.8 ps for a self-aligned 0.35 μm process [77].

The uncertainty in making representative technology comparisons for logic circuits from the published data makes the prediction of future trends difficult. The HBT is perhaps the most promising contender for high-speed SSI circuitry, having demonstrated high speed from such modest dimensions and with the potential for good off-chip drive capability. For larger scale circuits, however, the high-power dissipation may prove inhibitive. For these circuits, the HEMT, with its higher f_T, is likely to have a performance edge over the MESFET. The higher electron mobility of the HEMT also leads to lower parasitic resistances which should result in a lower power dissipation.

New variants on the HEMT and HBT basic device topologies are likely to lead to further performance advantages. The two principal HEMT based variants are the SISFET (semiconductor insulator semiconductor FET) or MISFET (metal insulator semiconductor FET) and the pseudomorphic HEMT. The SISFET and MISFET, which are analogous to the Si MOSFET, typically use an n$^+$ GaAs (SISFET) or metal (MISFET) gate to create an inversion layer at an undoped AlGaAs ('insulator') – undoped

GaAs ('semiconductor') interface ([112], [113]). The principal advantage of this device over a conventional HEMT is that the threshold voltage is set by material potential discontinuities and is not determined by any doping thickness product. The uniformity of this technology may therefore be expected to be better than for conventional HEMT or MESFET technologies. In addition, complementary logic [114] is possible in this technology. The SISFET is expected to be useful for LSI/VLSI applications where device uniformity and low power are of paramount importance.

The pseudomorphic HEMT is similar to a conventional HEMT but incorporates a layer of InGaAs between the undoped GaAs channel and the AlGaAs [111]. In conventional HEMT devices, a large conduction band discontinuity is required to produce a high sheet carrier concentration and to inhibit real space transfer of hot electrons from the 2DEG into the AlGaAs. This requirement is met by an aluminium mole fraction ratio of around 43%. However, the aluminium mole fraction is normally restricted to lower values to reduce the PPC effect due to electron trapping in the AlGaAs (see Section 7.2.4). Since InGaAs has a smaller bandgap than GaAs, a layer of InGaAs between the GaAs and the AlGaAs results in a large conduction band discontinuity, even at lower mole fraction ratios. An additional improvement in device performance results from the fact that InGaAs has superior transport properties to GaAs [115]. Indeed a recent comparison for 0.25 μm gate length devices yielded f_T values of 70 and 80 GHz for HEMT and pseudomorphic HEMT respectively [111]. The use of pseudomorphic HEMTs should result in performance improvements in analogue and digital applications. In addition, the material growth requirements for the pseudomorphic device are essentially the same as in the conventional HEMT but with a thin ($\sim$ 150 Å) InGaAs layer added. The processing requirements of conventional and pseudomorphic devices are identical.

The main variant on the basic GaAs/AlGaAs HBT is the replacement of the p-GaAs base with p-InGaAs. In analogy with the pseudomorphic HEMT, the use of InGaAs results in a larger conduction band discontinuity and in improved transport properties over conventional GaAs material.

In addition to analogue and digital ICs, applications for heterojunction devices exist in the field of optoelectronic integrated circuits (OEICs). These applications generally involve the integration of circuit functions such as amplifiers and mux/demux combinations with optical sources and detectors for use in optical fibre communication systems. Monolithic integration of these components should lead to performance and cost improvements over conventional hybrid technologies by removing the need for off-chip interconnection. HBTs are suitable candidates for integration with laser diodes due to their high current capability and high switching speed [116]. Incorporation of double heterojunction bipolar transistors (DHBTs) would have the added advantage of a similar epitaxial layer structure to that of the laser diode. The integration of InGaAsP/InP DHBT devices with laser diodes has been demonstrated and functional laser/ transistor operation achieved [117].

On the receiver side, a low noise detector – amplifier combination is required. The HEMT is a suitable device for this application due to its low noise and high gain performance at microwave frequencies. A monolithically integrated p-i-n photodiode – HEMT or Schottky photodiode – HEMT [118] would therefore be expected to form an ideal high frequency receiver, possibly followed by additional amplifier or logic functions which could be integrated on the same chip.

REFERENCES

1. Anderson, R. L. (1960) 'Germanium-gallium arsenide heterojunctions' *IBM J. Res. and Develop.* **44** (July) 283–287.
2. Esaki, L. and Tsu, R. (1970) 'Superlattice and negative differential conductivity in semiconductors' *IBM J. Res. and Develop.* **14** 61.
3. Dingle, R., Stormer, H. L., Gossard, A. C. and Wiegmann, W. (1978) 'Electron mobilities in modulation-doped semiconductor heterojunction super-lattices' *Appl. Phys. Lett* **33** (7) (Oct.) 665–667.
4. Mimura, T., Hiyamizu, S., Fujii, T. and Nanbu, K. (1980) 'A new field-effect transistor with selectively doped GaAs/n–$Al_xGa_{1-x}As$ heterojunctions' *Japan J. Appl. Phys.* **19** (5) (May) L225–L227.
5. Delagebeaudeuf, D., Delescluse, P., Etieene, P., Laviron, M., Chaplart, J. and Linh, N.T. (1980) 'Two-dimensional electron gas MESFET structure' *Elect. Lett.* **16** (17) (Aug.) 667–668.
6. Wira, S. J., Wang, W. I., Chao, P. C., Wood, C. E. C., Woodard, D. W. and Eastman, L. F. (1981) 'Modulation-doped MBE GaAs/n–$Al_xGa_{1-x}As$ MESFETS' *IEEE Elect. Dev. Lett.* **EDL–2** (1) (Jan.) 14–15.
7. Mimura, T., Hiyamizu, S., Joshin, K. and Hikosaka, K. (1981) 'Enhancement-mode high electron mobility transistors for logic applications' *Japan J. Appl. Phys.* **20** (5) (May) L317–L319.
8. Kuroda, S., Mimura, T., Suzuki, M., Kobayashi, N., Nishiuchi, K., Shibatomi, A. and Abe, M. (1984) 'New device structure for 4 Kb HEMT SRAM' *Tech. Dig. 1984 GaAs IC Symp.* Boston, MA USA.
9. Drummond, T. J., Morkoc, H., Su, S. L., Fischer, R. and Cho, A. Y. (1981) 'Enhanced mobility in inverted $Al_xGa_{1-x}As$/GaAs heterojunctions: binary on top of ternary' *Elect. Lett.* **17** (23) (Nov.) 870–871.
10. Kinoshita, H., Nishi, S., Akiyama, M. and Kaminishi, K. (1985) 'High speed low power ring oscillators using inverted-structure modulation-doped GaAs/n–AlGaAs field effect transistors' *Japan, J. Appl. Phys.* **24** (8) (Aug.) 1061–1064).
11. Drummond, T. J., Klem, J., Arnold, D., Fisher, R., Thorne, R. E., Lyons, W. G. and Morkoc, H. (1983) 'Use of a superlattice to enhance the interface properties between two bulk heterolayers' *Appl. Phys. Lett.* **42** (7) (April) 615–617.
12. Cirillo, N. C., Shur, M. S. and Abrokawah, J. K. (1986) 'Inverted GaAs/AlGaAs modulation doped field effect transistors with extremely high transconductances' *IEEE Elect. Dev. Lett.* **EDL–7** (2) (Feb.) 71–74.
13. Shah, N. J., Pei, S. S., Tu, C. W. and Tiberio, R. C. (1986) 'Gate-length dependence of the speed of SSI circuits using submicrometer selectively doped

heterostructure transistor technology' *IEEE Trans. Elect. Dev.* **ED–33** (5) (May) 543–547.

14. Camnitz, L. H., Maki, P. A., Tasker, P. J. and Eastman, L. F. (1984) 'Sub-micrometer quantum well HEMT with $Al_{0.3}Ga_{0.7}As$ buffer layer' *11th Int. Conf. GaAs and related compounds* Biarritz.

15. Camnitz, L. H., Tasker, P. J., Lee, H., Van Der Merwe, D. and Eastman, L. F. (1984) 'Microwave characterisation of very high transconductance MODFET' *IEDM* San Francisco, CA, USA.

16. Inoue, K. and Sakaki, H. (1984) 'A new highly-conductive (AlGa)As/GaAs/(AlGa)As selectively doped double-heterojunction field-effect transistor (SD–DH–FET)' *Japan J. Appl. Phys.* **23** (2) (Feb.) L61–L63.

17. Sovero, E., Gupta, A. K., Higgins, J. A. and Hill, W. A. (1986) '35 GHz performance of single and quadruple power heterojunction HEMT's' *IEEE Trans. Elect. Dev.* **ED–33** (10) (Oct.) 1434–1438.

18. Hiyamizu, S., Saito, J., Nanbu, K. and Ishikawa, T. (1983) 'Improved electron mobility higher than $10^6 cm^2/Vs$ in selectively doped GaAs/N–AlGaAs hetero-structures grown by MBE' *Japan J. Appl. Phys.* **22** (10) (Oct.) L609–L611.

19. Walukiewicz, W., Ruda, H. E., Lagowski, J. and Gatos, H. C. (1984) 'Electron mobility limits in a two-dimensional electron gas: GaAs–GaAlAs heterostructures' *Phys. Rev. B* **29** (8) (April) 4818–4820.

20. Stormer, H. L. (1983) 'Electron mobilities in modulation-doped GaAs–(AlGa)As heterostructures' *Surface Science* **134** 519–526.

21. Rode, D. L. (1970) 'Electron mobility in direct band gap semiconductors' *Phys. Rev. B* **2** 1012–1024.

22. Stormer, H. L., Pinczuk, A., Gossard, A. C. and Wiegmann, W. (1981) 'Influence of an undoped (AlGa)As spacer on mobility enhancement in GaAs–(AlGaAs) superlattices' *Appl. Phys. Lett.* **38** (90) (May) 691–693.

23. Witkowski, L. C., Drummond, T. J., Stanchak, C. M. and Morkoc, H. (1981) 'High mobilities, in $Al_xGa_{1-x}As$–GaAs heterojunctions' *Appl. Phys. Lett.* **37** (11) (Dec.) 1033–1035.

24. Delescluse, P., Laviron, M., Chaplart, J., Delagebeaudeuf, D. and Linh, N. T. (1981) 'Transport properties in GaAs–$Al_xGa_{1-x}As$ heterostructures and MESFET application' *Elect. Lett.* **17** (10) (May) 342–344.

25. Wallis, R. H. (1982) 'Effect of free carrier screening on the electron mobility of GaAs: a study by field-effect measurements' *Proc. 16th Conf. Phys. Semicon.* Montpellier, France.

26. Mori, S. and Ando, T. 'Electronic properties of a semiconductor superlattice II. Low temperature mobility perpendicular to the superlattice' *J. Phys. Soc. Jpn.* **48** (3) (March) 865–873.

27. Hiyamizu, S., Fujii, T., Mimura, T., Nanbu, K., Saito, J. and Hashimoto, H. (1981) 'The effect of growth temperature on the mobility of two-dimensional electron gas in selectivity doped GaAs/N–AlGaAs heterostructures grown by MBE' *Japan. J. Appl. Phys.* **20** (6) (June) L455–L458.

28. Wiesbuch, C., Dingle, R., Petroff, P. M., Gossard, A. C. and Wiegmann, W. (1981) 'Dependence of the structural and optical properties of GaAs–$Ga_{1-x}Al_xAs$ multiquantum-well structures on growth temperature' *Appl. Phys. Lett.* **38** (11) (June) 840–842.

29. Drummond, T. J., Kopp, W., Morkoc, H. and Keever, M. 'Transport in modulation-doped structures ($Al_xGa_{1-x}As$/GaAs) and correlations with Monte Carlo calculations (GaAs)' *Appl. Phys. Lett.* **41** (3) (Aug.) 277–279.

30. Fawcett, W. A., Boardman, A. D. and Swan, S. (1970) 'Monte Carlo determination of electron transport properties in gallium arsenide' *J. Phys. Chem. Solids* **31** (9) (Sept.) 1963–1990.

31. Masselink, W. T., Kopp, W., Henderson, T. and Morkoc, H. (1985) 'Measurement of the electron velocity-field characteristic in modulation-doped structures using the geometrical magnetoresistance method' *IEEE Elect. Dev. Lett.* **EDL–6** (10) (Oct.) 539–541.

32. Inouc, M., Inayama, M., Hiyamizu, S. and Inuishi, Y. (1983) 'Parallel electron transport and field effects of electron distributions in selectively-doped GaAs/n–AlGaAs' *Japan J. Appl. Phys.* **22** (4) (April) L213–L215.

33. Lee, K., Shur, M. S., Drummond, T. J. and Morkoc, H. (1984) 'Parasitic MESFET in (AlGa)As/GaAs modulation doped FETs and MODFET characterisation' *IEEE Trans. Elect. Dev.* **ED–31** (1) (Jan.) 29–35.

34. Hirano, M., Takanashi, Y. and Sugeta, T. (1984) 'Current-voltage characteristics of an AlGaAs/GaAs heterostructure FET for high gate voltages', **EDL–5** (11) (Nov.) 496–499.

35. Drummond, T. J., Su, S. L., Lyons, W. G., Fischer, R., Kopp, W., Morkoc, H., Lee, K. and M. S. Shur (1982) 'Enhancement of electron velocity in modulation-doped (AlGa)As/GaAs FETs at cryogenic temperatures' *Elect. Lett.* **18** (24) (Nov.) 1057–1058.

36. Cappy, A., Versnayen, C., Vanoverschelde, A., Salmer, G., Delagebeaudeuf, D., Linht, N. T. and Laviron, M. (1982) 'A novel model of two-dimensional electron gas field effect transistors', *Proc. 1982 GaAs IC Symp.* New Orleans, USA.

37. Drummond, T. J., Fischer, R. J., Kopp, W. F., Morkoc, H., Lee, K. and Shur, M. S. (1983) 'Bias dependence and light sensitivity of (AlGa)As/GaAs MODFETs at 77 K' *IEEE Trans. Electron. Dev.* **ED–30** (12) (Dec.) 1806–1811.

38. Fischer, R., Drummond, T. J., Klem, J., Kopp, W., Henderson, T. S., Perrachione, D. and Morkoc, H. (1984) 'On the collapse of drain I–V characteristics in modulation-doped FETs at cryogenic temperatures' *IEEE Trans. Electron. Dev.* **31** (8) (Aug.) 1028–1032.

39. Nelson, R. J. (1977) 'Long lifetime photoconductivity effect in n-type GaAlAs' *Appl. Phys. Lett.* **31** 351–353.

40. Fischer, R., Masselink, W. T., Klem, J., Henderson, T. and Morkoc, H. (1984) 'The elimination of drain IV collapse in MODFETS through the use of a thin n–GaAs/AlGaAs superlattice' *Elect. Lett.* **20** (18) (Aug.) 743–744.

41. Delagebeaudeuf, D. and Linh, N. T. (1982) 'Metal–(n) AlGaAs–GaAs two-dimensional electron gas FET' *IEEE Trans. Elect. Dev.* **ED–29** (6) (June) 955–959.

42. Drummond, T. J., Morkoc, H., Lee, K. and Shur, M. (1982) 'Model for modulation doped field effect transistor' *IEEE Elect. Dev. Lett.* **EDL–3** (11) (Nov.) 338–341.

43. Loret, D., Baets, R., Snowden, C. M. and Hughes, W. A. (1986) 'Two-dimensional numerical models for the high electron mobility transistor' *Proc. Int. Conf. on Simulation of Semiconductor Devices and Processes* Swansea, UK.

44. Radaioly, U. and Ferry, D. K. (1986) 'MODFET ensemble Monte Carlo model including the quasi two-dimensional electron gas' *IEEE Trans. Elect. Dev.* **ED–33** (5) (May) 677–681.

45. Ando, T., Fowler, A. B. and Stern, F. 'Electronic properties of two-dimensional systems' *Rev. Mod. Phys.* **54** (2) (April) 437–672.

46. Pengelly, R. S. (1982) *Microwave field-effect transistors – theory, design and applications* Research Studies Press.

47. Fukui, H. (1979) 'Optimal noise figure of microwave GaAs MESFETS' *IEEE Trans. Elect. Dev.* **ED–26** (7) (July) 1032–1037.

48. Laviron, M., Delagebeaudeuf, D., Rochette, J. F., Jay, P. R., Delecluse, P., Chevrier, J. and Linh, N. T. (1984) 'Ultra low noise and high frequency microwave operation of FETs made by MBE' *11th Int. Symp GaAs and Related Compounds* Biarritz, France.

49. Wood, C. E. C. (1986) 'GaAs materials for monolithic microwave integrated circuits' *GEC J. Res.* **4** (2) 72–90.

50. Millter, D. L. (1984) 'MBE materials considerations for HEMT circuits' *GaAs IC Symp.* Boston, MA, USA.

51. Laviron, M., Delagebeaudeuf, D., Delescluse, P., Chaplart, J. and Linh, N. T. (1981) 'Low-noise two-dimensional electron gas FET' *Elect. Lett.* **17** (15) (July) 536–537.

52. Mimura, T., Joshin, K., Hiyamizu, S., Hikosaka, K. and Abe, M. (1981) 'High electron mobility transistor logic' *Japan J. Appl. Phys.* **20** (8) (Aug.) L598–L600.

53. Niori, N. and Saito, T. (1983) 'A 20 GHz high electron mobility transistor amplifier for satellite communications', *1983 IEEE Int. Solid State Circuits Conf. Dig.* New York, USA.

54. Duh, K. H. G., Chao, P. C., Smith, P. M., Lester, L. F., Lee, B. R. and Hwang, J. C. M. (1986) 'Millimeter wave low noise HEMTs' *Dev. Res. Conf.* (June).

55. Camnitz, L. H., Tasker, P. J., Lee, H., van der Merwe, D. and Eastman, L. F. (1984) 'Microwave characterisation of very high transconductance MODFET' *IEDM, San Francisco.*

56. Linh, N. T., Laviron, M., Delescluse, P., Tung, P. N., Delagebeaudeuf, D., Diamand, F. and Chevrier, J. (1983) 'Low-noise performance of two-dimensional electron gas FETS' *Proc. 9th IEEE Cornell Biennial Conf.* Cornell University, Ithaca, NY, USA 187–193.

57. Joshin, K. (1983) 'Noise performance of microwave HEMT' *1983 IEEE MTT–S Digest* (June) 563–565.

58. Mishra, U. K., Palmateer, S. C., Chao, P. C., Smith, P. M. and Hwang, J. C. M. (1985) 'Microwave performance of 0.25 μm gate length high electron mobility transistors' *Elect. Dev. Lett.* **EDL–6** (3) (March) 142–145.

59. Smith, P. M., Chao, P. C., Duh, K. H. G., Lester, L. F. and Lee, B. R. (1986) '94 GHz transistor amplification using an HEMT' *Elect. Lett.* **22** (15) (July) 780–781.

60. Tanaka, K., Ogawa, M., Togashi, K., Takakuwa, H., Ohke, H., Kanazawa, M., Kato, Y. and Watanabe, S. (1986) 'Low-noise HEMT using MOCVD' *IEEE Trans. Electron Devices* **ED–33** (12) (Dec.) 2053–2058.

61. Kamei, K., Hori, S., Kawasaki, H. and Shibata, K. (1984) 'Low noise high electron mobility transistor' *Gallium Arsenide and Related Compounds Eleventh International Symp.* Biarritz, France 545–550.

62. Laviron, M., Delagebeaudeuf, D., Rochette, J. F., Jay, P. R., Delescluse, P., Chevrier, J. and Linh, N. T. (1984) 'Ultra low noise and high frequency

operation of TEGFETS made by MBE' *Proc. 14th European Solid State Device Research Conf.* Lille, France 376–379.

63. Joshin, K., Yamashita, Y., Niori, M., Saito, J., Mimura, T. and Abe, M. (1984) 'Low noise HEMT with self-aligned gate structure' *Extended Abstracts 16th Internat. Conf. Solid State Devices and Materials* Kobe, Japan 347–350.

64. Gupta, A. K., Sovero, E. A., Pierson, R. L., Stein, R. D., Chen, R. T., Miller, D. L. and Higgins, J. A. (1985) 'Low-noise high electron mobility transistors for monolithic microwave integrated circuits' *IEEE Electron Device Lett.* **EDL–6** (2) (Feb.) 81–82.

65. Sovero, E. A., Gupta, A. K. and Higgins, J. A. (1986) 'Noise figure characteristics of 1–2 μm gate single heterojunction high-electron-mobility FETs at 35 GHz' *IEEE Electron Device Lett.* **EDL–7** (3) (March) 179–181.

66. Chao, P. C., Palmateer, S. C., Smith, P. M., Mishra, U. K., Duh, K. H. G. and Hwang, J. C. M. (1985) 'Millimeter-wave low-noise high electron mobility transistors' *IEEE Electron Device Lett.* **EDL–6** (10) (Oct.) 531–533.

67. Chye, P. W. and Huang, C. (1982) 'Quarter micron low noise GaAs FETS' IEEE *Electron Device Lett.* **EDL–3** (12) (Dec.) 401–403.

68. Laviron, M., Delagebeaudeuf, D., Rochette, J. F., Jay, P. R., Delescluse, P., Chevrier, J. and Linh, N. T. (1984) 'Ultra low noise and high frequency microwave operation of FETs made by MBE' *Gallium Arsenide and Related Compounds, Eleventh Internat. Symp.* Biarritz, France 539–543.

69. Abe, M., Mimura, T., Yokoyama, N. and Ishikawa, H. (1982) 'New technology towards GaAs LSI/VLSI for computer applications' *IEEE Trans. Electron Devices* **ED–29** (7) (July) 1088–1093.

70. Kiehl, R. A., Feuer, M. D., Handle, R. H., Hwang, J. C. M., Keramidas, V. G., Allyn, C. L. and Dingle, R. (1985) 'Selectively doped heterostructure frequency dividers' *IEEE Elect. Dev. Lett.* **EDL–4** 377–379.

71. Nishiuchi, K., Mimura, T., Kuroda, S., Hiyamizu, S., Nishi, H. and Abe, M. (1983) 'Device characteristics of short channel high electron mobility transistors (HEMT)' *41st Device Research Conf.* Burlington, Vermont, USA. *IEEE Trans Elect. Dev* **ED–30** (11) 1569.

72. Pei, S. S., Shah, N. J., Hendel, R. H., Tu, C. W. and Dingle, R. (1984) 'Ultra high speed integrated circuits with selectively doped heterostructure transistors' *Tech Dig. 1984 GaAs IC Symp.* Boston, MA, USA.

73. Lee, S. L., Lee, C. P., Hou, D. L., Anderson, R. J. and Miller, D. L. (1984) 'Static random access memory using high electron mobility transistors' *IEEE Elect. Dev. Lett.* **EDL–5** (4) (April) 115–117.

74. Nishiuchi, K., Kobayashi, N., Kuroda, S., Notomi, S., Mimura, T., Abe, M. and Kobayashi, M. (1984) *Tech Dig. 1984 Int. Solid State Circuits Conf.* 48.

75. Kuroda, S., Mimura, T., Suzuki, M., Koybayashi, N., Nishiuchi, K., Shibatomi, A. and Abe, M. (1984) 'New deive structure for 4 Kb HEMT SRAM' *Tech Dig. 1984 GaAs Symp.* Boston, MA, USA.

76. Yamasaki, K., Kato, N. and Hirayama, M. (1984) 'Below 10ps/gate operation with buried p-layer SAINT FETs' *Elect. Lett.* **20** (25/26) (Dec.) 1029–1031.

77. Sakai, T., Konaka, S., Yamamoto, Y. and Suzuki, M. (1985) 'Prospects of SST technology for high-speed LSI' *IEDM Tech. Dig.* Washington DC. USA 18–21.

78. Chang, M. F., Asbeck, P. M., Wang, K. C., Sullivan, G. J. and Miller, D. L. (1986) 'AlGaAs/GaAs heterojunction bipolar transistor circuits with improved high-speed performance' *Elect. Lett.* **22**, (22) (Oct.) 1173–1174.

79. Sugeta, T., Mizutani, T., Ino, M. and Horiguchi, S. (1986) 'High speed

technology comparison – GaAs vs Si' *GaAs IC Symp.* Grenelefe, Florida, USA 3–6.

80. Schockley, W. (1981) US Patent 2569347 (filed 26th June 1948, issued 26 Sept. 1951).

81. Chang, M. F., Asbeck, P. M., Wang, K. C., Sullivan, G. J. and Miller, D. L. (1986) 'AlGaAs/GaAs heterojunction bipolar transistor circuits with improved high-speed performance' *Elec. Lett.* **22** (22) 1173.

82. Kroemer, H. (1957) *Proc. IRE* **45** 1535.

83. Sze, S. M. (1969) *Physics of semiconductor devices* Willey Inter.

84. Dunke, W. I., Woodall, J. M. and Rideout, V. L. (1972) *Solid-state electronic* **15** 1329.

85. Konagai, M. and Takahashi, K. (1975) 'GaAlAs/GaAs heterojunctions transistor with high injection efficiency' *J. Appl. Phys.* **46** 2120.

86. Kressel, H. (1980) *Materials for Heterojunction Devices* **10** Academic, NY, USA. 287.

87. Hayes, J. R., Capasso, F., Malik, R. J., Gossard, A. C. and Wiegman, W. (1983) 'Optimum emitter grading for heterojunction bipolar transistor' *Appl. Phys. Lett.* **43** (10) 949.

88. Yamauchi, Y. and Ishibashi, T. (1986) 'Equivalent circuit and ECL ring oscillators of graded-bandgap base HBTs' *Electron. Lett.* **22** (1) 18.

89. Levine, B. F., Tsung, W. T., Bethea, C. G. and Capasso, F. (1982) 'Electron drift velocity measurement in compositionally graded Al_xGa_{1-x} As by time-resolved optical picosecond reflectivity' *Appl. Phys. Lett.* **41** 470.

90. Kroemer, H. (1982) 'Heterojunction bipolar transistors and circuits' *Proc. IEEE* **70** (1) 13.

91. Dubon, C., Azoulay, R., Desrousseaux, P., Dangha, J., Duchenois, A. M., Hountonkji, M. and Ankri, D, (1983) 'Double heterojunction GaAs–GaAlAs bipolar transistors grown by MOCVD for emitter coupled logic circuits' *IEEE, IEDM* 689.

92. Su, S. L., Tejayadi, D., Drummond, T. J., Fisher, R. and Merkoc, H. (1983) 'Double heterojunction AlGaAs/GaAs bipolar transistor by MBE' **4** (5) 130.

93. Arnold, S. R. and Pritchett, R. L. (1965) *Bell Labs. Rep. 20, Contract DA 36–639 AMC–02227 (E)*.

94. Nakamura, T., Miyazaki, T., Takahashi, S., Kure, T., Okabe, T. and Gaguta, M. (1982) 'Self-aligned transistor with sidewall base electrode' *IEEE Trans. Electron. Devices* **ED–29** 596.

95. Ning, T. H., Isaac, R. D., Solomon, P. M., Tang, D. D. and Yu, H. N. (1980) 'Self-aligned npn bipolar transistors' *IEDM Tech. Dig.* 823.

96. Asbeck, P. M., Miller, D. L., Anderson, R. D. and Eisen, F. H. (1984) 'GaAs/GaAlAs heterojunction bipolar transistors with buried oxygen-implanted isolation layers' *IEEE Electron. Device Lett.* **EDL–5** (8) 310.

97. Konaka, A., Yamamoto, Y. and Sakai, T. (1984). 'A 30 ps bipolar IC using super self-aligned process technology' *16th Conf. Solid-State Devices and Materials* Kobe, Japan 209.

98. Tiwari, S., Kuan, T. S. and Tierney, E. (1983) 'Ohmic contacts to n-GaAs with germanide overlayers' *IEEE IEDM* 115.

99. Lin, H.-H. and Lee, S. (1985) 'Super-gain AlGaAs/GaAs heterojunction bipolar transistors using an emitter edge-thinning design' *Appl Phys. Lett.* **47** (8) 839.

100. Rezazadeh, A. A., Barnard, J. A. and Kerr, T. M. (1985) 'The role of

electron traps in the DC current gain of GaAs/GaAlAs heterojunction bipolar transistors prepared by MBE' *Proc. 12th SYmp. on GaAs and Related Compounds* Karaziawa, Japan (79).

101. Frost, M. S., Riches, M. and Kerr, T. (1985) 'A pnp AlGaAs heterojunction bipolar transistor for high temperature operation' *J. Appl. Phys.* **60** (5) 2149.

102. Rezazadeh, A. A., Frost, M. S., Kerr, T. M. and Wood, C. E. C. (1986) 'npn and pnp GaAs/GaAlAs heterojunction bipolar transistors for high speed and high temperature applications prepared by MBE' *J. Vac. Sci. Tech.* **B4** (3) (May/June) 773.

103. Asbeck, P. M. *et al.*, (1985) 'Microwave performance of GaAs–(Al,Ga)As heterojunction bipolar transistors' *IEEE IEDM, Digest Tech.* San Francisco, USA (19) 9.

104. Zipperian, T. E. and Dawson, L. K. (1983) 'GaP/Al$_x$ Ga$_{1-x}$ P heterojunction transistors for high temperature electronic applications' *J. Appl. Phys.* **54** 6019.

105. Asbeck, P. M., Chang, M. F., Wang, K. C., Sullivan, G. J. and Miller, D. L. (1986) 'GaAlAs/GaInAs/GaAs heterojunction bipolar transistor technology for sub–35ps current-mode logic circuits' *IEEE Bipolar Circuits and Technology Meeting* 25.

106. Asbeck, P. M., Miller, D. L., Anderson, R. J., Hou, L. D., Deming, R. and Eisen, F. (1984) 'Non threshold logic ring oscillators implemented with GaAs/GaAlAs heterojunction bipolar transistors' *IEEE Electron. Device Lett.* **EDL–5** (5) 181.

107. Asbeck, P. M., Miller, D. L., Anderson, R. J., Deming, R. N., Chen, R. T., Liechti, C. A. and Eisen, F. H. (1984) 'Applications of heterojunction bipolar transistors for high-speed, small-scale digital integrated circuits' *GaAs IC Symp. IEEE* 133.

108. Asbeck, P. M., Miller, D. L., Anderson, R. J., Deming, R. N., Hou, L. D., Liechi, C. A. and Eisen, F. H. (1984) '4.5 GHz frequency dividers using GaAs/GaAlAs heterojunction bipolar transistors' *ISSCC 84/Wed.* (Feb. 22) 50.

109. Sakai, T., Konaka, S., Yamamoto, Y. and Suzuki, M. (1985) 'Prospects of SST technology for high-speed LSI' *IEDM Tech. Dig.* 18.

110. Ashburn, P., Rezazadeh, A. A., Chor, E. F. and Brunnschweiler, A. (1987) 'Comparison of silicon bipolar and GaAs HBT for high speed ECL circuits' *Bipolar Circuits and Technology Conference, Minnesota, USA* (21–22 Sept.) 61.

111. Pustai, J. (1987) 'MM-wave Transistors: the key to advanced systems' *Microwaves and RF* **26** (3) (March) 125–177.

112. Matsumoto, M., Ogura, M., Wada, T., Hashizume, N., Yao, T. and Hayashi, Y. (1984) 'n+–GaAs/undoped GaAlAs/undoped GaAs field effect transistor' *Elect. Lett.* **20** (11) (May) 462–363.

113. Maezawa, K., Ito, H. and Mizutani, T. (1987) 'An AlGaAs/InGaAs/GaAs strained channel MISFET' *Japan. J. Appl. Phys.* **26** (1) (Jan.) L74–L76.

114. Mizutani, T., Fujita, S., Hirano, M. and Kondo, N. (1986) 'Circuit performance of complementary heterostructure MISFET inverter using high mobility 2DEG and 2DHG' *GaAs IC Symp.* Grenelefe, Florida, USA 107–110.

115. Henderson, T., Masselink, W. T., Kopp, W. and Morkoc, H. (1986) 'Determination of carrier saturation velocity in high-performance

In $_y$Ga$_{1-y}$As/Al$_x$Ga$_{1-x}$As modulation doped field effect transistors [^{y}0≤y≤0.2]' *IEEE Elect. Dev. Lett.* **EDL-7** 288–290.

116. Katz, J., Barchaim, N., Chen, P. C., Margalit, S., Urg, I., Wilt, D., Yust, M. and Yarito, Y. (1980) *Appl. Phys. Lett.* **37** 211–213.

117. Su, L. M., Grote, N., Kaumanns, R. and Katzschner, W. (1985) *IEEE Elect. Dev. Lett.* **EDL-6** (1) 14.

118. Hughes, W. A. and Parker, D. G. (1986) 'Operation of a high-frequency photodiode–HEMT hybrid photoreceiver at 10 GHz' *Elect. Lett.* **22** (1) (May) 509–510.

Chapter 8

Applications of GaAs Integrated Circuits

Sven ROOSILD and Allen FIRSTENBERG

8.1 INTRODUCTION

When silicon replaced germanium in the early 1960s as the semiconductor of choice for solid state devices, it was able to convert the entire industry in just a few years because of two important characteristics. First, silicon has a higher energy band-gap, which permits silicon based devices to operate over a wider temperature range – a feature that is especially important to military requirements. Second, and more important, silicon has a native oxide, which provided initially for improved stability and planar, rather than mesa, type devices. The planar technology soon spawned the integrated circuits. The intergrated circuit in turn brought on the electronics revolution, allowing the complexity of circuits to increase by a factor of two every year (Moore's Law) and brought us from single transistors to megabit memory chips.

As a semiconductor gallium arsenide (GaAs) has one of the characteristics that proved to be so important to the rapid development of silicon technology. GaAs has an even higher band-gap than silicon, allowing operation over an even wider temperature range. However, instead of a native oxide, GaAs provides a semi-insulating substrate, an attribute that neither silicon nor geranium possess. This semi-insulating substrate is ideal for an all ion implanted planar device technology. Furthermore, despite the absence of a useful native oxide, the GaAs surface is stable; stabilized by surface defect states rather than passivated by native oxide. This is a plus for the development of GaAs radiation hardened circuits that are of vital interest to many military systems, since radiation-induced charging of native or deposited insulators will not cause shifts in device characteristics as it does in silicon devices. Unfortunately, the surface properties of GaAs are a definite minus for commercial production of high-complexity, low-cost chips: the lack of passivation eliminates the possibility of MOS-type circuits that form the basis for the widest segment of present day silicon VLSI.

In the early 1960s a lot of effort was spent on developing bipolar transistors in GaAs. These efforts were frustrated by the low lifetime of minority carriers in this direct band-gap semiconductor, the difficulties of introducing dopants by diffusion, and the thermal conversion (i.e. becoming electrically conductive after heat treatments) of the semi-insulating substrates. Real progress in GaAs-based circuits began when the principal cause of thermal conversion – allowing crystal growth to occur even slightly on the gallium-rich side of the binary phase diagram – was uncovered. Ion implantation replaced diffusion as the means for introducing dopants, and MESFETS and JFETS emerged as devices suitable for building integrated circuits in GaAs; these devices being majority carrier devices for which the short lifetime of GaAs is not a difficulty. Once a workable integrated circuit fabrication technology became available, additional impetus to the development of a GaAs based semiconductor technology came from the high electron mobility inherent to GaAs; that translates into an improved speed-power product. Coupling this advantage of GaAs to its demonstrated excellent total dose radiation hardness made it a natural area for military research.

8.2 SYSTEM IMPACT: THE ROLE OF GaAs

The potential of lower cost microwave systems is a major reason for the large investment by the Department of Defense and a long list of companies to develop an alternative to hybrid MIC technology by integrating active microwave transistor circuits onto a single semiconductor chip. GaAs microwave integrated circuits, with their superb microwave characteristics in terms of low noise and medium power amplification, wide bandwidth capability, and RF signal switching, have become the standard active electronic device for analog microwave circuits above about 1 to 2 GHz. The types of circuitry realisable with monolithic microwave integrated circuit (MMIC) technology range from simple functional modules at the lowest level of integration; i.e. low noise, power, and broadband amplifiers, mixers, oscillators and phase shifters, to system building blocks such as receiver front ends, modulators and demodulators, and transceivers at higher levels of integration.

However, when considering the system impact of GaAs integrated circuits, one generally thinks of improved system performance: with systems having higher processing rates, providing higher throughput rates compatible with higher input rates. The higher speed can permit new and more complex algorithms to be performed, improving system functional capabilities. On the other hand, many wideband systems are distributed, solving the system requirements with a great deal of parallelism. In these distributed parallel systems, for a given throughput or given capability, higher speed electronics can offer greatly improved architectures, reducing parallelism, and reducing system complexity, saving in many important areas; size, weight, power, reliability and cost.

What is the role of GaAs digital integrated circuits in systems? Considering the complex trade-offs to be made relative to speed, power, complexity, reliability, availability, radiation hardness, etc., this is indeed a complex question. To help answer this question, a generic, wideband digital signal processing system, comprised of a preprocessor, programmable signal processor and data processor, is shown in Fig. 8.1 [1]. By definition of a wideband digital system, we are referring to systems characterized by wideband analogue waveform sources. Typical wideband sensors are: electronic warfare (EW) warning receivers, smart communication receivers, infra-red imaging sensors, radar receivers, sonar arrays, time division multiple access (TDMA) receivers etc. At the preprocessor, the types of functions performed are sampling, analogue-to-digital conversion, shift register delay, digital filtering, pulse parameterization, demultiplexing, error decoding and correction, convolving, synchronization, etc. These functions are flow form continuous, where the processing is fixed format and is not dependent upon the input data stream. In the programmable signal processor signal transformation and categorization takes place. Here, the primary elements are Fast Fourier Transforms (FFT), high speed arithmetic elements such as multiplier/accumulators and arithmetic logic units (ALU), address generators, code generators, high-speed gate arrays, serial memories, RAMs/ROMs and content addressable memories. In the data processor/computer, signal interpretation and system level decisions are reached, such as fire control and system display. This is where microprocessors, bit slice arithmetic logic units, large semicustom logic, lower speed gate arrays, EPROMs and bus interface logic find wide utilization. As the signal goes from left to right within Fig. 8.1, the amount of processing programmability goes from low to extensive; the signal throughput rate and associated chip clock rate goes from very high, to less than 50 million operations per second (MOPS); the required complexity goes from medium scale to very large scale and beyond.

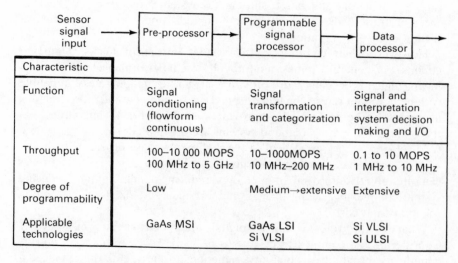

Characteristic	Pre-processor	Programmable signal processor	Data processor
Function	Signal conditioning (flowform continuous)	Signal transformation and categorization	Signal and interpretation system decision making and I/O
Throughput	100–10 000 MOPS 100 MHz to 5 GHz	10–1000MOPS 10 MHz–200 MHz	0.1 to 10 MOPS 1 MHz to 10 MHz
Degree of programmability	Low	Medium→extensive	Extensive
Applicable technologies	GaAs MSI	GaAs LSI Si VLSI	Si VLSI Si ULSI

Fig. 8.1 Requirements of wideband digital signal processing systems.

If one thinks of IC throughput as a product of device complexity and speed, where distributed, parallel processing is envisioned to replace raw speed, then silicon, at least in the near term, has a definite advantage due to its large dominance in complexity. Where, then, does GaAs apply in near term?

There are a number of processing areas where parallel processing is not efficient. Data acquisition and sampling front ends of wide bandwidth signal processors are important examples. Another is bit stream processing, which includes error detection and correction and data correlation. In these types of processing, GaAs high speed capabilities are needed today.

GaAs digital integrated circuits will play an important role in pre-processors, where the highest speeds are needed but where chip complexity is not a dominating factor. GaAs chips will hand over to silicon chips as soon as possible, that is, as soon as the processing speeds are low enough for silicon to handle, due to silicon's lower price, greater availability and maturity. In the programmable signal processor, silicon will have the dominant role, however, in processing tasks such as FFTs, requiring multiplication, GaAs will also be present. Signal transformations and digital filtering requiring complex multiply and add operations are dominated by the multiply time since multiplication requires many more gate delays than addition. For that reason, a GaAs multiplier would be useful to achieve high programmable signal processor throughput rates. In the data processor, where speed is not the major factor, but chip complexity is, silicon VLSI and ULSI chips will be predominantly used. In certain space based and tactical military applications, where radiation hardness beyond silicon levels is important, systems made entirely of GaAs components are envisioned. Shown in Table 8.1 are the primary military applications of GaAs digital integrated circuits. The super high speed of GaAs devices point particularly to electronic warfare and communication applications whereas the radiation hardness advantages of GaAs point toward space based applications [2].

Table 1.   GaAs military applications

- Electronic warfare
- Communications (secure, satellite)
- Space-based surveillance
- Intelligence
- Distributed processing

8.3   ELECTRONIC WARFARE APPLICATIONS [3]

The world of electronic warfare has experienced rapid technical growth, and that trend is continuing. The microwave environment is becoming increasingly cluttered as a result of increased emitter/beam densities and higher frequencies of operation. New sophistication in radar equipment is making it

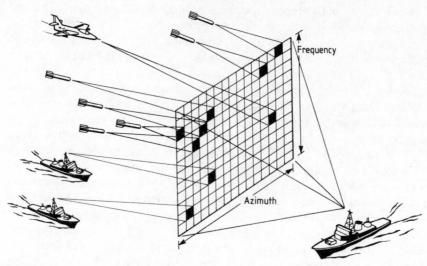

Fig. 8.2 Electronic warfare systems under operating conditions.

common to have systems that are agile in frequency, pulse repetition interval and pulse-width. In addition, advanced radar systems are using spread spectrum techniques and complex intrapulse modulations. These features all place new and demanding performance requirements on electronic warfare processing, with the need for more processing in less time (Fig. 8.2).

To keep pace with projected EW requirements, both electronic support measure and electronic countermeasure systems will need to be designed with improved algorithms and processing techniques. They then must be implemented with higher speed electronic components. The performance advantages of GaAs integrated circuits, particularly in the area of speed, place this technology as the prime candidate to fulfil wideband signal and data processing system requirements.

One example of GaAs integrated circuit application in EW equipment is in radar warning receivers (RWR) (Fig. 8.3). Whereas GaAs monolithic microwave integrated circuits will find utilization in the receiver portion of the system, digital electronics in the preprocessor will permit signal sorting (clustering) at maximum input rates up to 100 million pulses per second. Advanced architectures performing two-dimensional signal clustering (frequency and position) designed to handle low signal strength, spread spectrum, and burst-modulated emitters, require chip clock rates of approximately 2.5 GHz to accommodate input pulse rate to 100 million pulses. These clock rates are beyond the capability of 1 μm silicon devices, but within the performance levels of GaAs devices. The main RWR processor, providing system control and display, is being addressed by the VHSIC program and will be accomplished with silicon devices.

A second example of GaAs application to EW systems is in the area of

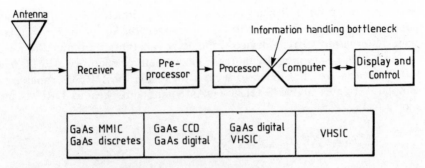

Fig. 8.3 GaAs application to radar warning receiver.

electronic counter measurement (ECM). Coherent digital RF memories (DRFM) with significantly enhanced capabilities are achievable in GaAs (Fig. 8.4). The system is envisioned as part of a repeat jammer where RF would be sampled, digitized and stored in memory. At a selectable delay, the information is read from memory, converted to analogue form, and retransmitted, for jamming purposes. Since the sampling of 2 to 18 GHz signals requires sampling rates in excess of 36 GHz, channelling to parallel, lower bandwidth signals is needed.

The 'data acquisition system' could also be implemented as a hybrid containing both GaAs and ECL digital integrated circuits. System resolution of 6 bits and sample rates of 1 to 2 GHz are feasible. This may be obtained in a very straightforward manner since the analogue-to-digital conversion may be performed within 1 nS using GaAs. Current silicon systems in production utilize multiphase sampling schemes to achieve high sampling rates, but 500 MHz bandwidth systems, even with only one-bit

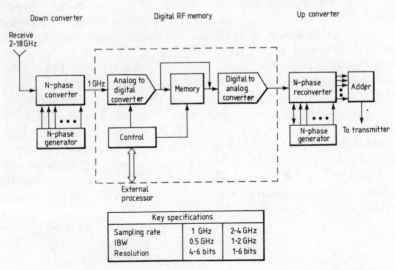

Key specifications		
Sampling rate	1 GHz	2-4 GHz
IBW	0.5 GHz	1-2 GHz
Resolution	4-6 bits	1-6 bits

Fig. 8.4 GaAs application in repeat jammer.

resolution, have been difficult to achieve on a production basis. In GaAs, sampling rates and analogue-to-digital conversion to 1 to 2 GHz are achievable within current device technology capabilities without the need for multiphase sampling techniques. GaAs heterostructure (HEMT and HBT) devices are envisioned to yield 1 bit resolution DRFMs capable of sampling at rates above 5 GHz. GaAs components key to DRFMs are shown in Fig. 8.5.

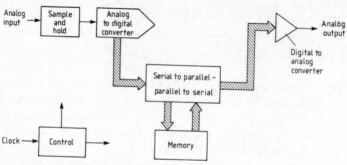

Fig. 8.5 Digital RF memory (DRFM) key components.

8.4 COMMUNICATIONS APPLICATIONS

In order to meet the increasing demands placed on satellite communications traffic, satellite communications systems will move into the millimeter wave band of operation. These satellites will use time division multiple access (TDMA) to route from point-to-point the five-fold increase in communication traffic expected to occur between 1990 and 2000. Satellites will be required to route simultaneously 100 up-link signals to 100 down-line stations. The switching as envisioned will be perfomed as IF (6 GHz) with a 100 × 100 matrix switch with 10 000 cross-point switch amplifiers. These amplifiers represent a sizeable market which can only be implemented affordably using GaAs monolithic microwave integrated circuits (MMICs.)

The specific attributes of MMICs that make this an attractive technology are small size, light weight, potential for low cost, reproducible performance, broad bandwidth capability, and multifunctional performance. These attributes are derived directly from the GaAs technology utilized to design and construct the circuits. Small size is intrinsic to the monolithic approach and light weight is an adjunct to the small size and integrated functionality of the monolithic circuitry. Low cost potential is anticipated from the batch fabrication of many circuits on each wafer and is directly impacted by wafer and chip size and yield.

Reproducible performance of MMICs is a result of the high resolution capability of the lithography used to define the active device and circuit geometries, in conjunction with the highly reproducible doping profiles attainable with ion implantation doping. Capability of broadband performance in monolithic circuitry arises from the elimination of parasitic elements, such as wire bond inductances and bonding pad capacitances, which tend to limit the bandwidth in conventional hybrid microwave integrated circuits. Finally, one of the most important features of MMICs is the capability to provide several circuit functions integrated into a single chip, e.g. an entire receiver on a chip, which offers extraordinary advantages in terms of size, weight, cost and reliability. The United States MIMIC program is addressing all of these issues with its goal of establishing a production base in GaAs MMIC technology.

GaAs digital multiplexers and demultiplexers are key components in serial-to-parallel and parallel-to-serial conversion for time division multiplexed communication systems. The high speed of GaAs digital integrated circuits can accommodate several silicon channels with no compromise to individual channel bandwidth. Integrated electro-optical devices where lasers, detectors and drive electronics can all be monolithically integrated with multiplexers and demultiplexers to provide multi-gigahertz transceiver capability are in the development stage (Fig. 8.6).

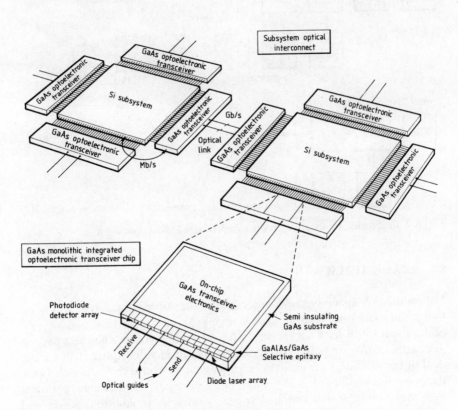

Fig. 8.6 GaAs integrated opto-electronic transceiver for subsystem optical interconnect.

Another area of importance for GaAs in the communication arena is frequency synthesis. In Fig. 8.7 is shown a typical IF generator at 1242 to 1479 MHz. Since ECL prescalers cannot operate at 1479 MHz, an intermediate frequency must be generated to mix against the output signal to present a lower frequency signal to the prescalar. In the example shown, a 348 MHz signal is generated and mixed with the output signal to generate a difference signal in the 261 and 498 MHz range. This signal can then be handled by an ECL divide by two circuit and subsequent variable modulus divider providing the required signal to the phase detector completing the phase locked loop. In contrast, with a GaAs variable modulus divider/ prescalar that can operate directly at the output frequency, all of the components shaded in Fig. 8.7 can be replaced with a single GaAs component providing significant savings in cost and power dissipation while providing a phase-locked loop circuit with enhanced performance.

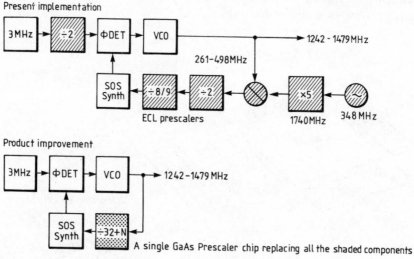

Fig. 8.7 Frequency synthesiser product improvement with GaAs.

8.5 SPACE APPLICATIONS

Future military space systems, especially in the surveillance area, will need to be autonomous, have at least a five-year operating life, and remain operational for at least six months after a strategic conflict has pumped up the earth's radiation belts. Autonomous operation will require that a great deal of the satellite's sensor data be processed on-board the satellite. Thus, these satellites will have a large number of electronic components that have high radiation tolerance and are extremely reliable. In addition, power and weight are always at a premium in space and therefore space applications are also seeking the technology that consumes the least amount of power while still accommodating the required processing speeds. The fundamental physical properties of GaAs (i.e. mobility, band-gap, pinned surface states) are such that it holds great promise for providing the electronic technology that satisfies the unique requirements of space based applications. It is only the technological immaturity of the GaAs technologies (i.e. low yield, low production volume) that keeps us from having this technology in hand.

Internally, GaAs integrated circuits operate at lower power than the lowest power silicon technology, i.e. complementary metal-oxide-semi-conductor (CMOS), for systems with clock speeds above approximately 40 MHz, and always exhibit superior total dose radiation tolerance. This advantage can be lost if the integration level is too low because of the number of interconnects needed; input/output drivers require a lot of power essentially independent of the technology that is used. Furthermore, reliability is also affected by the integration level; thus despite the fact that GaAs's larger bandgap should improve its reliability, if integration levels comparable to silicon are not reached, the overall system's reliability will suffer. In the following paragraphs radiation tolerance, power dissipation, complexity and reliability of GaAs ICs will be discussed.

Satellites must be immune to several types of radiation damage in the space environment. Initially, the main concern of a satellite system was vulnerability to radiation damage caused by high energy electrons that become trapped by the earth's magnetic field. However, as future military systems become more and more dependent on space assets, the effects of nuclear weapons must also be considered in the selection of electronic components. Finally, all space-based systems must take into account single event upset (SEU) – both from cosmic rays in any orbit or from high energy protons that are also trapped in the earth's magnetic field. The primary and secondary effects of the various threat scenarios are listed in Table 8.2.

Table 8.2. Radiation effects on semiconductor electronics

The threat	Primary effects	Secondary effects	Remarks
High energy-trapped electrons	Total dose		Low altitude only
Ballistic ASAT	Total dose	Dose rate Neutrons	
Guided ASAT	Dose rate Total dose	Neutrons	
Particle beams – wide divergence – narrow divergence	Total dose Dose rate	Dose rate Total dose	
Natural space radiation	SEU		Protons ($> 20\,MeV$) Cosmic rays

The conventional system's solution to the radiation effects problem is shielding. However, for some orbits that solution may be inadequate in the case of a nuclear conflict and thus it is instructive to consider worse case conditions in the earth radiation belts (Fig. 8.8). Shielding can reduce total dose effects only by about a factor of 500 due to brehmstrahlung re-radiation. For synchronous orbits (right-hand scale) the maximum exposure in ten days is a few times 10^6 Rads and shielding can reduce that to a level where specially processed silicon technologies can survive. For low altitude satellite orbits (left-hand scale), however, there is very little likelihood that silicon circuits can be made tolerant enough to survive, even with massive shielding. In addition, while some shielding will always be present, requiring maximum effective shielding results is a heavy weight penalty. Natural decay of the radiation after a nuclear burst in space is such that in six months the total absorbed ionizing dose will be four times the dose accumulated in the first ten days. Therefore, for the worst case scenario, utilizing maximum effective shielding but considering the case where the belts are just saturated once and allowed to undergo natural decay thereafter one concludes that six-month survival necessitates the electronics to remain operational past an exposure of at least 40 megarads of ionizing radiation.

The same need for high tolerance to ionizing radiation exists for electronics used in the control systems of nuclear reactors. This is especially true for those that may be employed in space. Super-hard electronics can reduce the shielding requirements to such an extent that the savings in lift-off cost for one system alone can pay for a substantial fraction of the cost of setting up a GaAs pilot production facility. As an example, elecronics

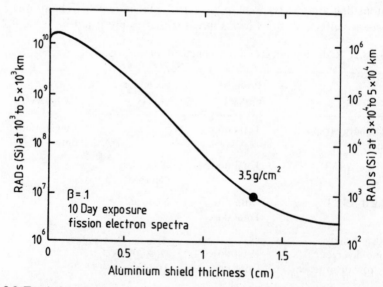

Fig. 8.8 Total dose hardness *vs* aluminium shield thickness requirement for two altitudes.

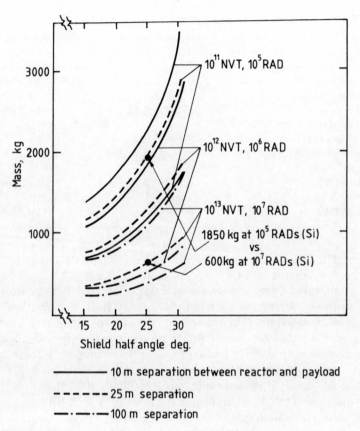

Fig. 8.9 Shield mass required for space-based nuclear reactor.

capable of surviving an accumulation of 10 megarads rather than 0.1 megarads would reduce the shield weight of a 7 year, 1600 kilowatt space-based nuclear reactor by 1250 kilograms (Fig. 8.9) [4]. With present-day launch costs at \$50 000 per kg for synchronous orbit or \$4000 per kg for shuttle orbits this reduced shielding requirement translates to a multi-million dollar cost reduction.

As seen earlier in Table 8.2, total dose figures prominently in most of the threat scenarios. The fundamental materials properties of GaAs account for the superior performance in the total dose environment when compared to silicon-based electronics. As already mentioned, GaAs differs from silicon in that rather than having a passivating oxide on its surface, the GaAs surface is stabilized by such a high density of mid-gap surface states that the surfaced Fermi-level can only be moved by tens of millivolts. This feature has prevented the fabrication of devices analogous to silicon MOS, but for the development of total dose tolerant circuits this constitutes a major plus. Radiation induced charging of any insulator that is placed on the GaAs surface cannot cause large shifts in surface potential as is the case for silicon

Table 8.3. Total dose ionizing radiation tests of GaAs ICs

Circuit	Number	Failed	10^7 rad (GaAs)	5×10^7 rad (GaAs)
Ring-oscillator (JFET)	14	0	$\Delta t_{pd} = 0$	$\Delta t_{pd} = 5\%$
Gate-chain R-load (JFET)	1	0	$\Delta t_{pd} = 0$	$\Delta t_{pd} = 2\%$
Divide-by-two circuit D-JFET load (JFET)	1	0	$\Delta f = 0$	$\Delta t = 0$
Current limiter (JFET)	20	0	$\Delta I_{DS} = 0$	$\Delta I_{DS} = 4\%$
Divide-by-eight (MESFET)	3	0	$\Delta t_{pd} = 0$	$\Delta t_{pd} = 0$
8 by 8 multiplier (MESFET)	1	0	–	$\Delta t_{pd} = 19\%$

devices. The actual data taken on present experimental GaAs circuits, that have been irradiated in a Co^{60} cell, is shown in Table 8.3.

At 10 megarads, one cannot see any degradation. Upon measuring the most sensitive parameter, namely pulsed delay (pd), one sees only small changes at 50 megarads. The changes at 50 megarads are believed to be due to displacement damage. Such damage results in carrier removal; a damage mechanism akin to damage caused by dielectric charging. Recent tests on 256–bit and 1024–bit memory chips have shown them to remain fully operational even after exposure greater than 100 megarads [5]. A comparison with silicon technologies is shown in (Fig. 8.10).

Transient radiation hardness on the other hand, depends on the interplay among material properties, device design and circuit design. The low lifetime of a direct band-gap semiconductor like GaAs provides some relief

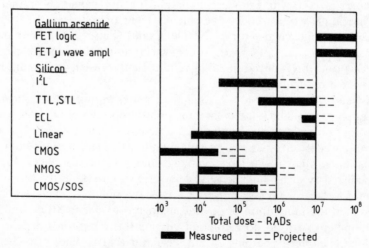

Fig. 8.10 Total dose tolerance of semiconductor integrated circuits.

in reducing the primary photocurrents, the semi-insulating substrate insures latch-up prevention. However, circuit or cell noise margins still determine whether upset occurs or does not occur in a transient environment. The complementary-junction field effect transistor (C–JFET) member cell has emerged as the prime candidate for achieving transient radiation hard memory technology (Fig. 8.11). Using a p-channel JFET as a nonlinear load, the noise margins in the memory cells are enhanced; furthermore, the standby power dissipation is decreased for the same reasons that make silicon CMOS such a low-power technology. Unfortunately, the low mobility of p-type GaAs makes p-channel JFETs only acceptable in memory cells and not in the peripheral decode circuits or in random logic circuits. Data taken on 256–bit memory chips build with C-JFET memory cells shows

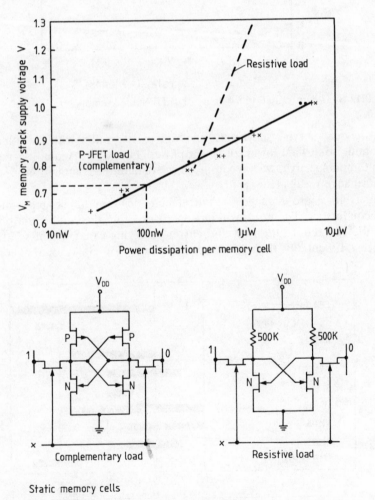

Fig. 8.11 Power consumption in JFET static memory.

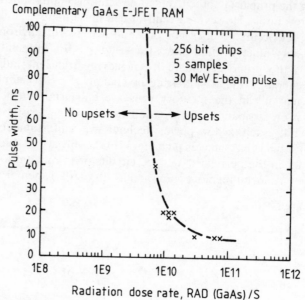

Fig. 8.12 Transient radiation hardness of JFET static memory.

that they can survive short pulses as high as 10^{11} rad/s, with the tolerance becoming somewhat lower for larger pulses (Fig. 8.12). Again, placing this result in perspective, we have inserted the data into a compilation of silicon performance results (Fig. 8.13).

Neutrons cause damage by introducing new, mid-band-gap states in semiconductors. These states remove carriers from the conduction process and thereby reduce the gain and current carrying capability of solid state devices. Essentially the carrier removal rates for GaAs and silicon are

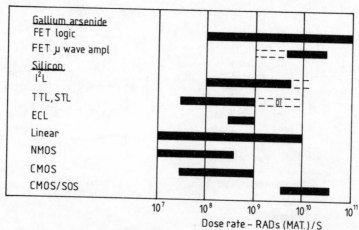

Fig. 8.13 Reported (solid) and projected (dotted) short pulse upset threshold.

similar, thus components made from either material degrade similarly. In both cases the primary hardness is obtained by keeping the doping level high so that a carrier removal rate between 5 and 10 per neutron per cm^2 does not significantly affect the total carrier density in the devices.

Single event upset (SEU) is a naturally occurring radiation phenomena that has become increasingly important as integration levels have increased and device dimensions have decreased. It was first observed in dynamic random access memories (DRAMS) as a result of alpha particle emission by impurity elements present in the DRAM packaging materials. However, in the space environment, SEU is the result of both cosmic rays and, high energy (20 MeV) protons, in the radiation belts. Akin to the transient radiation hardness problem, SEU immunity depends on noise margins; the charge that is collected onto a critical circuit node, as a result of ionizing by the primary particle or collision induced secondaries, must not cause that node to undergo a change of state. Initial tests on GaAs memories gave disappointing results, but again the increased noise margin of the C-JFETs reduced this vulnerability. The C-JFET memory cell is also amenable to the same SEU hardening methodology as is done for CMOS – inclusion of cross coupling resistors in the memory cell. This procedure has eliminated upset by 40 Mev protons and is expected to increase greatly the hardness to upset from cosmic rays. Work is underway to determine the optimum relationship between additional SEU immunity and the speed penalty that GaAs memories suffer due to the inclusion of these resistors. Silicon CMOS circuits do not suffer a speed penalty since the memory cell transition times are fast compared to the peripheral circuit speed; in GaAs these times are comparable.

It should be noted that the effort to achieve increased SEU immunity for GaAs memories is the first time that any effort and funds have been spent to increase the radiation hardness of GaAs components. All the excellent radiation results obtained in the areas of total dose and dose rate were obtained on circuits that had been built to demonstrate either the high speed or the low power potential of GaAs ICs and then tested for radiation hardness as an afterthought. Further, one should note, that if the SEU frequency is not too great, system's solutions are possible at the cost of system complexity. The research need in the SEU area is to reduce its frequency without seriously effecting component performance.

Having covered the various radiation effects and the various degrees of improvement that GaAs components have to offer over silicon ones, a word of caution must be inserted. Good semi-insulating GaAs starting material is vital. If the starting material contains a large number of traps, as did the heavily chromium-doped material that was prevalent a few years back, then transient pulses can cause long-term external logic level shifts that have prohibitively long recovery times even though internal circuit functions are not upset. Today, properly qualified, undoped semi-insulating material allows devices to be fabricated whose recovery is essentially prompt (Fig. 8.14).

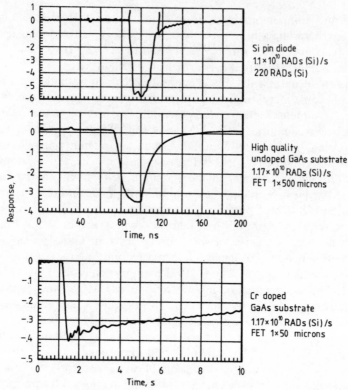

Fig. 8.14 Material influence on radiation induced long term transients.

Future satellite systems will be designed to be autonomous and to accomplish a great percentage of the signal processing of sensor inputs on-board the satellite. These characteristics drive both the need for signal processing throughput (operation/s) and mass memory capacity to increase by orders of magnitude (Fig. 8.15). However, it is well known that for large systems it is the memory size that determines the system's power requirements which in turn determines the weight of the system. For modern digital electronics, in silicon as well as GaAs, the dominant fraction of the chip power is consumed in the output buffers. Consequently, a dramatic reduction in system power, weight and ultimately cost results from pushing memory chip capacity to the highest limits of a given technology. A specific example was worked out for silicon CMOS (Fig. 8.16), but qualitatively these results hold for any technology. The importance of this result is further illustrated when one observed that some of the contemplated future space systems cannot be launched by today's launch vehicle if they are built with low complexity chips of less than 16k (Fig. 8.17).

The above described need for high density chips presents the greatest difficulty to the utilization of GaAs technology in future space systems. The silicon technologies have reached their level of complexity through years of

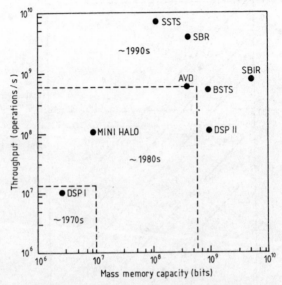

Fig. 8.15 Projected growth in space-borne computer requirements.

		CMOS computer power (watts)				Weights savings (kg)*			Cost savings† (millions of dollars per spacecraft)		
	k-bits/chip	4	16	64	256	4→16	4→64	4→256	4→16	4→64	4→256
Mission	SBR 4 × 10⁸ bits	1 110	115	35	4	1 300	1 400	1 450	100	108	112
	DSP II 10⁹ bits	1 300	1 000	300	100	950	1 550	1 750	73	119	135
	SBIR 7 × 10⁹ bits	11 000	8 000	4 000	900	11 000	13 000	15 500	847	1 000	1 190

* Computer plus power supply weight savings; 1 w power equals 1 lb wt
† 1 lb launch weight costs $35000

Fig. 8.16 Silicon CMOS example for synchronous orbit to illustrate increased component density on-chip results in large savings for space-based systems.

moving up the 'learning curve' – an empirical relationship that predicts a 30% increase in yield every time the total quantity of fabricated parts has been doubled. GaAs has no such production base.

Yield results for high density circuits fabricated in an R&D environment of 1 to 2 wafers/week throughput are disastrous, and it is not until throughput has been increased to the 10–20 wafers/week level that some hope of achieving required complexities emerges (Fig. 8.18).

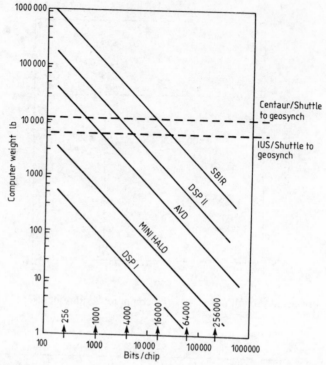

Fig. 8.17 System computer weight *vs* chip capacity to illustrate low component density may limit space system capabilities.

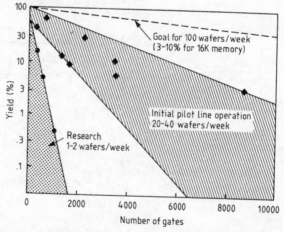

Fig. 8.18 Present (solid) and projected (dotted) GaAs IC yield *vs* gate complexity.

The challenge to achieve a minimally acceptable yield (1%) for the 16–kbit memory chips led to the DARPA program for GaAs pilot production lines that have a minimum throughput of 100 wafers/week and are expected to tilt the yield vs complexity curve substantially upward.

8.6 DISTRIBUTED PROCESSING

The need for high-speed processing is clearly established by the United States' desire to develop a broad line of intelligent machines. Machines that are experts in the solution of various problems, capable of natural language and vision, and with the ability to plan and reason are currently in the early stages of concept definition and development. The processing rate required just for the task of limited domain image recognition alone is approximately 10E9 operations per second (OPS) and a similar rate of processing is required for the task of speech recognition. State-of-the-art computers currently under development do not begin to approach the processing rates required for complex, intelligent machines.

The system responsible for strategic defense will be faced with critical real-time decision making, performing complex processing tasks under conditions of extremely high input rates, and with extraordinarily large databases. Performance multipliers of many orders of magnitude over current computer capabilities are required, and for certain defense applications, these super-systems must be space-based, compatible with the power dissipation, size, weight, volume and radiation survivability constraints of space-based systems.

The giant steps forward needed in processing throughput are not anticipated exclusively through the development of improved electronic components. Whereas, the very high speed integrated circuit (VHSIC) programme and associated GaAs integrated circuit programmes are addressing improved device performance through improved circuit designs, improved wafer processing techniques, smaller and smaller (sub-micron) minimum lithographic dimensions, higher electron mobility and radiation hard materials (GaAs, InP), larger wafers, higher speed packaging and interconnection methods, complex heterostructures for enhanced mobility devices, improved computer-aided design techniques, etc., factors of hundreds or thousands will remain to be resolved through other avenues such as improved algorithms, improved processing methods (i.e. systolic processing) and through the utilization of distributed systems.

Many projects are currently underway for the development of local area networks and distributed systems where literally thousands of processors can work together, simultaneously, in parallel. Efficient operating systems, and fault-tolerant architectures incorporating the highest possible speed micro-processors, are under development. Supporting programs to develop state-of-the-art microprocessors (i.e. Si and GaAs reduced instruction set computers) are also in progress. Whereas, much attention has been placed on improving the processing speed at each node, an important use of very high speed GaAs electronics will be to speed up the communication between processors.

8.7 THE US DARPA GaAs IC PROGRAMME

One of the first systems applications that sought to use digital GaAs to achieve a radiation hard signal processor was the advanced on-board signal processor (ADSP) project. The United States Defence Advanced Research Projects Agency (DARPA) was investigating spaceborne signal processing for a broad range of military missions and it concluded that a need existed for radiation-resistant, low-power, highly reliable and programmable signal processing systems. To meet this need, the AOSP programme was initiated to develop a suitable architecture. One of the goals of this programme was to fabricate an AOSP brassboard. After careful investigation, the best candidate technology was found to be GaAs.

The first ingredient of reliable electronic systems is a reduced total chip count. Studies showed that a minimum of 16 K static random access memory (SRAM) was necessary with a strong desire for 64 K SRAM. Additionally, a large configurable gate array (6 K gates) was necessary. Considering that the most complex GaAs chip successfully fabricated at that time (1980s) was a laboratory fabricated 8 by 8 multiplier composed of 1008 gates, the task ahead was formidable and of extremely high risk – just the type of technical challenge for which DARPA was created. In 1982, DARPA started the development of an all-GaAs prototype of the AOSP. Both complexity and yield of GaAs integrated circuits had to be increased by orders of magnitude – simultaneously! Only by phenomenally increasing yield could a sufficient number of memory and gate array chips of the desired complexity be produced to ever contemplate the possibility of realising the GaAs AOSP prototype. Clearly, it was necessary to emulate the same 'learning curve' improvement in yield and complexity that had propelled silicon technology to the VLSI era – and this had to be done in a shorter time period than silicon had required. Fortunately, much of the methodology and equipment used by the silicon industry is transferrable to GaAs circuit fabrication with but minor changes. Lines processing a minimum of 100 three-inch wafers a week were contractually required, based on surveys of the practices in the silicon industry, that such a throughput is the minimum necessary to ensure the manufacturing discipline essential to realising a reasonable yield of complex chips. The first pilot line programme was begun in 1983, the next in 1985, and a third in 1987. In addition to developing the chips for the AOSP, these pilot lines will provide the foundation for military and commercial usages of GaAs digital circuits in advanced computers, instrumentation and communication systems.

In setting the specifications for components manufactured on the GaAs pilot line, it was necessary to keep in mind the primary concerns for spaceborne systems – weight and power. Consequently, not raw speed but millions of instructions per second per unit of power (MIPS per watt) became the critical figure-of-merit. Despite pressure from the contractor community, DARPA maintained this commitment and set the specification of the SRAM at one microwatt per bit for static power with a 10 nS access time. Similarly, for the gate arrays the requirements were 400 micro-watts for a D-flip-flop toggling at 50 MHz. These low-power goals require operations at low-threshold voltage with extremely tight control on uniformity to achieve adequate noise margins. An IC fabrication process developed for these requirements can always be relaxed to accommodate the higher drive needs of fast circuits. The reverse, however, would not necessarily be true. (Furthermore, an increase of two in frequency, often requires an order of magnitude increase in power.)

Because of the multiplicity of GaAs pilot lines being developed under the auspices of DARPA contracts, and the resulting legacy for military use, it was recognized that some mechanism would be required which would allow defence contractors not intimately familiar with the details of these lines to be able to exploit their manufacturing potential. Each pilot line uses a different GaAs chip technology. Since each technology possesses different strengths, it was recognized that the capability to fabricate the same chip design(s) in more than one of these technologies could be useful for system optimization. In the commercial silicon VLSI world, in general, the design of each integrated circuit must be targeted from the outset for a particular chip technology (i.e., 2 micron CMOS, 1.25 micron NMOS, ECL, etc.). If it is desired later to refabricate the same chip design in a different silicon chip technology, nearly the entire design process must be repeated. To minimize this problem across the several GaAs pilot lines and technologies, DARPA has sponsored the development at the Mayo Foundation of a comprehensive computer-aided design (CAD) package, which allows technology-independent chip designs in a straightforward manner.

Hosted on Digital Equipment Corporation VAX computers, this large software system reduces dramatically the amount of redesign effort required to change from one GaAs technology to another. Every integrated circuit, or in fact the entire processor system, is designed initially using a 'generic' library of gate array and/or standard cell 'functional components'. The CAD system also supports a number of technology-specific libraries of functional components. Every 'functional component', or 'macro', in every technology library has a corresponding macro component in the generic library. This matching between generic and technology-specific 'functional components' permits an entire chip or system design to be retargeted from its initial generic form into a technology-specific form, and then laid out on the specific integrated circuits presenting the selected GaAs technology. The retargeting process can be fully automatic, fully manual, or a combination of both, depending upon desired chip packing density (efficiency) and/or desired turnaround time. Simply by selecting different technology-specific libraries, the same generic design can be retargeted for any one or more of the emerging GaAs technologies. It is also intended that at least one silicon technology will be targeted so that functionality can be tested at minimum cost.

In the spring of 1984 it was decided that, in addition to the gate array and the RAM, a custom GaAs microprocessor was needed as an engine to drive the AOSP. While it is possible to construct a microprocessor from gate arrays, performance would clearly suffer. A need for a high performance 32-bit machine added a further complication.

At the time a 10 K custom chip was considered the outer limit to the level of complexity achievable in the program's timeframe. Since there was also an interest in high performance, a desire existed to build the microprocessor on a single chip. An exhaustive search was made of the known micro-processor architectures. The only single-chip design that met all the needs outlined above, was a DARPA developed reduced instruction set computer (RISC) design at Stanford called microprocessor without interlocked pipeline stages (MIPS).

In the autumn of 1984 three Phase I contracts were competitively awarded to Texas Instruments, RCA, and McDonnell-Douglas to start the construction of a single chip, all GaAs, MIPS microprocessor with a 200 MHz clock frequency. In the autumn of 1985, two of the three contractors, Texas Instruments and McDonnell-Douglas, were selected to proceed to a full development of the MIPS microprocessor, with a floating point coprocessor and the necessary bus and memory interface chips.

As part of the first Phase I effort, each of the contractors has undertaken to produce portions of the final design constituting 2000–3000 gates. Completion of the GaAs MIPS microprocessor is schedule for late 1987. Current projections for the microprocessor performance are 5 nS for a fixed point add, 30 nsec floating point add and 90 nS for a floating point multiply. In order to obtain these speeds a small (1 K word) cache memory with a 1 nS access time will be needed. Thus, the final design will include working chips for both fast and slow memory access, as well as chips for bus and sensor interfaces. Both contractor designs will use the same core instruction set to which several software compilors (PASCAL, Ada, C) and translators (1750A, 68000) have already been targeted. McDonnell Douglas is also designing a vector processor utilizing the MIPS chips. This will allow an orderly transition of the planned AOSP demonstration onto the GaAs brassboard.

The United States Strategic Defense Initiative (SDI) was formed in 1984. DARPA's GaAs Pilot Program was transitioned to SDI as a fundamental part of the radiation-resistant space signal processor program. By mutual consent, DARPA has continued to manage the GaAs program for the SDI [6]. The SDI office (SDIO) has in turn been a staunch supporter of both GaAs and the AOSP. The all-GaAs AOSP system is currently scheduled for completion in 1990 as one of the major technology milestones in the SDI program. Although the specifications have not been set, it is likely that this system will operate at processing rates up to several GFLOPS and occupy no more than half a standard rack.

Having identified high speed, radiation-hard, low-power circuitry as a unique nitch for GaAs digital technology, and having identified the AOSP architecture for implementation with these circuits, there is every reason to expect that GaAs-based digital semiconductor technology will become firmly established at the end of this DARPA/SDIO programme.

The challenge facing us is to achieve manufacturability for GaAs circuits of VLSI complexity. To accomplish this, it will take a manufacturing science and technology programme that will exploit as much as possible of the processing rigour and equipment that has been so successful in silicon IC manufacturing. Specifically, four assumptions underlie the GaAs Pilot Line effort:

(1) state-of-the-art GaAs semi-insulating substrate material is adequate for some types of digital VLSI circuits;

(2) low yield of present GaAs LSI circuits is due primarily to random processing defect, not lack of reproducibility in device parameters;

(3) strict process control, as practice in silicon-based VLSI manufacturing, will result in achieving comparable 'learning curve' type yield improvements, and

(4) 100 wafer/week throughput is a necessary minimum to achieve a pilot manufacturing discipline.

In general, these precepts have been acquiesced to by workers in the field. The first two result from the observation that the yield of integrated circuits is nowhere near that which can be calculated from the standard deviations in device parameters that are measured on large arrays of single transistors. The last two assumptions are based on the fact that a great deal of the production experience and equipment that is used in the silicon industry is directly applicable to GaAs IC production.

The first digital LSI GaAs pilot production implementing depletion-mode MESFET technology began wafer processing in September 1985, at the Rockwell facility in Newbury Park, California and both 4 K SRAMs and 6 k gate arrays have been designed, fabricated, and demonstrated to be fully functional. A 3.5% yield on the 4 K SRAMs represents the best result reported to date for a GaAs chip of equivalent complexity. Essentially, the technology gap in digital GaAs that had developed between US capability and Japanese capability, has been closed by this SDIO programme. The Rockwell pilot line has systematically increased the wafer throughput during the programme. Initially starting at 10 wafer per week, they progressed to 20 wafers per week and at the time of writing (mid 1987) are running at 60 wafers per week. At the same time, the complexity of the circuits being fabricated has now been extended to include the 16 K SRAM that has a 10 ns access time and will function even after an expose of 100 megarads of total accumulated ionizing radiation.

Based on the excellent radiation tolerance and 'speed–power' figure of merit of the C-JFET technology, a second pilot production programme was awarded to McDonnell-Douglas. In addition to the low power, transient radiation hard memory, the pilot line is fabricating demonstration circuits such as a 32-bit 200 megahertz microprocessor based on the MIPS version of a RISC architecture. As part of the initial demonstration a 32-bit arithmetic logic unit (ALU) was fabricated on this pilot line. This 1500 gate part provides a full 34-bit adder, three 32-bit registers as well as multiplexers and built-in-test circuits. The time for a full 32-bit add was 9.7 ns. The next step is to add a full register stack and three 32-bit bus circuits to the ALU.

The initial operation of the two pilot line programs, operating still at only 20–60 wafers per week, has proven the validity of the assumptions that were made at the start of the programme. The yield of all circuits fabricated on the pilot lines has increased markedly to the point where a clear separation from the realm of research result is evident (Fig. 8.18). It is equally clear however that for the massive memory requirements of autonomous space systems of the 1990s even the yield goals of the present GaAs pilot line programmes have to be exceeded. Consequently, there will be a continuation of the successful pilot line programmes to further improve yield and eventually proceed to 64 K SRAMs and 10 K to 12 K gate arrays. Equally important, the signal processing throughput will eventually have to be increased. Therefore, a third pilot line effort, to fabricate custom circuits in a GaAs technology with better speed performance than required in the first two programmes, was started in 1987 at the AT&T facilities in Reading, Pennsylvania.

REFERENCES

1. Firstenberg, A. (1984) 'GaAs digital integrated circuits' *Defense Science & Electronics Magazine* (Aug.).
2. Firstenberg, A. and Roosild, S. (1985) 'GaAs ICs for new defense systems offer speed and radiation hardness benefits' *Microwave Journal* (March) 145–163.
3. Firstenberg, A. (1985) 'GaAs digital ICs applied to EW systems' *J. Electronic Defense* (Oct.) 119–134.
4. Angelo, J. Jr. and Buden, D. (1982) *Shielding considerations for advanced space nuclear reactor systems* Los Alamos Publication, LA-VR-82-2002.
5. Simons, M. (1983) 'Radiation effects in GaAs integrated circuits: A comparison with silicon' *Technical Digest 1983, GaAs IC Symposium.*
6. Karp, S. and Roosild, S. (1985) 'DARPA, SDI, and GaAs' *Computer* (Oct.) 17–19.

Index

Index